AF686289

Bioengineering of the Motor System

Bioengineering of the Motor System

Guest Editor

Carlo Albino Frigo

MDPI

Basel • Beijing • Wuhan • Barcelona • Belgrade • Novi Sad • Cluj • Manchester

Guest Editor
Carlo Albino Frigo
Movement Biomechanics and
Motor Control Lab, DEIB
Politecnico di Milano
Milan
Italy

Editorial Office
MDPI AG
Grosspeteranlage 5
4052 Basel, Switzerland

This is a reprint of the Special Issue, published open access by the journal *Bioengineering* (ISSN 2306-5354), freely accessible at: https://www.mdpi.com/journal/bioengineering/special_issues/WS36YXZC41.

For citation purposes, cite each article independently as indicated on the article page online and as indicated below:

Lastname, A.A.; Lastname, B.B. Article Title. *Journal Name* **Year**, *Volume Number*, Page Range.

ISBN 978-3-7258-3843-1 (Hbk)
ISBN 978-3-7258-3844-8 (PDF)
https://doi.org/10.3390/books978-3-7258-3844-8

© 2025 by the authors. Articles in this book are Open Access and distributed under the Creative Commons Attribution (CC BY) license. The book as a whole is distributed by MDPI under the terms and conditions of the Creative Commons Attribution-NonCommercial-NoDerivs (CC BY-NC-ND) license (https://creativecommons.org/licenses/by-nc-nd/4.0/).

Contents

About the Editor

Carlo Albino Frigo

Carlo Albino Frigo, formerly of the Department of Electronics, Information, and Bioengineering, Politecnico di Milano (now retired), has worked in movement analysis and motor rehabilitation since his graduation in 1976. He was among the founders of the Italian Society of Clinical Movement Analysis (SIAMOC), of which he was the First Secretary and then President. He also helped to found the Italian branch of ISPO (the International Society for Prosthetics and Orthotics), of which he was Vice-President. As a member of the ESMAC (the European Society for Movement Analysis in Adults and Children) committee, he organised the ESMAC-SIAMOC joint congress in 2001. He has been a member of several scientific societies and editorial boards. In 2012, he was invited to deliver the Honorary Baumann Lecture at the ESMAC annual congress (Stockholm, Sweden). His interests include movement analysis, musculoskeletal modelling, motor rehabilitation, biomechanics, and motor control. In these areas, he has participated in several European projects and educational courses. He has been Guest Editor of three Special Issues of MDPI journals: Movement Biomechanics and Motor Control (2020), Musculoskeletal Models in a Clinical Perspective (2021), and Bioengineering of the Motor System (2025). His scientific output consists of 45 contributions to multi-author scientific books and more than 120 articles published in international scientific journals (H index 38, 7432 Citations in Google Scholar).

bioengineering

Editorial

Bioengineering of the Motor System

Carlo Albino Frigo

Department of Electronics, Information and Bioengineering, Politecnico di Milano, 20133 Milan, Italy; carlo.frigo@polimi.it

About fifty years ago, which seems very recent, new technologies for motion analysis were being developed, promising a more detailed and precise study of the human motor system. These technologies allowed the measurement of biomechanical variables and the quantification of phenomena that were previously difficult to assess through clinical observation. In the 1980s, the challenge from a bioengineering perspective was to make these systems as reliable and user-friendly as possible, enabling comprehensive movement analysis across various settings, including clinical environments.

Over the years, extensive research has been conducted to validate these technologies and develop application protocols, demonstrating their potential for research and clinical investigations. Today, well-established technologies such as stereophotogrammetric systems, dynamometric platforms, surface electromyography, and portable metabolimeters are commonly used in dedicated movement analysis laboratories. However, as technology has evolved, new demands and opportunities have emerged.

One of the most significant developments has been the increasing need for a more "ecological" approach to movement analysis, made possible by portable motion sensors. Research in this area has been highly active, leading to the widespread use of wearable systems based on miniaturized accelerometers, gyroscopes, and magnetometers, commonly known as Inertial Measurement Units (IMUs). Although IMUs have lower accuracy compared to complex stereophotogrammetric systems, they offer key advantages such as affordability, ease of use, and, most importantly, the ability to analyze individuals in natural settings such as gyms, sports fields, and workplaces. Additionally, the integration of these sensors into Virtual Reality systems has expanded their applications, enhancing our understanding of motor control mechanisms.

Another major advancement has been the improvement of computational power, which has facilitated the development of musculoskeletal modeling. Initially, these models provided representations of bones, joints, muscles, and ligaments, simulating the body's degrees of freedom. The next step involved simulating movement by applying forces to the model, solving what is known as the forward dynamics problem. Once considered an unattainable challenge, this approach is now widely used by researchers to analyze the behavior of internal structures and assess the effects of modifying their mechanical properties.

These advances have led to a vast body of literature covering both technological developments and applications in diverse fields such as motor rehabilitation, neurology, orthopedics, sports science, ergonomics, and psychophysics. This Special Issue, "Bioengineering of the Motor System", was conceived to collect research that exemplifies current interests in this field, and it has successfully fulfilled this objective.

An analysis of the twelve papers in this issue reveals that technological advancements now focus on developing algorithms to exploit data from existing systems [1,2] and incorporating sensorized devices [3,4]. Other studies explore applications involving Virtual Reality and exoskeletons [5–7]. Musculoskeletal modeling is represented in a study by [8] which

Received: 5 February 2025
Revised: 13 February 2025
Accepted: 17 February 2025
Published: 18 February 2025

Citation: Frigo, C.A. Bioengineering of the Motor System. *Bioengineering* **2025**, *12*, 199. https://doi.org/10.3390/bioengineering12020199

Copyright: © 2025 by the author. Licensee MDPI, Basel, Switzerland. This article is an open access article distributed under the terms and conditions of the Creative Commons Attribution (CC BY) license (https://creativecommons.org/licenses/by/4.0/).

highlights its potential for understanding internal anatomical structures during movement. Additional contributions focus on motor function recovery using electromyography [9], electrical stimulation [10], and specialized algorithms for identifying muscle synergies [11]. Finally, an intriguing study on human behavior examines psychological aspects using a motion capture system with a bioengineering approach [12].

These contributions illustrate how this Special Issue, entitled "Bioengineering of the Motor System", has successfully assembled research that represents the state of the art and key areas of interest in this domain. Looking ahead, we can expect rapid advancements in methodologies based on current and emerging technologies. These developments will likely be directed toward clinical applications, significantly enhancing our understanding of the human motor system and driving progress in both research and practical implementations.

Conflicts of Interest: The author declares no conflict of interest.

References

1. Du, G.; Ding, Z.; Guo, H.; Song, M.; Jiang, F. Estimation of Lower Limb Joint Angles Using sEMG Signals and RGB-D Camera. *Bioengineering* **2024**, *11*, 1026. [CrossRef] [PubMed]
2. Aviles, M.; Alvarez-Alvarado, J.M.; Robles-Ocampo, J.-B.; Sevilla-Camacho, P.Y.; Rodríguez-Reséndiz, J. Optimizing RNNs for EMG Signal Classification: A Novel Strategy Using Grey Wolf Optimization. *Bioengineering* **2024**, *11*, 77. [CrossRef] [PubMed]
3. Abdollahi, M.; Rashedi, E.; Kuber, P.M.; Jahangiri, S.; Kazempour, B.; Dombovy, M.; Azadeh-Fard, N. Post-Stroke Functional Changes: In-Depth Analysis of Clinical Tests and Motor-Cognitive Dual-Tasking Using Wearable Sensors. *Bioengineering* **2024**, *11*, 349. [CrossRef] [PubMed]
4. De Pasquale, P.; Bonanno, M.; Mojdehdehbaher, S.; Quartarone, A.; Calabrò, R.S. The Use of Head-Mounted Display Systems for Upper Limb Kinematic Analysis in Post-Stroke Patients: A Perspective Review on Benefits, Challenges and Other Solutions. *Bioengineering* **2024**, *11*, 538. [CrossRef] [PubMed]
5. Adamo, P.; Longhi, G.; Temporiti, F.; Marino, G.; Scalona, E.; Fabbri-Destro, M.; Avanzini, P.; Gatti, R. Effects of Action Observation Plus Motor Imagery Administered by Immersive Virtual Reality on Hand Dexterity in Healthy Subjects. *Bioengineering* **2024**, *11*, 398. [CrossRef] [PubMed]
6. Wang, L.; Li, X.; Peng, Y.; Han, J.; Zhang, J. Balance Evaluation Based on Walking Experiments with Exoskeleton Interference. *Bioengineering* **2024**, *11*, 386. [CrossRef] [PubMed]
7. Kreipe, S.; Helbig, T.; Witte, H.; Schumann, N.-P.; Anders, C. Comparison of sEMG Onset Detection Methods for Occupational Exoskeletons on Extensive Close-to-Application Data. *Bioengineering* **2024**, *11*, 119. [CrossRef] [PubMed]
8. Donno, L.; Galluzzo, A.; Pascale, V.; Sansone, V.; Frigo, C.A. Walking with a Posterior Cruciate Ligament Injury: A Musculoskeletal Model Study. *Bioengineering* **2023**, *10*, 1178. [CrossRef] [PubMed]
9. Park, J.; Lee, K.; Kim, J.; Song, C. Electromyography-Triggered Constraint-Induced Movement Cycling Therapy for Enhancing Motor Function in Chronic Stroke Patients: A Randomized Controlled Trial. *Bioengineering* **2024**, *11*, 860. [CrossRef] [PubMed]
10. Veldema, J.; Steingräber, T.; von Grönheim, L.; Wienecke, J.; Regel, R.; Schack, T.; Schütz, C. Direct Current Stimulation over the Primary Motor Cortex, Cerebellum, and Spinal Cord to Modulate Balance Performance: A Randomized Placebo-Controlled Trial. *Bioengineering* **2024**, *11*, 353. [CrossRef] [PubMed]
11. Borzelli, D.; De Marchis, C.; Quercia, A.; De Pasquale, P.; Casile, A.; Quartarone, A.; Calabrò, R.S.; D'avella, A. Muscle Synergy Analysis as a Tool for Assessing the Effectiveness of Gait Rehabilitation Therapies: A Methodological Review and Perspective. *Bioengineering* **2024**, *11*, 793. [CrossRef] [PubMed]
12. Tanabe, H.; Yamamoto, K. The Psychological Nature of Female Gait Attractiveness. *Bioengineering* **2024**, *11*, 1037. [CrossRef] [PubMed]

Disclaimer/Publisher's Note: The statements, opinions and data contained in all publications are solely those of the individual author(s) and contributor(s) and not of MDPI and/or the editor(s). MDPI and/or the editor(s) disclaim responsibility for any injury to people or property resulting from any ideas, methods, instructions or products referred to in the content.

bioengineering

MDPI

Article

Estimation of Lower Limb Joint Angles Using sEMG Signals and RGB-D Camera

Guoming Du [1], Zhen Ding [2], Hao Guo [1], Meichao Song [1] and Feng Jiang [1,*]

[1] School of Computer Science and Technology, Harbin Institute of Technology, Harbin 150001, China; 19b903025@stu.hit.edu.cn (G.D.); guohao@hit.edu.cn (H.G.); 23b936020@stu.hit.edu.cn (M.S.)
[2] College of Computer and Control Engineering, Northeast Forestry University, Harbin 150040, China; dingzhen@nefu.edu.cn
* Correspondence: fjiang@hit.edu.cn

Abstract: Estimating human joint angles is a crucial task in motion analysis, gesture recognition, and motion intention prediction. This paper presents a novel model-based approach for generating reliable and accurate human joint angle estimation using a dual-branch network. The proposed network leverages combined features derived from encoded sEMG signals and RGB-D image data. To ensure the accuracy and reliability of the estimation algorithm, the proposed network employs a convolutional autoencoder to generate a high-level compression of sEMG features aimed at motion prediction. Considering the variability in the distribution of sEMG signals, the proposed network introduces a vision-based joint regression network to maintain the stability of combined features. Taking into account latency, occlusion, and shading issues with vision data acquisition, the feature fusion network utilizes high-frequency sEMG features as weights for specific features extracted from image data. The proposed method achieves effective human body joint angle estimation for motion analysis and motion intention prediction by mitigating the effects of non-stationary sEMG signals.

Keywords: joint angle estimation; sEMG; RGB-D camera; dual-branch convolutional network

Citation: Du, G.; Ding, Z.; Guo, H.; Song, M.; Jiang, F. Estimation of Lower Limb Joint Angles Using sEMG Signals and RGB-D Camera. *Bioengineering* **2024**, *11*, 1026. https://doi.org/10.3390/bioengineering11101026

Academic Editor: Carlo Albino Frigo

Received: 13 September 2024
Revised: 8 October 2024
Accepted: 11 October 2024
Published: 15 October 2024

Copyright: © 2024 by the authors. Licensee MDPI, Basel, Switzerland. This article is an open access article distributed under the terms and conditions of the Creative Commons Attribution (CC BY) license (https://creativecommons.org/licenses/by/4.0/).

1. Introduction

Human–machine interaction (HMI) technology has consistently been a promising domain in automation, healthcare, and other fields of human activities for decades. To conduct a quantitative analysis of human motions and activities during HMI, human body joint angles and positions are utilized as essential metrics. Human activities can be captured and analyzed using wearable devices [1–4] and vision-based sensors [5].

In-depth analysis of human activity allows for joint angles to be extracted during movements, providing key information for motion pattern detection, movement recognition, and prediction. For example, Da Silva et al. [6] utilized an FBG sensing glove to achieve continuous hand posture reconstruction through joint angles. Kim J. S. et al. [7] enhanced the accuracy and applicability of fiber Bragg grating (FBG) sensors and proposed a real-time tracking algorithm for hand joints. Although strain-based sensors can provide low-latency data for posture, similar applications for larger body parts such as the knee joint and elbow joint are less convenient. Joint axis and position identification has proven effective for joint angle measurement using inertial measurement unit (IMU) sensors [8–10], leading to improved accuracy and flexibility in higher-level applications such as gait analysis [8] and activity recognition [11]. Signals captured by surface electromyographic (sEMG) equipment are more sensitive to variations in muscle actions. Consequently, motion models for primary human limbs, such as arms [12] and legs [13,14], yield less invariant results across individuals. Furthermore, the time-advanced features of sEMG signals in human motion provide opportunities for human motion prediction [15–18].

Compared to wearable devices, sensors such as optical cameras, laser-based cameras, and electromagnetic field sensing devices offer greater flexibility. Gu et al. [19] used a single

RGB camera to conduct gait analysis via 2D joint angular information. More detailed 3D joint features can be achieved using red green blue and depth (RGB-D) cameras, yielding outstanding results [20,21]. Radar-based sensing and analysis have achieved promising outcomes in human posture detection [22–24], but the relatively higher errors in joint location estimation limit their precise application. As convolutional neural networks become increasingly integrated with image processing technologies, predicting and generating redundant data for joint estimation have become more feasible. Mehta et al. [25] presented a novel approach for 3D motion sensing using a single RGB camera. This approach extends to multi-individual analysis due to the wide sensing field of wear-less sensors. Wear-less sensors have achieved excellent results in joint angle estimation and motion analysis, but disadvantages such as occlusions and insufficient resolutions remain challenges that need to be mitigated.

Attempts to use fused data for angle estimation have been made to improve the accuracy of results and alleviate misrecognition caused by sensor limitations. Filter-based algorithms function effectively with the classical combination of IMU and vision systems [26], while Bo et al. [27] aided in rehabilitation using a combination of depth cameras and inertial sensors. With network techniques, different types of sensing fusion methods offer more options for pose estimation [28–30]. To address real-time online processing, synchronization of multiple sensors is a critical issue. The hysteresis in capture timing and varying signal window durations may cause cumulative regression. Data fusion plays a key role in training and processing. Cippitelli et al. [31] achieved relatively precise synchronization of depth cameras and IMUs through auxiliary communication nodes. Processing embedded data remains a challenging issue in computer science, as multiple data sources are handled in both raw and processed forms [32,33]. In human motion recognition applications, rapid information fusion is required without losing detailed semantic-level features.

Achieving accurate human joint angle estimation necessitates addressing individual variations as a key challenge. These variations are caused by differences in physical body structure, movement behaviors, sensor placements, environmental conditions, and other factors. Optimal sensor utilization is an effective approach to improving robustness and adaptability in these complex conditions. Vibrations in sEMG signals introduce noise patterns, which can have significant effects during the processing stage. Wavelet transform techniques on a timescale can mitigate this effect [33], but the loss of detailed information may also decrease the accuracy of joint angle regression and motion pattern recognition. Introducing vision-based sensors and processing techniques enhances data diversity, thereby strengthening stability across different scenarios. The speed of computation is critical in human motion-related applications; tools like encoder–decoder [34,35] and fast convolutional [36,37] operations can provide timely and effective feature extraction.

This paper aims to improve the accuracy and stability of human joint angle estimation for applications in human motion analysis. The contributions of this work can be summarized as follows:

1. Our work presents a novel dual-branch framework for human joint angle estimation that leverages sEMG signals and RGB-D camera data. The multi-dimensional feature extraction block offers a more comprehensive representation of the human joint model across different scenarios.
2. This paper proposes an enhanced sEMG feature extraction method that utilizes multiple channels and scales of sEMG signals through an online adaptive convolutional autoencoder structure, coupled with feature enhancement based on correlation principles.
3. We employ a feature extraction and fusion pipeline that combines sEMG signals and RGB-D data, which has proven effective and efficient in convolutional networks by leveraging multi-scale input and a sensitive data fusion block.

The structure of this paper is as follows. The design of human joint angle estimation is presented in Section 2. The experimental verification of the framework in this paper is described in Section 3. The discussion appears in Section 4 and conclusions are presented in Section 5.

2. Materials and Methods

2.1. Overview

The proposed dual-branch framework is briefly described in Figure 1. The processing system mainly consists of three components: the sEMG signal processing branch, the depth image processing branch, and the feature fusion and angle estimation module.

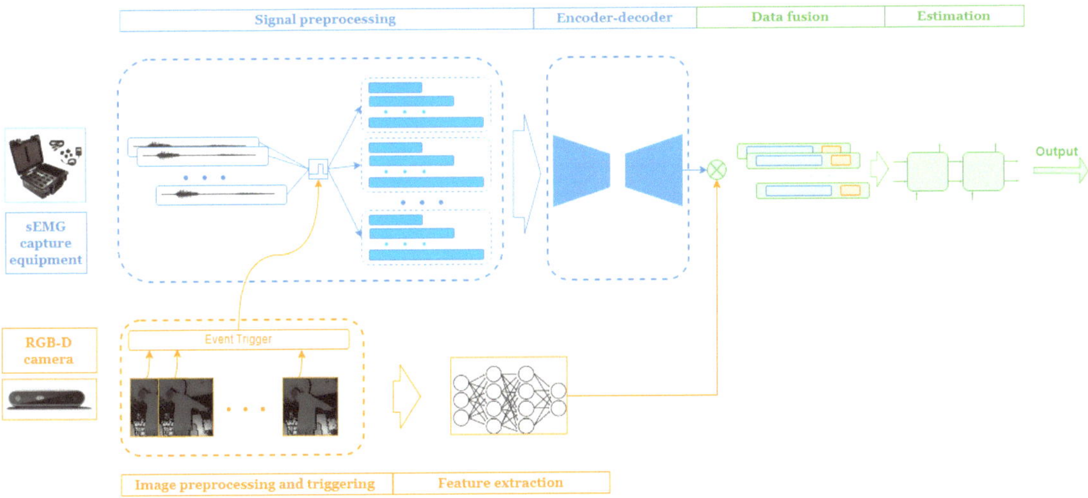

Figure 1. Global framework.

The surface EMG signal processing branch consists of two main operations: data preprocessing and feature extraction. The preprocessing stage employs standard data filtering techniques to mitigate distributional changes caused by the non-stationarity of sEMG signals. A time-triggered windowing method synchronizes sEMG signals with the lower-rate depth image data and provides multi-scale raw data channels to generate more reliable feature vectors for the training process. An encoder–decoder framework with shared weights provides a higher level of feature compression vectors for angle estimation.

Considering the lower data acquisition rate of the RGB-D camera compared to the sEMG equipment, a synchronization trigger signal is created for each frame. Using 3D information provided by the RGB-D camera, a CNN-based feature extraction block simultaneously captures spatial characteristics of 3D human joint positions.

To adapt to different dimensions and scales of the extracted features, the proposed framework employs a feature fusion block through a series of feature concatenations. To expedite and enhance the effectiveness of information acquisition, multiple down-sampling and pooling layers are introduced. The final joint angle regression is performed using an LSTM network.

2.2. sEMG Signal Processing

2.2.1. Preprocessing

The noise in sEMG signals can significantly affect the process of data acquisition and feature extraction. Considering the baseline energy density spectrum of EMG signals, band-pass filters are applied to eliminate noise and artifacts. Due to the potential loss of information when standardizing the data sequence, no additional preprocessing is applied to the raw data.

Typically, sEMG signals are segmented using a sliding window to obtain multiple samples over different time intervals, but in our case, the sliding window may increase the complexity of data asynchrony. Triggered by the acquisition of a depth image, a timestamp is generated to create a series of fixed windows for sEMG signals, each containing a different length of sEMG signal. The varying window lengths provide different timescales of digitized sEMG data, enhancing the diversity and redundancy of the sEMG signal information.

As shown in Figure 2, each frame captured by the RGB-D camera generates a time-triggered event, and this timestamp automatically forms a base window for each sEMG channel. The interval of the base window is fixed to 300 cycles of sEMG capturing periods. To achieve different scales of sEMG signals, processing windows are refactorized to 3 scales, and each signal scale ends precisely when the depth image is captured with a duration of T, $\frac{2}{3}T$, and $\frac{1}{3}T$.

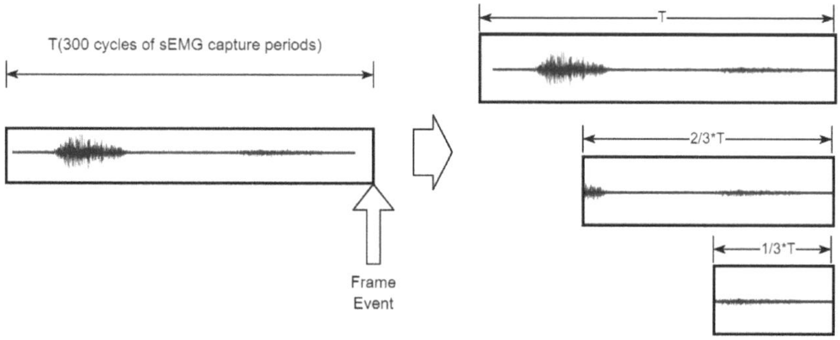

Figure 2. Windowing different scales.

2.2.2. Multi-Scale Feature Extraction Block

The convolutional neural network (CNN) in feature extraction of sEMG signals has been verified as an efficient and robust method. Based on the CNN structure and the capabilities of the autoencoder network, this paper proposes a novel feature extraction block to handle multiple scales of digitized sEMG data, yielding more compressed and reconstructive information. The reconstruction of the sEMG image can better adapt to varying data distributions in raw sEMG data by utilizing reconstruction error. The encoder–decoder network has proven effective in extracting higher-dimensional features. We redesigned the network to be more sensitive to differently scaled windows of sEMG data by using shared weight vectors. By updating shared weights, the extraction block maintains both training efficiency and information sensitivity.

As shown in Figure 3, separate encoders and decoders are applied to each sEMG data channel. The encoder and decoder have symmetric architectures with feature outputs, although their corresponding parameters differ. The input of the encoder network is a serial combination of different channels of sEMG signals, each representing a different source of signal. Depending on the varying timescales of the sEMG data sequences, different network parameters are applied. Using the horizontal and vertical indices of the sEMG data, the discrete sEMG data are combined in the form of an image. The convolutional layers of the encoder consist of single-kernel and multi-kernel components. The single-kernel component compresses only the temporal information, while the multi-kernel component applies 2D convolution to a mixture of temporal and spatial information. A shuffle technique is applied between the single-kernel and multi-kernel components to capture the interrelationships and interactional impacts between related muscle groups.

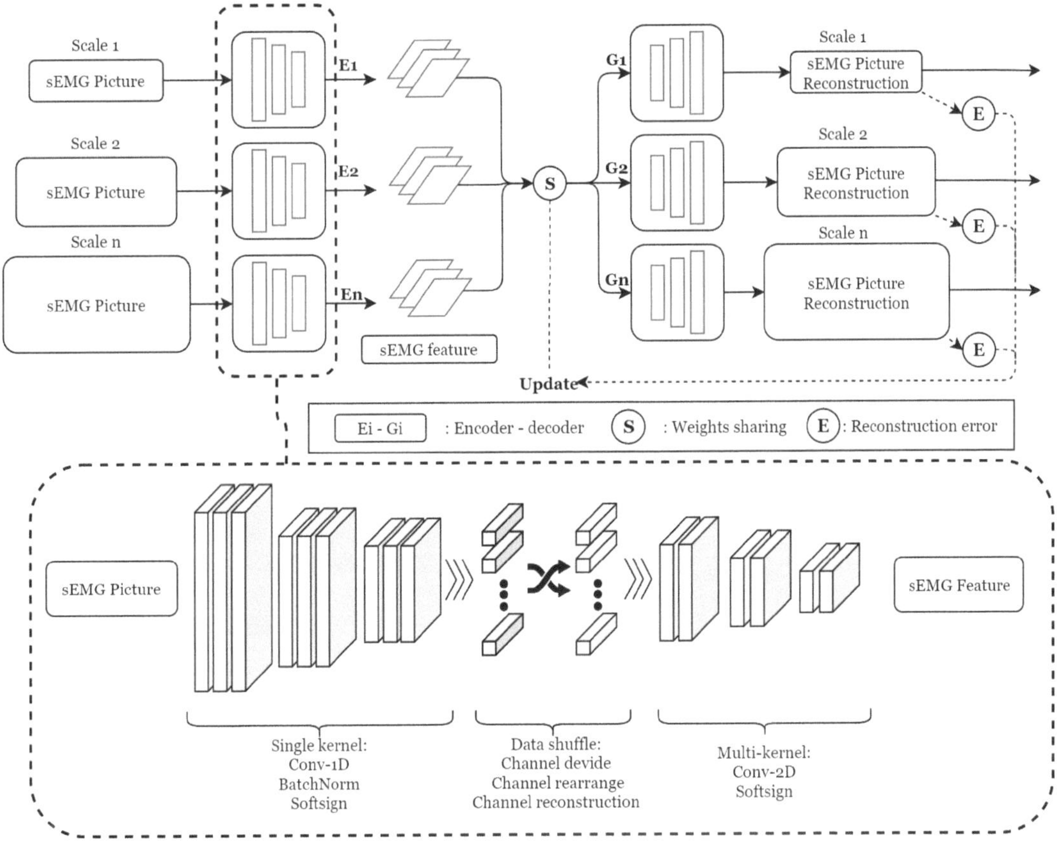

Figure 3. Multi-scale processing and updating.

2.3. Vision-Based Feature Extraction

To estimate human body joint angles, the processing system should construct a skeleton model that represents human body posture using the appearance in the image. In this paper, we use a human pose estimation method based on a convolutional neural network (CNN) using depth images acquired from an RGB-D camera. Although depth images exhibit less detailed information than optical images, the feature extraction block faces a trade-off between spatial reliability and locational precision of image features. By eliminating background noise and lighting effects through 3D imaging techniques, a more accurate silhouette of the human body can be extracted from the depth image. The idea is to employ a series of small, lightweight networks to perform feature extraction for depth images. The initial convolutional block uses small kernel sizes, with batch normalization and ReLU applied after each convolutional layer. Several convolutional layers are combined to form a residual block. We also implemented average pooling layers at the beginning, middle, and end of the feature extraction block.

The aim of the depth image feature extraction block is to form a high-precision heatmap of joint 3D locations. The feature vectors are then delivered to the data fusion block along with multi-scale sEMG features.

2.4. Data Fusion and Angle Estimation

To obtain a reliable and stable outcome from the estimation network, attention must be paid to both global and local information. Motivated by the architecture of the feature pyramid, we designed a fusion network as shown in Figure 4. The input of the data fusion block is a concatenation of sEMG features and depth image features. Extracted image features are copied to each scale of the sEMG feature vector. The combined data are subjected to a series of down-sampling operations. After four iterations of the convolutional layer with a stride of 2, the number of channels doubles each time. The convolutional layer consists of three parts: a convolutional layer with a 3×3 kernel and padding of 1, batch normalization, and Leaky ReLU. To account for the loss of superficial semantic information in the first four layers, three global average pooling layers are applied to the outputs of the 1st, 4th, and 7th convolutional layers to generate the final fused feature map. As multi-layered semantic information is formed by three feature vectors, concatenation is applied to obtain the fused feature map of sEMG signals and depth images.

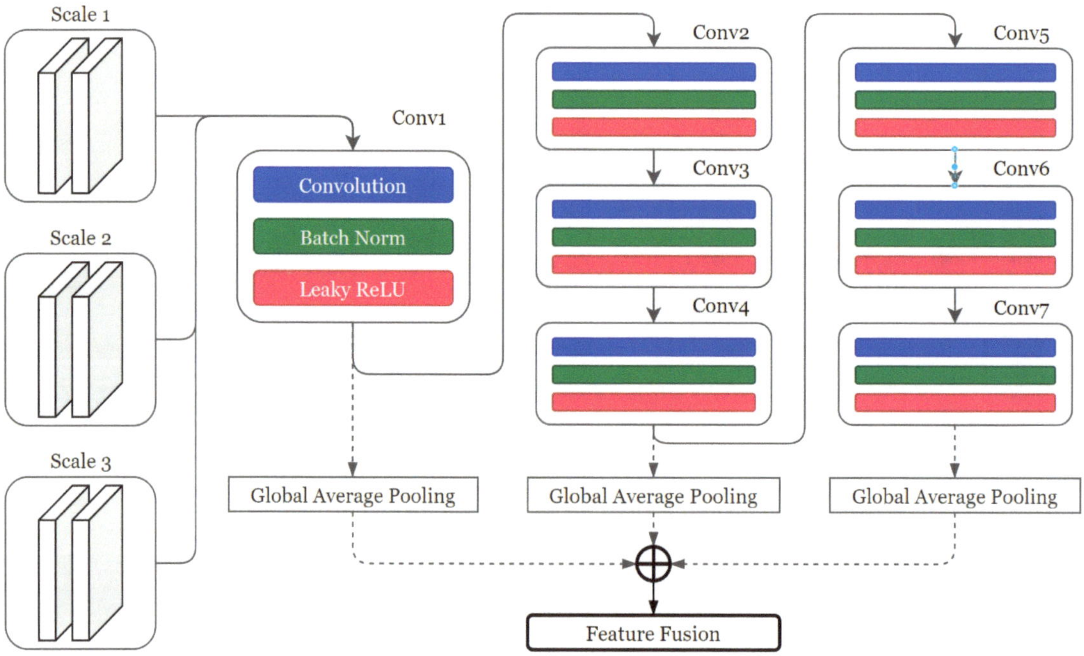

Figure 4. Feature fusion network.

To leverage the strong temporal correlation among joint angles, sEMG signals, and visual signals, we fed the fused features into an LSTM network for joint angle prediction. LSTM is well suited to capturing the dynamic characteristics of time-series data and can more effectively handle multimodal temporal features. This approach improves the model's generalization ability and leads to enhanced performance.

2.5. Experimental Environments

Since no current public dataset contains simultaneously captured data of sEMG signals and depth images of body movements, we created a custom dataset using sEMG signal capture equipment and an RGB-D camera, with motion capture system markings. To evaluate the performance of the proposed framework, we performed a horizontal analysis between our method and state-of-the-art methods on our custom dataset to conduct the following experiments and achieve corresponding results.

1. We compared single-modal sensor capture and processing of sEMG signals and RGB-D camera data. By highlighting the disadvantages of a single-sensor system, we demonstrated the improved accuracy of joint angle estimation using the combination of sEMG signals and RGB-D camera data.

2. Considering subject diversity, we performed cross-comparisons between different subjects and analyzed the errors by accounting for scenario effects and human behavior effects.
3. Noting that ground obstacles may influence visual data capture, we introduced a series of occlusions in our test of knee angle prediction.
4. We compared a series of state-of-the-art methods with ours using our dataset and analyzed the performance from different perspectives.

Under laboratory experimental conditions, the hardware equipment used in this study includes the following:

- Computer: Windows 10 64-bit (DELL, Beijing, China)and Ubuntu 18.04.6 LTS 64-bit double operating system, carrying a CPU of Intel® Core™ i7-8700K with 16 GB RAM and an Nvidia GEFORCE GTX 1660 Ti graphics processing card.
- RGB-D camera: Intel RealSense Depth Camera D435i with specifications of 1920×1080 colored resolution and 1280×720 depth resolution @30 fps, 69° horizontal and 42° vertical field-of-view wide-angle lens.
- sEMG signal capture equipment: Delsys wearable sensors with analog output in range of 11 mV and sampling rate of 1111 Hz. The average noise of overall channel is inferior to 3 uV RMS @ 10–850 Hz.
- Motion capture equipment: VICON motion capture system with 1.3 MP resolution @ 250 Hz. The capture system has 98.1° horizontal and 50.1° vertical field of view.

2.5.1. Participants

Motivated by the hand gesture dataset created by putEMG [38], we created a dataset containing a total of 12 participants aged 24 to 32 years. The average height of our participants is 173.8 ± 6.2 cm, and the average weight is 72.5 ± 8.9 kg. None of the participants had motor dysfunction. Participants were instructed to walk on a treadmill at a constant speed of 4.5 km/h with a zero-degree incline. Each participant completed the procedure eight times, ensuring that the movements of the hip, knee, and ankle joints were consistently and accurately captured by the camera.

This study was conducted in accordance with the Declaration of Helsinki. Written informed consent was obtained from all participants.

2.5.2. Data Acquisition

The main goal of establishing the dataset is to obtain synchronized multi-channel sEMG signals and RGB-D camera images, the dataset capture environment is illustrated in Figure 5. Due to the low latency and high accuracy of the motion capture system, we utilized the VICON motion capture system, as mentioned above, to establish ground truth for human joint angles. Under laboratory conditions, we established the hardware setup as follows. In this study, a total of nine sensors were used to monitor the lower limb muscle groups. These muscles include the rectus femoris (RF), vastus medialis (VM), vastus lateralis (VL), tibialis anterior (TA), soleus (SOL), semitendinosus (SEM), biceps femoris (BF), medial gastrocnemius (MG), and lateral gastrocnemius (LG). The placement of the sensors was approximate and adjusted through palpation and EMG recordings for different subjects. The skin of each subject was cleaned with rubbing alcohol prior to electrode placement to ensure optimal conditions.

A crucial aspect of data acquisition is the synchronization of the depth video stream and the sEMG signal stream. The capture moment of each depth image can be used as a time stamp to generate a synchronization signal. To achieve real-time performance in the data acquisition system, threading techniques were implemented by creating parallel tasks consisting of a system-level hard real-time timer, an image capture task, and a multi-channel sEMG capture task. Maximizing the utilization of system RAM improves performance by reducing latency between sEMG signals and image capture events through pipelined procedures for storing data in RAM and writing data to the hard disk.

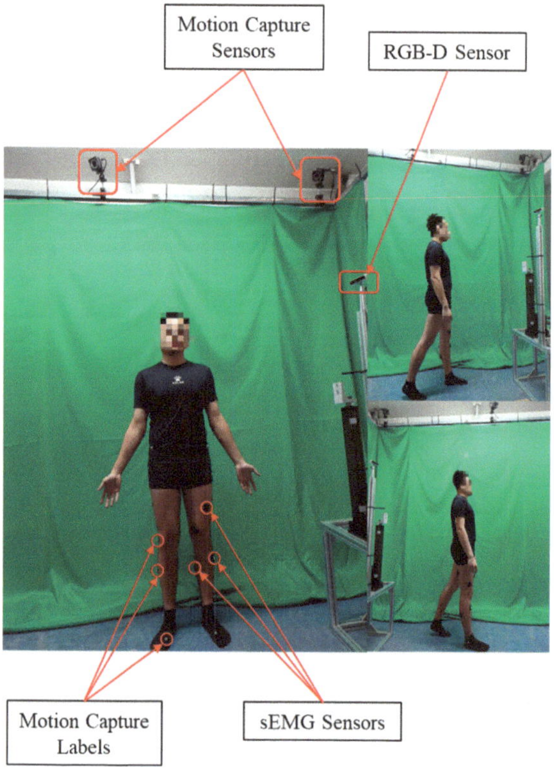

Figure 5. Dataset capture environment.

3. Results

Through a single depth image containing the 3D locational information of each pixel, we can relatively easily reconstruct the human body with 3D joint locations; however, the consistency and stability of the regression output cannot be guaranteed. Vision data processing allowed us to extract 19 human body joint 3D locations. The data were recorded indoors with a single person at two different ranges. Considering the application scenario, participants were recorded from different angles. As shown in Figure 6, the joints could be accurately extracted when facing the depth camera and in the absence of occlusion. Despite the accuracy of joint locations, different angles significantly affect the feature extraction stage, leading to misrecognitions and altering the relationships between neighboring joints. To estimate the prediction error with occlusions, we conducted predictions using blocked inputs in our experimental series.

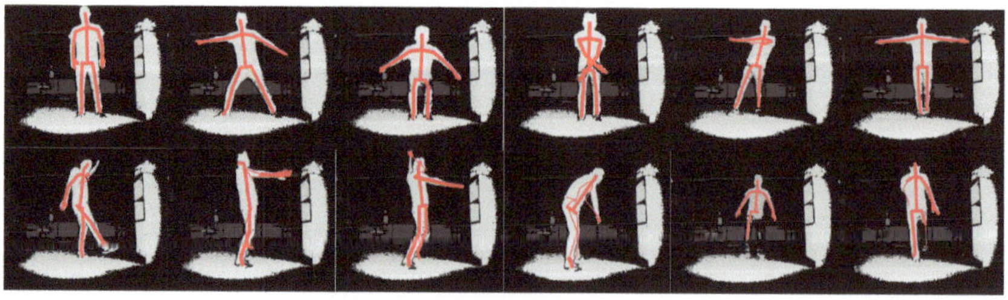

Figure 6. Depth camera-processed data: the upper row shows the correct recognition scenarios in short (1.5 m) range and long range (3 m); the lower row shows the misrecognition scenarios.

The joint angles were calculated from 3D locations. We used a convolutional neural network on depth images [39] to calculate the knee angle during walking. Conversely, we used a BP neural network [40] on sEMG signals to obtain knee angles. Using 10 sets of unrelated data, we obtained three averaged curves of knee angle variation over time.

Figure 7 presents the predictions corresponding to different algorithms, with the gray curve representing the standard knee joint angle results computed by VICON. The average maximum prediction error for the sEMG-based methods is 2.72 ± 1.32°, while for the visual methods, it is 2.67 ± 0.69°. The fusion of visual and EMG-based methods results in a lower prediction error of 1.04 ± 0.52°.

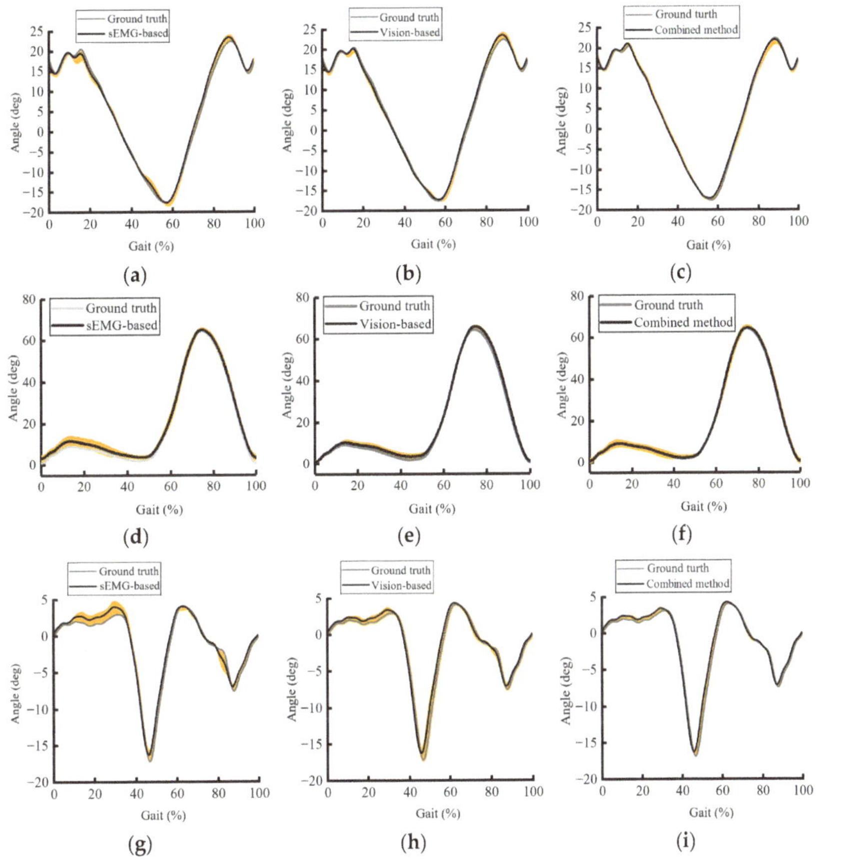

Figure 7. Knee angle, hip angle, and ankle angle prediction while walking on the treadmill based on different methods, the yellow shaded area represents the variance of the estimation results: (**a**) sEMG-based hip angle prediction; (**b**) vision-based hip angle prediction; (**c**) combined hip angle prediction; (**d**) sEMG-based knee angle prediction; (**e**) vision-based knee angle prediction; (**f**) combined knee angle prediction; (**g**) sEMG-based ankle angle prediction; (**h**) vision-based ankle angle prediction; (**i**) combined ankle angle prediction.

Individual variations may cause significant differences in predictions; therefore, we applied different datasets with varying distributions. During the stance phase, the sEMG-based method achieved a prediction error of 4.42 ± 1.71°, the vision-based method achieved a prediction error of 1.71 ± 0.99°, and the combined method achieved a prediction error of 1.17 ± 0.76°. During the swing phase, the sEMG-based method achieved a prediction error of 3.82 ± 1.63°, the vision-based method achieved a prediction error of 2.89 ± 0.93°, and the combined method achieved a prediction error of 1.26 ± 0.82°. To assess the model's adaptability, we also validated the hip and ankle joints, and the results are presented in Figure 8 and Table 1 below.

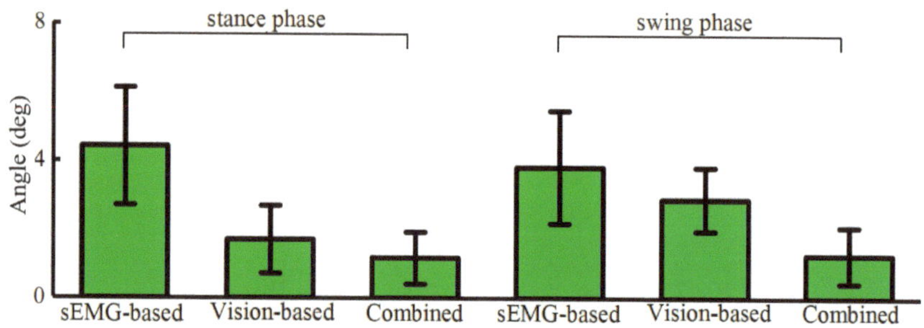

Figure 8. Knee angle predictions via different participants.

Table 1. Estimation errors in degrees for hip angle, knee angle, and ankle angle via different participants.

		sEMG-Based	Vision-Based	Combined
Hip	Stance phase	3.97 ± 1.35	2.15 ± 0.87	1.47 ± 0.63
	Swing phase	3.64 ± 1.11	2.57 ± 0.86	1.89 ± 0.67
Knee	Stance phase	4.42 ± 1.71	1.71 ± 0.99	1.17 ± 0.76
	Swing phase	3.82 ± 1.63	2.89 ± 0.93	1.26 ± 0.82
Ankle	Stance phase	4.21 ± 1.45	2.23 ± 0.89	1.52 ± 0.65
	Swing phase	3.97 ± 1.18	2.98 ± 0.89	2.12 ± 0.55

By combining sEMG signals and vision signals, we achieved promising results with partial blocks. To simulate obstacles on the ground that may cause loss of visual capture of the ankles, we implemented blocks of four different heights.

Figure 9 presents knee angle predictions with four different blocks implemented at the ankle. Without blocks, we achieved a prediction error of 1.04 ± 0.26°, while with increasing obstacle height, the mean errors rose from 1.31 degrees to 3.67 degrees, with variances ranging from 0.79 degrees to 1.25 degrees.

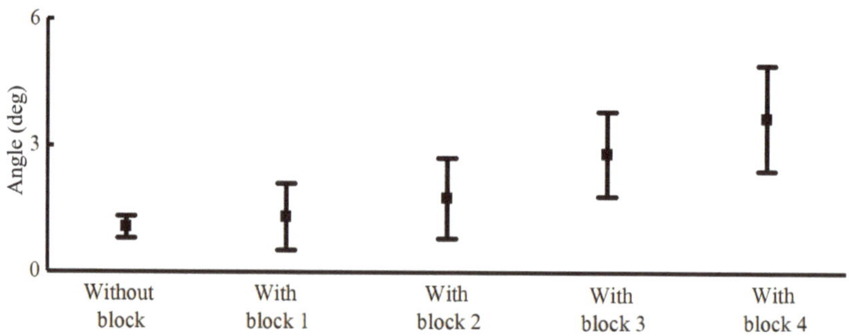

Figure 9. Knee angle prediction with partial blocks.

Using deep belief networks [41] and long short-term memory [42,43], we calculated the root mean square errors for each method. As shown in Table 2, our method achieves significant improvements compared to baseline methods.

Table 2. Comparisons of RMSE between different methods.

	DBN [41]	LSTM [42]	LSTM [43]	Ours
Hip	3.58	1.31–2.48 *	1.33	0.68
Knee	3.96	1.37–1.74 *	2.16	1.68
Ankle	2.45	1.31–1.56 *	1.73	0.73

* The results are calculated in three phases; we note the RMSE range in the table.

4. Discussion

In knee angle prediction, both surface electromyographic (sEMG) and visual methods can achieve a certain level of accuracy. These results indicate an inherent correlation between EMG, visual information, and joint angles. This implicit mapping relationship can be constructed using an end-to-end approach.

Compared to visual methods, EMG-based methods show inferior performance in both mean and variance. The average maximum prediction error for EMG-based methods is $2.72 \pm 1.32°$, while for visual methods, it is $2.67 \pm 0.69°$. On the one hand, this is because EMG signals are inherently related to motion, making their relationship less explicit compared to direct visual observation. On the other hand, EMG signals are characterized by high-amplitude noise, which can decrease algorithm performance due to interference. The maximum error for EMG-based methods occurs during the stance phase of the lower limb, where significant muscle vibrations increase EMG signal noise, leading to higher prediction errors. In contrast, during the relatively stable swing phase, EMG-based methods yield better results. For visual-based predictions, the maximum error occurs during the swing phase due to continuous changes in multiple limbs and background variations, impacting predictions and decreasing performance. Pure visual methods exhibit good stability and strong result repeatability. The fusion of visual and EMG-based methods effectively compensates for the shortcomings of both, leading to improved accuracy with a maximum angle error of $1.04 \pm 0.52°$.

In practical applications, data often have distributions different from our training set. When the distributions of training and test data differ, prediction errors increase. Particularly in EMG-based methods, individual differences significantly impact predictions, as variations in EMG distribution due to individual differences lead to a rapid increase in errors. Conversely, methods based on visual information are relatively less affected by individual differences. Furthermore, relationships between sparse sEMG signals of cooperative muscles can be extracted using multi-scale feature maps.

Similar phenomena were observed in further experiments on hip and ankle angle predictions. However, we observed a slight increase in variation in ankle angle predictions, which may be due to the flexibility of movement patterns. Our model is designed to combine sEMG signal feature maps and vision features to address this issue, but the insufficient resolution of the RGB-D camera introduces drifting of foot posture. The results from ablation experiments show significant improvements using the combined method.

Ground obstacles can cause feature loss in visual capture, leading to prediction errors. To test the robustness of our prediction model, we introduced occlusions in the knee angle prediction scenario. Four categories of occlusions were introduced in this research, and as the level of occlusion increased, the prediction error for EMG-based methods also increased gradually. Notably, in occlusion category 1, which involves relatively low occlusion height, the visual impact was minimal. Overall, prediction error with the highest level of occlusion still shows better performance compared to the EMG-based method.

Compared to joint angle prediction methods based on sEMG signals using DBN and LSTM, our method demonstrated more competitive results in our test cases as shown in Table 2. By incorporating visual depth data, the spatial positioning of joints becomes more precise. Higher-dimensional data sequences, such as depth image data, offer a more robust representation of human skeletal postures. However, 3D imaging techniques typically require more time for sensing and preprocessing, which leads to temporal sparsity in the data sequence. In critical real-time HMI scenarios, the reaction and adjustment phases of the decision–reaction cycle are not as competitive due to the slower capture cadence of 3D imaging data. For this reason, human exoskeleton control benefits from the use of high-frequency sensors like sEMG and IMU, which provide faster and more frequent data. By combining sEMG signals and depth image data to extract distinct physical features, we developed a muscle–skeleton model that establishes a temporal–spatial link, enabling a feature fusion pattern for more accurate movement prediction. By incorporating 3D visual data as spatial markers and calibration references within the proposed framework, the robustness and accuracy of lower limb joint angle predictions are significantly improved.

In future work, we plan to focus on the long-term stability of our proposed human joint angle prediction method, ensuring that it remains robust and reliable over extended periods and across various conditions. This will involve continuous refinement of the algorithm to adapt to real-world challenges, such as signal drift and variations in sensor data. Additionally, we aim to implement our method in exoskeleton systems, where accurate joint angle estimation is critical for providing responsive and precise assistance to users. By integrating our approach into exoskeletons, we hope to enhance their effectiveness in applications such as rehabilitation and mobility support, ultimately contributing to improved user outcomes.

5. Conclusions

In this paper, we propose a novel dual-branch network framework for estimating human joint angles, leveraging the strengths of both sEMG signals and RGB-D camera data. By integrating a convolutional autoencoder for high-level sEMG feature extraction with a vision-based joint regression network, the proposed method addresses the challenges associated with non-stationary sEMG signals and the limitations of vision-based approaches, such as latency and shading. The enhanced feature extraction and fusion pipeline effectively combines multi-scale sEMG and RGB-D data, resulting in a reliable and accurate estimation of human joint angles. The proposed framework not only advances the field of motion analysis and gesture recognition but also demonstrates the potential for more precise motion intention prediction by overcoming the inherent challenges of sEMG signal variability and vision data acquisition.

Author Contributions: Conceptualization, G.D.; methodology, G.D.; software, G.D.; validation, G.D., Z.D., M.S., and H.G.; formal analysis, G.D.; investigation, G.D.; resources, G.D. and Z.D.; data curation, G.D., M.S., and H.G.; writing—original draft preparation, G.D.; writing—review and editing, G.D., Z.D., and F.J.; visualization, G.D.; supervision, F.J.; project administration, F.J.; funding acquisition, F.J. All authors have read and agreed to the published version of the manuscript.

Funding: This work is partly funded by Sichuan Local technological development Program (NO: 24YRGZN0010), the Technique Program of Jiangsu (No.BE2021086) and the National Natural Science Foundation of China (No.62076080).

Institutional Review Board Statement: This study is conducted according to the Declaration of Helsinki, and the protocol of Registering Clinical Trials (ChiECRCT20200319) is approved by the Chinese Ethics Committee.

Informed Consent Statement: Not applicable.

Data Availability Statement: The data are available from the corresponding author on reasonable request.

Conflicts of Interest: The authors declare no conflicts of interest.

Abbreviations

The following abbreviations are used in this manuscript:

HMI	Human–machine interaction
sEMG	Surface electromyography signal
FBG	Fiber Bragg grating
IMU	Inertial measurement unit

CNN	Convolutional neural network
RGB-D	Red green blue and depth
RAM	Random-access memory

References

1. Biagetti, G.; Crippa, P.; Falaschetti, L.; Orcioni, S.; Turchetti, C. Human activity monitoring system based on wearable sEMG and accelerometer wireless sensor nodes. *Biomed. Eng. Online* **2018**, *17*, 132. [CrossRef] [PubMed]
2. Supuk, T.G.; Skelin, A.K.; Cic, M. Design, development, and testing of a low-cost sEMG system and its use in recording muscle activity in human gait. *Sensors* **2014**, *14*, 8235–8258. [CrossRef] [PubMed]
3. Panahandeh, G.; Mohammadiha, N.; Leijon, A.; Händel, P. Continuous hidden Markov model for pedestrian activity classification and gait analysis. *IEEE Trans. Instrum. Meas.* **2013**, *62*, 1073–1083. [CrossRef]
4. Zhang, M.; Sawchuk, A.A. A feature selection-based framework for human activity recognition using wearable multimodal sensors. In Proceedings of the BodyNets Conference, Beijing, China, 7–10 November 2011; pp. 92–98.
5. Zhang, S.; Wei, Z.; Nie, J.; Huang, L.; Wang, S.; Li, Z. A review on human activity recognition using vision-based methods. *J. Healthc. Eng.* **2017**, *2017*, 3090343. [CrossRef]
6. Da Silva, A.F.; Gonçalves, A.F.; Mendes, P.M.; Correia, J.H. FBG sensing glove for monitoring hand posture. *IEEE Sens. J.* **2011**, *11*, 2442–2448. [CrossRef]
7. Kim, J.S.; Kim, B.K.; Jang, M.; Kang, K.; Kim, D.E.; Ju, B.K.; Kim, J. Wearable hand module and real-time tracking algorithms for measuring finger joint angles of different hand sizes with high accuracy using FBG strain sensors. *Sensors* **2020**, *20*, 1921. [CrossRef]
8. Seel, T.; Raisch, J.; Schauer, T. IMU-based joint angle measurement for gait analysis. *Sensors* **2014**, *14*, 6891–6909. [CrossRef]
9. Chen, L.; Yan, X.; Hu, D. A deep learning control strategy for IMU-based joint angle estimation in a hip power-assisted swimming exoskeleton. *IEEE Sens. J.* **2023**, *23*, 15058–15070. [CrossRef]
10. Zhu, K.; Li, J.; Li, D.; Fan, B.; Shull, P.B. IMU shoulder angle estimation: Effects of sensor-to-segment misalignment and sensor orientation error. *IEEE Trans. Neural Syst. Rehabil. Eng.* **2023**, *31*, 4481–4491. [CrossRef]
11. Zhao, L.; Sukthankar, G.; Sukthankar, R. Robust active learning using crowdsourced annotations for activity recognition. In Proceedings of the Workshops at the Twenty-Fifth AAAI Conference on Artificial Intelligence, San Francisco, CA, USA, 7–11 August 2011.
12. Zhuojun, X.; Yantao, T.; Yang, L. sEMG pattern recognition of muscle force of the upper arm for intelligent bionic limb control. *J. Bionic Eng.* **2015**, *12*, 316–323. [CrossRef]
13. Tu, J.; Dai, Z.; Zhao, X.; Huang, Z. Lower limb motion recognition based on surface electromyography. *Biomed. Signal Process. Control* **2023**, *81*, 104443. [CrossRef]
14. Wang, J.; Dai, Y.; Si, X. Analysis and recognition of human lower limb motions based on electromyography (EMG) signals. *Electronics* **2021**, *10*, 2473. [CrossRef]
15. Aung, Y.M.; Al-Jumaily, A. sEMG based ANN for shoulder angle prediction. *Procedia Eng.* **2012**, *41*, 1009–1015. [CrossRef]
16. Ding, Z.; Yang, C.; Wang, Z.; Yin, X.; Jiang, F. Online adaptive prediction of human motion intention based on sEMG. *Sensors* **2021**, *21*, 2882. [CrossRef] [PubMed]
17. Tanaka, T.; Nambu, I.; Maruyama, Y.; Wada, Y. Sliding-window normalization to improve the performance of machine-learning models for real-time motion prediction using electromyography. *Sensors* **2022**, *22*, 5005. [CrossRef]
18. Fleischer, C.; Reinicke, C.; Hommel, G. Predicting the intended motion with EMG signals for an exoskeleton orthosis controller. In Proceedings of the 2005 IEEE/RSJ International Conference on Intelligent Robots and Systems, Edmonton, AB, Canada, 2–6 August 2005; IEEE: Piscataway, NJ, USA, 2005; pp. 2029–2034.
19. Gu, X.; Liu, J.; Zheng, Y. Markerless gait analysis based on a single RGB camera. In Proceedings of the 2018 IEEE 15th International Conference on Wearable and Implantable Body Sensor Networks (BSN), Las Vegas, NV, USA, 4–7 March 2018; IEEE: Piscataway, NJ, USA, 2018.
20. İnce, Ö.F.; Yüksel, M. Human activity recognition with analysis of angles between skeletal joints using an RGB-depth sensor. *ETRI J.* **2020**, *42*, 78–89. [CrossRef]
21. Wang, L.; Zhao, X.; Yu, T.; Wang, S.; Liu, Y. NormalGAN: Learning detailed 3D human from a single RGB-D image. In *Computer Vision—ECCV 2020, Proceedings of the 16th European Conference, Glasgow, UK, 23–28 August 2020*; Springer International Publishing: Cham, Switzerland, 2020; pp. 430–446.
22. Sengupta, A.; Maji, P.; Sinha, R. mm-Pose: Real-time human skeletal posture estimation using mmWave radars and CNNs. *IEEE Sens. J.* **2020**, *20*, 10032–10044. [CrossRef]
23. Wu, Q.; Xu, G.; Zhang, S.; Li, Y.; Wei, F. Human 3D pose estimation in a lying position by RGB-D images for medical diagnosis and rehabilitation. In Proceedings of the 2020 42nd Annual International Conference of the IEEE Engineering in Medicine & Biology Society (EMBC), Montreal, QC, Canada, 20–24 July 2020; IEEE: Piscataway, NJ, USA, 2020; pp. 5802–5805.
24. Zheng, J.; Shi, X.; Gorban, A.; Mao, J.; Song, Y.; Qi, C.R.; Liu, T.; Chari, V.; Cornman, A.; Zhou, Y.; et al. Multi-modal 3D human pose estimation with 2D weak supervision in autonomous driving. In Proceedings of the 2022 IEEE/CVF Conference on Computer Vision and Pattern Recognition (CVPR), New Orleans, LA, USA, 18–24 June 2022; pp. 4478–4487.

25. Mehta, D.; Singh, S.; Wray, S. XNect: Real-time multi-person 3D motion capture with a single RGB camera. *ACM Trans. Graph. (TOG)* **2020**, *39*, 82:1–82:17. [CrossRef]
26. Bo, A.P.L.; Hayashibe, M.; Poignet, P. Joint angle estimation in rehabilitation with inertial sensors and its integration with Kinect. In Proceedings of the 2011 Annual International Conference of the IEEE Engineering in Medicine and Biology Society (EMBC), Boston, MA, USA, 30 August–3 September 2011; IEEE: Piscataway, NJ, USA, 2011; pp. 3479–3483.
27. Huang, F.; Zeng, A.; Liu, M.; Lai, Q.; Xu, Q. DeepFuse: An IMU-aware network for real-time 3D human pose estimation from multi-view images. In Proceedings of the IEEE/CVF Winter Conference on Applications of Computer Vision (WACV), Snowmass, CO, USA, 1–5 March 2020; pp. 429–438.
28. Liu, L.; Yang, J.; Lin, Y.; Zhang, P.; Zhang, L. 3D human pose estimation with a single image and inertial measurement unit (IMU) sequence. *Pattern Recognit.* **2024**, *149*, 110175. [CrossRef]
29. Chen, M.; Tan, G. 3D human pose estimation based on wearable IMUs and multiple camera views. *Electronics* **2024**, *13*, 2926. [CrossRef]
30. Cippitelli, E.; Gasparrini, S.; Gambi, E.; Spinsante, S.; Wåhslény, J.; Orhany, I.; Lindhy, T. Time synchronization and data fusion for RGB-depth cameras and inertial sensors in AAL applications. In Proceedings of the 2015 IEEE International Conference on Communication Workshop (ICCW), London, UK, 8–12 June 2015; IEEE: Piscataway, NJ, USA, 2015; pp. 265–270.
31. Varshney, P.K. Multisensor data fusion. *Electron. Commun. Eng. J.* **1997**, *9*, 245–253. [CrossRef]
32. Liu, Z.; Meyendorf, N.; Mrad, N. The role of data fusion in predictive maintenance using digital twin. *AIP Conf. Proc.* **2018**, *1949*, 020023.
33. Phinyomark, A.; Phukpattaranont, P.; Limsakul, C. The usefulness of wavelet transform to reduce noise in the SEMG signal. In *EMG Methods for Evaluating Muscle and Nerve Function*; BoD—Books on Demand: Norderstedt, Germany, 2012; pp. 107–132.
34. Said, A.B.; Mohamed, A.; Elfouly, T.; Harras, K.; Wang, Z.J. Multimodal deep learning approach for joint EEG-EMG data compression and classification. In Proceedings of the 2017 IEEE Wireless Communications and Networking Conference (WCNC), San Francisco, CA, USA, 19–22 March 2017; IEEE: Piscataway, NJ, USA, 2017; pp. 1–6.
35. Dinashi, K.; Ameri, A.; Akhaee, M.A.; Englehart, K.; Scheme, E. Compression of EMG signals using deep convolutional autoencoders. *IEEE J. Biomed. Health Inform.* **2022**, *26*, 2888–2897. [CrossRef] [PubMed]
36. Lin, C.; Cui, Z.; Chen, C.; Liu, Y.; Jiang, N. A fast gradient convolution kernel compensation method for surface electromyogram decomposition. *J. Electromyogr. Kinesiol.* **2024**, *76*, 102869. [CrossRef]
37. Wen, Y.; Avrillon, S.; Hernandez-Pavon, J.C.; Kim, S.J.; Hug, F.; Pons, J.L. A convolutional neural network to identify motor units from high-density surface electromyography signals in real time. *J. Neural Eng.* **2021**, *18*, 056003. [CrossRef]
38. Kaczmarek, P.; Mańkowski, T.; Tomczyński, J. putEMG—A surface electromyography hand gesture recognition dataset. *Sensors* **2019**, *19*, 3548. [CrossRef]
39. Ye, M.; Wang, X.; Yang, R.; Ren, L.; Pollefeys, M. Accurate 3D pose estimation from a single depth image. In Proceedings of the 2011 International Conference on Computer Vision, Barcelona, Spain, 6–13 November 2011; IEEE: Piscataway, NJ, USA, 2011; pp. 731–738.
40. Zhang, F.; Li, P.; Hou, Z.G.; Lu, Z.; Chen, Y.; Li, Q.; Tan, M. sEMG-based continuous estimation of joint angles of human legs by using BP neural network. *Neurocomputing* **2012**, *78*, 139–148. [CrossRef]
41. Chen, J.; Zhang, X.; Cheng, Y.; Xi, N. Surface EMG based continuous estimation of human lower limb joint angles by using deep belief networks. *Biomed. Signal Process. Control* **2018**, *40*, 335–342. [CrossRef]
42. Mundt, M.; Thomsen, W.; Witter, T.; Koeppe, A.; David, S.; Bamer, F.; Potthast, W.; Markert, B. Prediction of lower limb joint angles and moments during gait using artificial neural networks. *Med. Biol. Eng. Comput.* **2020**, *58*, 211–225. [CrossRef]
43. Song, Q.; Ma, X.; Liu, Y. Continuous online prediction of lower limb joint angles based on sEMG signals by deep learning approach. *Comput. Biol. Med.* **2023**, *163*, 107124. [CrossRef]

Disclaimer/Publisher's Note: The statements, opinions and data contained in all publications are solely those of the individual author(s) and contributor(s) and not of MDPI and/or the editor(s). MDPI and/or the editor(s) disclaim responsibility for any injury to people or property resulting from any ideas, methods, instructions or products referred to in the content.

bioengineering

MDPI

Article

Electromyography-Triggered Constraint-Induced Movement Cycling Therapy for Enhancing Motor Function in Chronic Stroke Patients: A Randomized Controlled Trial

Jaemyoung Park [1], Kyeongjin Lee [2], Junghyun Kim [3,*] and Changho Song [1,*]

1 Department of Physical Therapy, Sahmyook University, Seoul 01795, Republic of Korea
2 Department of Physical Therapy, College of Health Science, Kyungdong University, Wonju 24764, Republic of Korea
3 Biomedical Research Institute, Seoul National University Hospital, Seoul 03080, Republic of Korea
* Correspondence: kiking0@naver.com (J.K.); chsong@syu.ac.kr (C.S.); Tel.: +82-2-3399-163 (C.S.)

Abstract: This single-blind randomized controlled trial investigated the effectiveness of surface electromyography (sEMG)-triggered constraint-induced movement cycling therapy (CIMCT) in improving balance, lower extremity strength, and activities of daily living in patients with chronic stroke. The participants included patients with chronic stroke-induced hemiplegia who had been diagnosed for more than 6 months, with a minimum score of 24 points on the Mini-Mental State Examination and above level 3 on the Brunnstrom stages. The trial lasted 4 weeks and participants were divided into a CIMCT group and a general cycling training (GCT) group. The CIMCT group (n = 20) used an sEMG-triggered constrained-induced movement therapy device, whereas the GCT group (n = 19) used a standard stationary bicycle. The primary outcome measures showed a significant increase in muscle strength on the affected side in the CIMCT group, as assessed by a manual muscle tester ($p < 0.05$), with a large effect size (d = 1.86), while no meaningful improvement was observed in the GCT group. Both groups demonstrated significant improvements in dynamic balance, as measured by the Timed Up and Go (TUG) test ($p < 0.05$), with the CIMCT group showing superior results compared to the GCT group, reflected by a large effect size (d = 0.96). Additionally, both groups showed significant improvements in balance as assessed by the Berg Balance Scale (BBS) and the Functional Reach Test (FRT). The CIMCT group exhibited more pronounced improvements than the GCT group, with large effect sizes of 0.83 for the BBS and 1.25 for the FRT. The secondary outcome measures revealed significant improvements in activities of daily living in both groups, as assessed by the modified Barthel index (MBI), with the CIMCT group achieving a substantial improvement ($p < 0.05$), accompanied by a large effect size (d = 0.87). This study concludes that sEMG-triggered CIMCT effectively improved muscle strength, postural balance, and activities of daily living in patients with chronic stroke.

Keywords: bioengineering; postural balance; stroke; electromyography; muscle strength

Citation: Park, J.; Lee, K.; Kim, J.; Song, C. Electromyography-Triggered Constraint-Induced Movement Cycling Therapy for Enhancing Motor Function in Chronic Stroke Patients: A Randomized Controlled Trial. *Bioengineering* **2024**, *11*, 860. https://doi.org/10.3390/bioengineering11090860

Academic Editor: Carlo Albino Frigo

Received: 25 July 2024
Revised: 20 August 2024
Accepted: 21 August 2024
Published: 23 August 2024

Copyright: © 2024 by the authors. Licensee MDPI, Basel, Switzerland. This article is an open access article distributed under the terms and conditions of the Creative Commons Attribution (CC BY) license (https://creativecommons.org/licenses/by/4.0/).

1. Introduction

Stroke remains one of the leading causes of death and disability worldwide, with over 12 million incident cases and 6.55 million deaths reported globally in 2019. Despite advancements in treatment, the burden of stroke has increased significantly, particularly in low- and middle-income countries [1]. Stroke imposes a significant financial burden on healthcare systems globally. The cost of stroke care varies considerably across countries, with the United States reporting the highest average per-patient annual cost at USD 59,900, followed by Sweden at USD 52,725 and Spain at USD 41,950. The lifetime cost per patient can also be substantial. For instance, in Australia, the lifetime cost per patient has been estimated at USD 232,100 across all identified definitions of stroke [2]. Stroke often results in long-term disability and profoundly affects the lives of affected individuals [3]. Approximately 80–90% of stroke patients experience impaired mobility [4]. Although 75% of these patients achieve independent gait within 3 months post-stroke, around 25% continue to require assistance for gait [5]. The primary challenge lies in the impediment to smooth gait due to limb paralysis on the affected side. Thus, enhancing gait capability is a critical goal in stroke rehabilitation [6].

Patients with stroke typically exhibit an asymmetric gait, often adopting compensatory movement patterns to counteract this [7]. Post-stroke gait is characterized by a shortened stance phase and an elongated swing phase on the affected side. This gait abnormality is primarily due to lower limb muscle weakness and impaired balance control. Spatiotemporal gait asymmetry is closely associated with muscle weakness [8,9]. Moreover, balance impairment exacerbates these gait issues, leading to decreased gait speed, reduced stride length, and further spatiotemporal asymmetry [10,11].

Recent technological advancements, such as robot-assisted therapy [12,13], virtual reality rehabilitation [14,15], and wearable assistive devices [16–18], have spurred various studies aimed at enhancing gait rehabilitation in patients with stroke. These cutting-edge technologies represent novel approaches that facilitate treatment and redefine the role of therapists. When integrated into clinically effective treatments, the aforementioned technologies can yield superior outcomes. One method garnering significant attention in the upper-extremity rehabilitation of patients with stroke is constraint-induced movement therapy (CIMT) [19]. This therapy involves restricting the use of the unaffected limb, thereby necessitating the use of the affected limb [19]. Originally aimed at activating the affected limb through enforced disuse of the unaffected side, CIMT has evolved to focus on the active use of the affected limb through repetitive and detailed movement training [20]. Although primarily applied to the upper extremities, recent research has begun to explore the application of the modality to the lower extremities. These include studies that incorporate lower extremity orthoses, shoe insoles [21,22], and stationary bicycles [23]. Surface electromyography (sEMG) is a technique used to measure and analyze the electrical signals of muscle activity through electrodes placed on the skin's surface [24,25]. By detecting and analyzing these electrical signals, which result from physiological changes in the muscle fibers, sEMG can provide valuable insights into muscle function and its role in movement [25]. A key application of EMG is in the use of electromyographic feedback to encourage specific movements; for instance, a previous study utilized EMG feedback to promote pedaling on the affected side in stroke patients, leading to improved muscle activation and gait on the paralyzed side [23]. However, there is still a significant gap in understanding how targeted interventions like sEMG-triggered pedaling specifically enhance muscle activation and promote functional recovery in the affected limbs of stroke patients.

In this study, we carefully selected and examined variables that are known to influence gait asymmetry, including muscle strength, static balance, and dynamic balance. Our objective was to explore how the use of sEMG feedback to encourage pedaling on the affected side could lead to improvements in these key areas, ultimately resulting in enhanced gait symmetry. Furthermore, we investigated whether this intervention could also improve daily living activities, offering a more comprehensive understanding of its potential benefits in stroke rehabilitation. We aimed to determine if this approach could indeed provide such advantages.

2. Materials and Methods

2.1. The Study Design

The study was structured as a single-blind randomized controlled trial spanning 4 weeks of intervention. The protocol, registered under the ClinicalTrials.gov identifier NCT06367140, was approved by the Ethics Committee of Kyungdong University (1041455-201706-HR-012-01) in accordance with the ethical standards outlined in the Declaration of Helsinki. All procedures adhered to the guidelines and regulations. Written informed consent was obtained from all participants after they were briefed about the research prior to the commencement.

2.2. Participants

This study included patients with stroke admitted to K Hospital in Seoul. The inclusion criteria for the participants were as follows: patients with chronic stroke-induced hemiplegia diagnosed with stroke more than 6 months prior to the start of the trial, achieving a minimum score of 24 points on the Mini-Mental State Examination [26], and demonstrating motor recovery at or above level 3 according to the Brunnstrom stages [27].

The exclusion criteria encompassed individuals with neurological damage unrelated to their stroke, orthopedic issues such as fractures or peripheral nervous system damage in the lower limbs, visual or auditory impairments, those who had experienced more than one stroke, and those with less than 80% participation rate in the study.

Prior to the commencement of the study, all participants were informed of its purpose and procedures, and only those who voluntarily signed the consent form were included in the study. The sample size for this study was determined based on the results of a pilot study. Before conducting the main experiment, a pilot study was performed with 5 participants in each group, using the same methods as the main study. The effect size was calculated to be 0.87 based on the Timed Up and Go (TUG) test, a dynamic balance variable. G-power 3.1.9 software was employed to compare the two groups' differences. The significance level was set at 0.05, and the power of the test was 0.8 [28], resulting in a required sample size of 18 participants per group. To account for a potential dropout rate of 10%, the final sample size was increased to 20 participants per group.

2.3. Experimental Procedures

We recruited 58 stroke patients receiving treatment at the K Rehabilitation Hospital in Seoul, and 40 patients who met the inclusion and exclusion criteria were selected. Before participating in the experiment, a screening test was conducted to determine whether the patients met the researchers' eligibility criteria. A total of 18 people who did not meet the inclusion criteria were excluded. After the pre-test, 20 participants were randomly allocated to the CIMCT group and 20 to the general cycling training (GCT) groups. To minimize selection bias, a computer program (random allocation software, version 1.0) was used [29], and random allocation was performed such that sex, side of paralysis, cause of stroke, disease duration, and cognitive ability were homogeneous. This task was carried out by researchers not involved in the training and assessment. The CIMCT group performed sEMG-triggered CIMCT for 50 min per session, five times a week for 4 weeks, and the GCT group performed general cycling training for 50 min per session, five times a week for 4 weeks. The training was conducted in alignment with the inpatient treatment schedule, which included two therapy sessions per day. Accordingly, training was administered once in the morning or afternoon, depending on each participant's schedule. Although the exact timing varied between participants, the procedures were executed with consistent time intervals between sessions to ensure uniformity across the study. As a pre-test, the participants' general characteristics, postural balance, lower limb strength, and activities of daily living were examined.

Those who could not participate in the program because of changes in their medical condition during the study were excluded from the final study. In the GCT group, 19 people participated in the study as one patient was discharged.

Ultimately, 20 participants in the CIMCT group and 19 in the GCT group took the same post-test as the pre-test and statistical analyses were performed on those who took the post-test without excluding anyone. Figure 1 presents the study process.

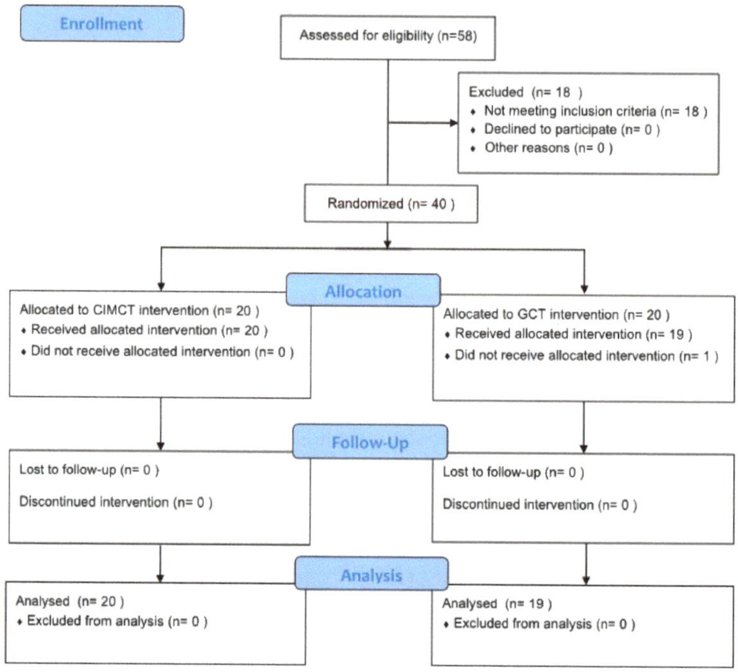

Figure 1. Flow chart.

2.4. Intervention

The sEMG-triggered CIMCT device used in this study represents a novel integration of bioengineering principles into rehabilitation. The device, comprising a myoelectric sensing unit, stationary bicycle, and control unit, uses four-channel sEMG sensors to capture muscle activity signals. The signals were processed into packets using an Arduino Pro Mini 328 (Arduino Pro Mini 328, SparkFun Electronics, Niwot, CO, USA), and then transmitted to the control unit via Bluetooth (Figure 2). These signals are transmitted via Bluetooth to a control unit that visualizes the data, providing patients with real-time feedback and enabling precise adjustments to therapy based on muscle activation. To minimize interference during the sEMG-triggered CIMCT study, we conducted the data collection in a space free from other electronic equipment. Additionally, to reduce interference caused by the sEMG cables touching each other, mesh coverings were applied to the cables.

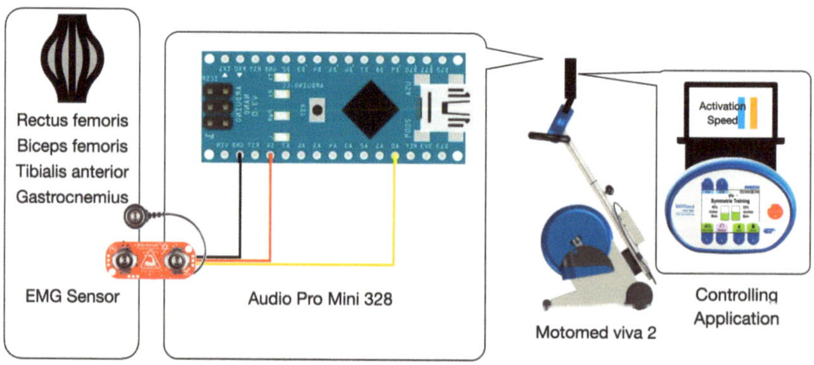

Figure 2. sEMG-triggered CIMCT system.

The muscle activation signals were collected using four sEMG sensors (MyoWare Muscle Sensor, SparkFun Electronics, Niwot, CO, USA), which have a high-input impedance of 110 GΩ. The sEMG electrode patches (Ag/AgCl surface electrode H2223H, 3M, Maplewood, MN, USA) were placed on the rectus femoris, biceps femoris, tibialis anterior, and gastrocnemius muscles. Each electrode was positioned according to the SENIAM project recommendations as follows:

Rectus Femoris: The electrodes were placed at 50% of the distance along the line from the anterior superior iliac spine to the superior part of the patella.

Biceps Femoris: The electrodes were placed at 50% of the distance along the line between the ischial tuberosity and the lateral epicondyle of the tibia.

Tibialis Anterior: The electrodes were placed at 1/3 of the distance along the line from the tip of the fibula to the tip of the medial malleolus.

Gastrocnemius: The electrodes were placed on the most prominent bulge of the muscle.

Although skin impedance was not directly measured, we minimized it by cleaning the skin with alcohol wipes before electrode application. These precautions helped ensure reliable sEMG signals and minimized any potential impact on the normalization process.

After applying the electrodes, participants pedaled a stationary bicycle (Motomed Viva 2, RECK-Technik GmbH & Co, Betzenweiler, Germany) at a comfortable speed for 30 s while sEMG data were collected. The average muscle activity during this period was calculated and used for normalization. A baseline threshold was defined as the average value of the normalized muscle's reference voluntary contraction (RVC) and was used to control the increase in bicycle speed.

When participants reached their threshold, the bicycle's speed increased by one step, and if they did not reach the threshold within 5 s, the speed decreased by one step. The amount of speed change per step followed the settings defined on the stationary bicycle. At the start of the exercise, the initial threshold was set at 150%, but this could be adjusted based on the therapist's judgment.

For sEMG-triggered CIMCT, the participants started with a 10-min warm-up at a comfortable speed, followed by therapist-guided acceleration and deceleration. The session comprised three 10-min periods interspersed with 1-min rests and concluded with a 5-min cool-down [30]. The therapist continuously monitored the participants for discomfort or dizziness, allowing breaks or cessation at any time.

For GCT, a regular stationary bicycle without an sEMG-triggered device was used, and the training time was the same as that for the CIMT cycling training. Moreover, the training was performed under the supervision of a guardian or caregiver and consisted of a 10-min warm-up, 30 min of main exercise, and 5 min of cool-down exercise. In the general pedaling training group, patients who experienced dizziness or complained of difficulty during the training were asked to take a break and train again. The participants were instructed to stop at any time according to their will. The evaluators were blinded to the intervention details of the participants.

2.5. Outcome Measurements

This study measured static balance using the GB300 postural measurement system (Metitur Ltd., Jyvaskyla, Finland). This system comprises a movable triangular platform and a scale that indicates the position of the feet. The GB300 system measures sway in the standing posture, recording medial–lateral (M–L) and anterior–posterior (A–P) movements. Sway is quantified by the velocity of movement and the area covered per second, referred to as the velocity moment. This system is widely used for assessing balance in various populations, including athletes, elderly individuals, stroke patients, and those with hemiplegia. The sampling frequency was set at 50 Hz. Participants were instructed to stand with their eyes open, facing forward, for 30 s while positioned on the device. This procedure was repeated three times. Subsequently, participants performed the same posture with their eyes closed and facing forward for an additional three measurements of 30 s each.

The Timed Up and Go (TUG) test was used to assess balance ability in this study. In the test, the participant sits in a chair with armrests, rises from the chair at the same time as the word "start" is spoken, walks at their most stable and comfortable speed to a point 3 m in front of them, and then returns to and sits down in the chair at which point the time is recorded. It has a high intra-rater reliability (r = 0.99) and inter-rater reliability (r = 0.98) [31]. The raters performed three measurements using a stopwatch and recorded the average value. The TUG test primarily assesses dynamic balance and mobility by measuring the time taken to stand up, walk a short distance, turn, and sit down.

The Berg Balance Scale (BBS) is used to assess functional balance in a wide range of participants, including older individuals at a high risk of falling and patients with acute and chronic diseases. Moreover, BBS is a functional balance test method that considers three aspects of functional balance: postural maintenance, postural control by manual exercise, and response to external perturbations. The test consists of 14 items that are common in daily activities, including getting up from a sitting position, standing without holding on, sitting without leaning back, sitting from a standing position, moving between chairs, standing with eyes closed, standing with feet together without holding on, reaching forward with arms from a standing position, picking up objects from the floor, turning left and right, spinning in place, alternating feet on a step stool, standing with one foot in front of the other, and standing on one leg; all 14 items are scored on a 5-point scale ranging from 0 to 4, with high scores indicating improved performance. The perfect score on this scale is 56, with a score below 45 indicating a risk of falling. This measure has high reliability and internal validity for assessing balance ability, with intra- and inter-rater reliabilities of r = 0.99 and r = 0.98, respectively [32]. The Berg Balance Scale (BBS) is a comprehensive measure that assesses both static and dynamic balance through a series of tasks, including sitting, standing, and transferring between positions.

The Functional Reach Test (FRT) assesses the limits of physical stability and measures dynamic balance and flexibility while the participant performs a functional task. The FRT measures the maximum distance a participant can extend their arm forward from a standing position while maintaining fixed support. The distance was measured in centimeters using a Laser Rangefinder (DLE50, BOSCH, Gerlingen, Germany). The results represent the averages of three consecutive measurements. The reliability of this test was 0.89 [33]. The Functional Reach Test (FRT) evaluates the limit of stability, specifically assessing how far an individual can reach forward without losing balance.

A manual muscle tester (Model 01163, Lafayette, LA, USA, 2003) was used to evaluate lower extremity muscle strength in this study. The knee extensors, knee flexors, dorsiflexors, and plantar flexors, which are primarily responsible for the pedaling motion, were assessed. Moreover, both sides were evaluated. Patients were seated on a fixed chair with a backrest. The knee extensors were measured with a pressure plate placed on the anterior aspect of the ankle, and the patient was instructed to extend the knee. The knee flexors were assessed with a pressure plate placed on the heel of the patient in the prone position. Ankle motion was gauged in a long sitting position with the knees extended. Dorsiflexors were measured with a pressure plate positioned on the distal aspect of the dorsum of the foot and plantar flexors were assessed on the distal aspect of the sole. The average of two measurements after each exercise was recorded.

The modified Barthel index (MBI) developed by was used to measure the perfor-263 mance of daily living behaviors. The modified Barthel index (MBI) developed by was used to measure the perfor-263 mance of daily living behaviors. The MBI consists of 10 items: self-care, bathing, feeding, climbing stairs, dressing, bowel control, bladder control, gait, and transferring. The scoring system ranged from 5 to 15, with a score of 100 if all items could be performed completely independently. The inter-rater reliability was 0.93–0.98, and the Cronbach's alpha value was 0.84 [34].

2.6. Statistical Analysis

Statistical analyses were performed using SPSS version 24 (IBM Corp., Armonk, NY, USA). The data were tested for normality using the Shapiro–Wilk test, and the mean and standard deviation were calculated. After the normality test, when the normal distribution assumption was satisfied, the sociodemographic characteristics of the participants were analyzed in real numbers, percentages, means, and standard deviations. The independent sample *t*-test and chi-square test were used to test the homogeneity between groups. The changes in the dependent variables before and after the intervention were analyzed using paired *t*-tests. For instances where significant differences were observed, repeated measures ANOVA was employed to compare the effects between the two groups and analyze the interaction effects between group and time. In addition, to determine the effect before and after training, the effect size of the training was investigated by dividing the difference between before and after training by the average deviation. The effect size was classified based on the minimal detectable change (MDC) of the measurements to complement related statistical methods. The MDC was adjusted for samples from two different measurements by multiplying the standard error of measurement (SEM) by 1.96, corresponding to a 95% confidence interval, and further multiplying by the square root of 2. Furthermore, SEM was estimated as the pooled standard deviation of the pre- and post-training assessments multiplied by the square root of $(1 - r)$, where r is the ICC. All statistical significance levels (α) of the data were set at 0.05.

3. Results

3.1. General Characteristics

The CIMCT and GCT groups were homogeneous in all general characteristics including age, height, weight, stroke type, and paralyzed side (Table 1).

Table 1. General characteristics of the subjects.

							$n =$	39
	CIMCT $n = 20$			GCT $n = 19$			χ^2/t	p
Age (year)	63.00	±	6.96	65.00	±	9.81	0.737	0.466
Height (cm)	162.15	±	7.74	162.58	±	8.69	0.163	0.871
Weight (kg)	58.89	±	8.32	60.17	±	8.69	0.470	0.641
BMI (point)	22.35	±	2.32	22.66	±	1.79	0.475	0.638
Duration of stroke (month)	13.85	±	6.09	17.00	±	5.83	1.648	0.108
MMSE-K	25.80	±	1.32	25.58	±	1.02	0.583	0.563
MBI	53.41	±	10.16	56.47	±	10.11	0.944	0.351
Gender (male/female)	9	/	11	10	/	9	0.634	0.227
Paretic side (right/left)	9	/	11	13	/	6	0.140	2.174
Stroke type (Infarction/hemorrhage)	13	/	7	12	/	7	0.905	0.014

Note. BMI = body mass index; MMSE-K = Mini-Mental State Examination-Korean; MBI = modified Barthel index. Values are expressed as mean ± standard deviation.

3.2. Muscle Strength

The strength of the knee flexor, extensor, dorsiflexor, and plantar flexor muscles on the affected side significantly increased after training in the CIMCT group ($p < 0.05$), with Cohen's d values indicating a large effect size ($d > 0.80$). However, no significant differences were observed in the GCT group. The muscles on the healthy side demonstrated no significant differences after training in either group (Tables 2 and 3).

Table 2. The changes in muscle strength on the affected side.

					n = 39	
		CIMCT *n* = 20	**GCT** *n* = 19	**t/F**	***p***	**MDC MDC%**	**Effect Size**
RF-A (*n*)	Pre	82.52 ± 35.98	98.50 ± 36.31	1.380	0.176		
	Post	92.10 ± 33.76	97.29 ± 34.29				
	Pre–Post	9.58 ± 7.49	−1.21 ± 3.19	33.643 †	0	4.64	1.86
	t	5.724 *	1.658			48.43	
	p	0	0.116				
BF-A (*n*)	Pre	50.03 ± 19.15	53.14 ± 21.41	0.635	3.449		
	Post	55.86 ± 21.37	53.40 ± 22.95				
	Pre–Post	5.83 ± 6.44	0.27 ± 2.88	11.898 †	0.001	3.99	1.1
	t	4.047 *	0.404			68.50	
	p	0.001	0.691				
TA-A (*n*)	Pre	59.17 ± 25.27	54.78 ± 21.11	0.588	0.56		
	Post	64.84 ± 22.22	54.80 ± 20.98				
	Pre–Post	5.67 ± 5.06	0.02 ± 3.54	16.180 †	0	3.13	1.29
	t	5.015 *	0.023			55.27	
	p	0	0.982				
Gastrocnemius-A (*n*)	Pre	92.84 ± 23.27	93.36 ± 21.26	0.073	0.942		
	Post	98.52 ± 23.92	93.87 ± 19.02				
	Pre–Post	5.67 ± 3.59	0.51 ± 4.25	16.891 †	0	2.23	1.32
	t	7.065 *	0.519			39.23	
	p	0	0.61				

Note. RF = rectus femoris muscle; BF = biceps femoris muscle; TA = tibialis anterior muscle; A = affected side; and MDC = minimal detectable change. Values are expressed as mean ± standard deviation (SD). * $p < 0.05$. † indicates that the *p*-value from the repeated measures ANOVA is less than 0.05.

Table 3. The changes in muscle strength on the non-affected side.

				n = 39	
		CIMCT *n* = 20	**GCT** *n* = 19	**t**	***p***
RF-NA (*n*)	Pre	159.42 ± 42.33	156.58 ± 34.68	0.229	0.82
	Post	161.68 ± 44.49	156.20 ± 35.24		
	Pre–Post	2.26 ± 5.37	−0.38 ± 5.50	1.521	0.137
	t	1.884	0.305		
	p	0.075	0.764		
BF-NA (*n*)	Pre	102.07 ± 32.07	102.83 ± 28.73	0.078	0.938
	Post	103.32 ± 32.12	103.07 ± 26.44		
	Pre–Post	1.25 ± 3.90	0.24 ± 5.56	0.655	0.516
	t	1.431	0.192		
	p	0.169	0.85		
TA-NA (*n*)	Pre	151.10 ± 38.12	150.88 ± 33.62	0.019	0.985
	Post	153.41 ± 32.25	151.89 ± 32.26		
	Pre–Post	2.31 ± 7.79	1.01 ± 4.77	0.625	0.536
	t	1.327	0.927		
	p	0.2	0.367		
Gastrocnemius-NA (*n*)	Pre	177.34 ± 30.15	173.17 ± 29.01	0.439	0.663
	Post	176.82 ± 28.50	174.98 ± 27.76		
	Pre–Post	−0.52 ± 7.29	1.81 ± 8.20	0.938	0.354
	t	0.321	0.961		
	p	0.752	0.35		

Note. RF = rectus femoris muscle; BF = biceps femoris muscle; TA = tibialis anterior muscle; and NA = non-affected side. Values are expressed as mean ± standard deviation (SD).

3.3. Static Balance

Changes in static balance ability, medial and lateral sway velocity, anteroposterior sway velocity, and moment velocity exhibited no significant differences after the intervention in either group, regardless of whether the eyes were open or closed (Table 4).

Table 4. The changes in static balance.

			CIMCT n = 20	GCT n = 19	n = 39 t	n = 39 p
EC	M-L speed (mm/s)	Pre	5.63 ± 2.19	5.41 ± 2.13	0.315	0.755
		Post	5.25 ± 3.14	5.66 ± 2.64		
		Pre–Post	0.38 ± 3.11	−0.25 ± 4.16	0.539	0.593
		t	0.546	0.264		
		p	0.591	0.795		
	A-P speed (mm/s)	Pre	7.49 ± 2.87	7.23 ± 3.67	0.249	0.805
		Post	6.88 ± 2.57	7.49 ± 2.80		
		Pre–Post	0.61 ± 3.12	−0.26 ± 3.83	0.775	0.443
		t	0.867	0.296		
		p	0.397	0.771		
	Velocity moment (mm^2/s)	Pre	6.50 ± 2.38	6.12 ± 2.55	0.489	0.628
		Post	6.29 ± 2.06	5.95 ± 1.69		
		Pre–Post	0.22 ± 2.61	0.17 ± 2.19	0.061	0.952
		t	0.369	0.336		
		p	0.716	0.741		
EO	M-L speed (mm/s)	Pre	4.16 ± 1.41	4.51 ± 1.20	0.845	0.404
		Post	4.32 ± 1.09	4.32 ± 1.12		
		Pre–Post	−0.16 ± 1.19	0.19 ± 1.58	0.781	0.440
		t	0.610	0.515		
		p	0.549	0.613		
	A-P speed (mm/s)	Pre	5.75 ± 1.78	5.68 ± 0.91	0.160	0.874
		Post	5.68 ± 0.91	5.39 ± 1.53		
		Pre–Post	0.30 ± 1.80	0.29 ± 1.69	0.017	0.987
		t	0.745	0.753		
		p	0.465	0.462		
	Velocity moment (mm^2/s)	Pre	4.89 ± 2.16	4.77 ± 2.14	0.174	0.863
		Post	4.34 ± 1.11	4.32 ± 1.12		
		Pre–Post	−1.27 ± 0.90	−0.66 ± 1.11	1.906	0.064
		t	1.131	0.660		
		p	0.272	0.518		

Note. M-L: Medial-Lateral; A-P: Anterior-Posterior; EO: Eyes Open; EC: Eyes Closed.

3.4. Dynamic Balance

For the TUG, BBS, and FRT, the CIMCT group demonstrated a significant reduction after training ($p < 0.05$), and the GCT group also displayed a significant decrease after training ($p < 0.05$). Additionally, Cohen's d values for BBS and FRT indicated a large effect size (d > 0.80). However, when comparing the differences between the groups by training method, the CIMCT group exhibited a significantly greater improvement compared to the GCT group ($p < 0.05$) (Table 5).

Table 5. The changes in dynamic balance.

		CIMCT n = 20	GCT n = 19	n = 39 t/F	n = 39 p	n = 39 MDC MDC%	n = 39 Effect Size
TUG (sec)	Pre	36.70 ± 4.89	35.50 ± 5.25	0.741	0.463		
	Post	31.82 ± 4.56	33.70 ± 4.51				
	Pre–Post	−4.88 ± 3.06	−1.79 ± 3.38	8.926 †	0.005	1.89 38.86	−0.96
	t	7.132 *	2.322 *				
	p	0	0.033				
BBS (point)	Pre	30.56 ± 7.88	29.00 ± 9.51	0.561	0.578		
	Post	35.05 ± 7.27	30.97 ± 9.12				
	Pre–Post	4.49 ± 2.94	1.97 ± 3.10	6.767 †	0.013	1.82 40.57	0.83
	t	6.832 *	2.793 *				
	p	0	0.012				
FRT (cm)	Pre	14.45 ± 4.33	13.17 ± 4.08	0.945	0.351		
	Post	17.83 ± 4.30	13.79 ± 4.59				
	Pre–Post	3.38 ± 2.85	0.62 ± 1.20	15.299 †	0	1.76 52.15	1.25
	t	5.315 *	2.254 *				
	p	0	0.038				

Note. TUG = Timed Up and Go; BBS = Berg Balance Scale; FRT = Functional Reach Test. Values are expressed as mean ± standard deviation (SD). * $p < 0.05$. † indicates that the p-value from the repeated measures ANOVA is less than 0.05.

3.5. Activities of Daily Living

For the MBI, both the CIMCT and GCT groups demonstrated a significant increase after training ($p < 0.05$), with Cohen's d values indicating a large effect size ($d > 0.80$). However, when comparing the differences between the groups according to the training method, the CIMCT group displayed a significantly greater improvement than the GCT group ($p < 0.05$) (Table 6).

Table 6. The changes in activities of daily living.

		CIMCT $n = 20$			GCT $n = 19$			t/F	p	MDC MDC% ($n = 39$)	Effect Size
MBI (score)	Pre	53.41	±	10.16	56.47	±	10.11	0.944	0.351		
	Post	62.40	±	14.77	58.34	±	9.69				
	Pre–Post	8.99	±	6.46	1.87	±	9.69	19.831 †	0	4.01	0.87
	t		6.224 *			3.058 *				44.53	
	p		0			0.007					

Note. MBI = modified Barthel index. Values are expressed as mean ± standard deviation (SD). * $p < 0.05$. † indicates that the p-value from the repeated measures ANOVA is less than 0.05.

4. Discussion

The present study investigates the effects of sEMG-triggered CIMCT, which focuses on constraining the movement of the unaffected side while inducing voluntary contractions on the affected side. The study also assesses muscle strength, postural balance, and activities of daily living in patients with chronic stroke.

Prior to the current study, we conducted a pilot study to investigate the muscle activation patterns of the affected side during the pedaling exercise. Based on the results of the pilot study, we developed an sEMG-triggered pedaling device that enabled pedaling exercises with muscle signals from the affected side. Continual monitoring of the four muscles and adjustment of their speed according to their activation provided feedback on muscle actions to patients, which motivated their participation. This method actively engaged the affected side, leading to improvements in gait function, which can be considered an advanced form of CIMT. In previous studies, the use of sEMG-triggered CIMCT was shown to improve spatiotemporal gait parameters and gait symmetry [23]. Building on these findings, the current study investigated changes in variables influencing spatiotemporal gait parameters and symmetry by evaluating muscle strength and balance as primary outcome measures. This approach aimed to address the significant gap in understanding how sEMG-triggered CIMCT contributes to enhancing spatiotemporal gait parameters and symmetry. Additionally, the study examined whether improvements in these variables could ultimately positively impact activities of daily living.

In the present study, training that induced voluntary contractions on the affected side enhanced muscle strength and function. The main muscles involved in pedaling are the knee extensors, knee flexors, dorsiflexors, and plantar flexors [35]. Before training, a significant difference in muscle strength between the affected and unaffected sides was noted, with approximately twice the strength in the upper leg and three times the strength in the lower leg. The effect sizes were notably large for knee extensors (1.86), knee flexors (0.98), dorsiflexors (1.29), and plantar flexors (1.16). These results satisfied the MDC criterion for muscle strength changes on the affected side, indicating statistical significance. The CIMCT group demonstrated significant improvements on the affected side and an enhancement in symmetry ratios after the intervention.

The proposition is that sEMG-triggered pedaling training, built upon electromyographic biofeedback and constraint-induced movement therapy (CIMT) principles for the lower extremities, induces improvements in muscle strength by engaging the motor cortex. This is achieved through the focused use of the affected muscles, promoting activity in the primary sensorimotor cortex (SMC) by leveraging neuroplastic mechanisms. The training combines volitional activation of the impaired side with continuous somatosensory feedback, which facilitates the reorganization of neural circuits. This process is more effective when the pyramidal tract (PT) is intact, leading to sustained long-term improvements. Even if the PT is compromised, consistent biofeedback training can still promote recovery through compensatory activation, though the effects may be less stable [36].

Improvement in postural balance is a secondary effect of pedaling training in patients with stroke. Lee [37] reported significant improvements in stability limits after 6 weeks of pedaling training in patients with stroke. However, Kim et al. [38] demonstrated no significant difference in BBS scores in patients with chronic stroke after pedaling training; although, significant differences were observed in TUG and gait speed.

In this study, static and dynamic postural balance was also assessed. Static balance and the ability to maintain posture demonstrated significant improvements in both groups; however, no significant differences were observed between them. The intervention in this study was more effective for dynamic balance than for static balance, as it induced reciprocal movements of both feet by inducing muscle contractions in the affected lower extremities. Dynamic balance was assessed using TUG, BBS, and FRT, and both groups exhibited significant improvements, with the sEMG-triggered training group displaying greater improvements than the general pedaling training group. The TUG time decreased by 12.93%. As the TUG test involves gait, sitting, standing, and turning, the test is likely to be related to improvements in gait. The BBS, which assesses the risk of falls, improved by 14.37%; however, the scores remained below 45, suggesting that the risk of falls was not eliminated. The FRT, which evaluates the limits of stability, demonstrated an improvement of 25.73%. Shen et al. [39] conducted a systematic review and meta-analysis on the effects of pedaling training on mobility and quality of life in patients with stroke and reported significant improvements in BBS, and [40] and Tang A, et al. [41] also reported increases in BBS in groups that used electronic gait training.

The improvements in postural balance observed in this study are thought to have been influenced by the improvement in muscle strength of the affected lower extremity. This, in turn, impacts the symmetry of the lower extremities. This aligns with previous research findings, suggesting that improvements in muscle tension, endurance, joint flexibility, and symmetry of the lower extremities enhance balance capabilities [42,43].

Another secondary effect observed in this study was the change in activities of daily living. Both groups demonstrated significant improvements in MBI scores, with the sEMG-triggered training group exhibiting a significant improvement (16.53%). The results are consistent with those of previous studies, indicating that task-oriented training [44] and voluntary movements of the affected lower extremities enhance cerebral reorganization and cortical restoration [45]. The aforementioned finding also aligns with research reporting a high correlation between balance and activities of daily living [46].

This study demonstrated that sEMG-triggered CIMCT effectively enhances lower limb muscle strength, postural balance, and activities of daily living in chronic stroke patients. The biofeedback device played a crucial role in engaging patients actively and improving motor function. Compared to previous studies, our research provides additional evidence supporting the practicality and effectiveness of sEMG-triggered devices. However, this study has certain limitations. Although the sample size was calculated, considering the complex pathologies of patients with stroke, representing all stroke populations is challenging. When cycling is used as an intervention, quantitatively comparing whether all patients are trained at the same intensity and duration within a given time is difficult. Additionally, the sustainability of the exercise effects could not be ascertained because follow-up observations were not conducted after the intervention.

5. Conclusions

This study confirms the effectiveness of sEMG-triggered CIMCT in activating voluntary muscle contractions in the lower extremities during chronic stroke rehabilitation. The findings suggest that sEMG-triggered CIMCT is a viable method for enhancing muscle strength, postural balance, functional activities, and activities of daily living in clinical settings. This approach offers a promising option for rehabilitation practitioners aiming to improve overall functional outcomes in stroke recovery. Future research should focus on integrating bioengineering technologies into rehabilitation programs and exploring their long-term benefits across different patient groups.

Author Contributions: C.S. and J.P. conceptualized the study; J.K. developed the methodology and provided the software; C.S., J.K. and K.L. validated the study; J.P. conducted the formal analysis and investigation and prepared the original draft; C.S. reviewed and edited the manuscript; K.L. was responsible for the visualization. All authors have read and agreed to the published version of the manuscript.

Funding: This research received no external funding.

Institutional Review Board Statement: This study was conducted in accordance with the Declaration of Helsinki and approved by the Institutional Review Board (or Ethics Committee) of Kyungdong University (1041455-201706-HR-012-01) and Clinical Trials (ClinicalTrials.gov identifier NCT06367140) for studies involving humans.

Informed Consent Statement: Informed consent was obtained from all subjects involved in the study.

Data Availability Statement: The original contributions presented in the study are included in the article; further inquiries can be directed to the corresponding author/s. https://osf.io/zmey3/, accessed on 21 August 2024 (DOI: 10.17605/OSF.IO/ZMEY3).

Conflicts of Interest: The authors declare no conflicts of interest.

References

1. Collaborators, G.S. Global, regional, and national burden of stroke and its risk factors, 1990–2019: A systematic analysis for the Global Burden of Disease Study 2019. *Lancet Neurol.* **2021**, *20*, 795.
2. Strilciuc, S.; Grad, D.A.; Radu, C.; Chira, D.; Stan, A.; Ungureanu, M.; Gheorghe, A.; Muresanu, F.-D. The economic burden of stroke: A systematic review of cost of illness studies. *J. Med. Life* **2021**, *14*, 606. [CrossRef] [PubMed]
3. Virani, S.S.; Alonso, A.; Benjamin, E.J.; Bittencourt, M.S.; Callaway, C.W.; Carson, A.P.; Chamberlain, A.M.; Chang, A.R.; Cheng, S.; Delling, F.N.; et al. Heart Disease and Stroke Statistics—2020 Update: A Report From the American Heart Association. *Circulation* **2020**, *141*, e139–e596. [CrossRef] [PubMed]
4. Duncan, P.W.; Zorowitz, R.; Bates, B.; Choi, J.Y.; Glasberg, J.J.; Graham, G.D.; Katz, R.C.; Lamberty, K.; Reker, D. Management of adult stroke rehabilitation care: A clinical practice guideline. *Stroke* **2005**, *36*, e100–e143. [CrossRef]
5. Kennedy, C.; Bernhardt, J.; Churilov, L.; Collier, J.M.; Ellery, F.; Rethnam, V.; Carvalho, L.B.; Donnan, G.A.; Hayward, K.S. Factors associated with time to independent walking recovery post-stroke. *J. Neurol. Neurosurg. Psychiatry* **2021**, *92*, 702–708. [CrossRef]
6. Kwakkel, G.; Stinear, C.; Essers, B.; Munoz-Novoa, M.; Branscheidt, M.; Cabanas-Valdés, R.; Lakičević, S.; Lampropoulou, S.; Luft, A.R.; Marque, P. Motor rehabilitation after stroke: European Stroke Organisation (ESO) consensus-based definition and guiding framework. *Eur. Stroke J.* **2023**, *8*, 880–894. [CrossRef]
7. Olney, S.J.; Richards, C. Hemiparetic gait following stroke. Part I: Characteristics. *Gait Posture* **1996**, *4*, 136–148. [CrossRef]
8. Hsu, A.-L.; Tang, P.-F.; Jan, M.-H. Analysis of impairments influencing gait velocity and asymmetry of hemiplegic patients after mild to moderate stroke. *Arch. Phys. Med. Rehabil.* **2003**, *84*, 1185–1193. [CrossRef]
9. Gonzalez-Suarez, C.B.; Ogerio, C.G.V.; dela Cruz, A.R.; Roxas, E.A.; Fidel, B.C.; Fernandez, M.R.L.; Cruz, C. Motor impairment and its influence in gait velocity and asymmetry in community ambulating hemiplegic individuals. *Arch. Rehabil. Res. Clin. Transl.* **2021**, *3*, 100093. [CrossRef]
10. Chen, G.; Patten, C.; Kothari, D.H.; Zajac, F.E. Gait differences between individuals with post-stroke hemiparesis and non-disabled controls at matched speeds. *Gait Posture* **2005**, *22*, 51–56. [CrossRef]
11. Middleton, A.; Braun, C.H.; Lewek, M.D.; Fritz, S.L. Balance impairment limits ability to increase walking speed in individuals with chronic stroke. *Disabil. Rehabil.* **2017**, *39*, 497–502. [CrossRef] [PubMed]
12. Xie, L.; Yoon, B.H.; Park, C.; You, J.S.H. Optimal Intervention Timing for Robotic-Assisted Gait Training in Hemiplegic Stroke. *Brain Sci.* **2022**, *12*, 1058. [CrossRef] [PubMed]
13. Choi, W. Effects of Robot-Assisted Gait Training with Body Weight Support on Gait and Balance in Stroke Patients. *Int. J. Environ. Res. Public Health* **2022**, *19*, 5814. [CrossRef] [PubMed]
14. Sana, V.; Ghous, M.; Kashif, M.; Albalwi, A.; Muneer, R.; Zia, M. Effects of vestibular rehabilitation therapy versus virtual reality on balance, dizziness, and gait in patients with subacute stroke: A randomized controlled trial. *Medicine* **2023**, *102*, e33203. [CrossRef]
15. Akinci, M.; Burak, M.; Yasar, E.; Kilic, R.T. The effects of Robot-assisted gait training and virtual reality on balance and gait in stroke survivors: A randomized controlled trial. *Gait Posture* **2023**, *103*, 215–222. [CrossRef]

16. Xie, R.; Zhang, Y.; Jin, H.; Yang, F.; Feng, Y.; Pan, Y. Effects of soft robotic exoskeleton for gait training on clinical and biomechanical gait outcomes in patients with sub-acute stroke: A randomized controlled pilot study. *Front. Neurol.* **2023**, *14*, 1296102. [CrossRef]
17. Gozzi, A. *ReAble: A Wearable Robotic Locomotor Trainer for Post-Stroke Individuals*; University of Illinois at Chicago: Chicago, IL, USA, 2022.
18. Kim, J.; Jung, S.; Song, C. The effects of auditory feedback gait training using smart insole on stroke patients. *Brain Sci.* **2021**, *11*, 1377. [CrossRef]
19. Kaneko, T.; Maeda, M.; Yokoyama, H.; Kai, S.; Obuchi, K.; Takase, S.; Horimoto, T.; Shimada, R.; Moriya, T.; Ohmae, H.; et al. Therapeutic effect of adjuvant therapy added to constraint-induced movement therapy in patients with subacute to chronic stroke: A systematic review and meta-analysis. *Disabil. Rehabil.* **2023**, 1–15. [CrossRef]
20. Hamzei, F.; Krüger, H.; Peters, M.; Ketels, G.; Blessin, J.; Ringer, T.; Liepert, J.; Taub, E. Shaping-Induced Movement Therapy for lower extremity (SIMT)—A pilot study. *Neurol. Und Rehabil.* **2012**, *18*, 236–241.
21. Numata, K.; Murayama, T.; Takasugi, J.; Oga, M. Effect of modified constraint-induced movement therapy on lower extremity hemiplegia due to a higher-motor area lesion. *Brain Inj.* **2008**, *22*, 898–904. [CrossRef]
22. Rodriguez, G.M.; Aruin, A.S. The effect of shoe wedges and lifts on symmetry of stance and weight bearing in hemiparetic individuals. *Arch. Phys. Med. Rehabil.* **2002**, *83*, 478–482. [CrossRef] [PubMed]
23. Lee, K. EMG-Triggered Pedaling Training on Muscle Activation, Gait, and Motor Function for Stroke Patients. *Brain Sci.* **2022**, *12*, 76. [CrossRef] [PubMed]
24. Reaz, M.B.I.; Hussain, M.S.; Mohd-Yasin, F. Techniques of EMG signal analysis: Detection, processing, classification and applications. *Biol. Proced. Online* **2006**, *8*, 11–35. [CrossRef]
25. Konrad, P. *A Practical Introduction to Kinesiological Electromyography*; Noraxon, Inc.: Scottsdale, AZ, USA, 2005.
26. Guzek, Z.; Dziubek, W.; Stefańska, M.; Kowalska, J. Evaluation of the functional state and mobility of patients after stroke depending on their cognitive state. *Sci. Rep.* **2023**, *14*, 1515.
27. Brunnstrom, S. Motor testing procedures in hemiplegia: Based on sequential recovery stages. *Phys. Ther.* **1966**, *46*, 357–375. [CrossRef]
28. Zabor, E.C.; Kaizer, A.M.; Hobbs, B.P. Randomized controlled trials. *Chest* **2020**, *158*, S79–S87. [CrossRef]
29. Saghaei, M. Random allocation software for parallel group randomized trials. *BMC Med. Res. Methodol.* **2004**, *4*, 26. [CrossRef] [PubMed]
30. Billinger, S.A.; Arena, R.; Bernhardt, J.; Eng, J.J.; Franklin, B.A.; Johnson, C.M.; MacKay-Lyons, M.; Macko, R.F.; Mead, G.E.; Roth, E.J. Physical activity and exercise recommendations for stroke survivors: A statement for healthcare professionals from the American Heart Association/American Stroke Association. *Stroke* **2014**, *45*, 2532–2553. [CrossRef]
31. Ng, S.S.; Hui-Chan, C.W. The timed up & go test: Its reliability and association with lower-limb impairments and locomotor capacities in people with chronic stroke. *Arch. Phys. Med. Rehabil.* **2005**, *86*, 1641–1647. [CrossRef]
32. Berg, K.O.; Wood-Dauphinee, S.L.; Williams, J.I.; Maki, B. Measuring balance in the elderly: Validation of an instrument. *Can. J. Public Health* **1992**, *83* (Suppl. S2), S7–S11.
33. Persad, C.C.; Cook, S.; Giordani, B. Assessing falls in the elderly: Should we use simple screening tests or a comprehensive fall risk evaluation? *Eur. J. Phys. Rehabil. Med.* **2010**, *46*, 249–259.
34. Hong, I.; Lim, Y.; Han, H.; Hay, C.C.; Woo, H.S. Application of the Korean Version of the Modified Barthel Index: Development of a keyform for use in Clinical Practice. *Hong Kong J. Occup. Ther.* **2017**, *29*, 39–46. [CrossRef] [PubMed]
35. Raasch, C.C.; Zajac, F.E. Locomotor strategy for pedaling: Muscle groups and biomechanical functions. *J. Neurophysiol.* **1999**, *82*, 515–525. [CrossRef] [PubMed]
36. Rijntjes, M.; Hamzei, F.; Glauche, V.; Saur, D.; Weiller, C. Activation changes in sensorimotor cortex during improvement due to CIMT in chronic stroke. *Restor. Neurol. Neurosci.* **2011**, *29*, 299–310. [CrossRef] [PubMed]
37. Lee, K. Speed-Interactive Pedaling Training Using Smartphone Virtual Reality Application for Stroke Patients: Single-Blinded, Randomized Clinical Trial. *Brain Sci.* **2019**, *9*, 295. [CrossRef]
38. Kim, S.J.; Cho, H.Y.; Kim, Y.L.; Lee, S.M. Effects of stationary cycling exercise on the balance and gait abilities of chronic stroke patients. *J. Phys. Ther. Sci.* **2015**, *27*, 3529–3531. [CrossRef]
39. Shen, C.; Liu, F.; Yao, L.; Li, Z.; Qiu, L.; Fang, S. Effects of MOTOmed movement therapy on the mobility and activities of daily living of stroke patients with hemiplegia: A systematic review and meta-analysis. *Clin. Rehabil.* **2018**, *32*, 1569–1580. [CrossRef]
40. Au, C.Y.; Mehra, P.; Leung, K.W.; Tong, R.K. Effects of Electromyographically-driven Neuromuscular Stimulatio Cycling System on the Lower-Limb of Stroke Survivors. *IEEE Int. Conf. Rehabil. Robot.* **2019**, *2019*, 300–304. [CrossRef]
41. Tang, A.; Sibley, K.M.; Thomas, S.G.; Bayley, M.T.; Richardson, D.; McIlroy, W.E.; Brooks, D. Effects of an aerobic exercise program on aerobic capacity, spatiotemporal gait parameters, and functional capacity in subacute stroke. *Neurorehabil. Neural Repair* **2009**, *23*, 398–406. [CrossRef]
42. Brouwer, B.; Parvataneni, K.; Olney, S.J. A comparison of gait biomechanics and metabolic requirements of overground and treadmill walking in people with stroke. *Clin. Biomech.* **2009**, *24*, 729–734. [CrossRef]
43. Ivey, F.M.; Hafer-Macko, C.E.; Macko, R.F. Task-oriented treadmill exercise training in chronic hemiparetic stroke. *J. Rehabil. Res. Dev.* **2008**, *45*, 249–259. [CrossRef]

44. Alsubiheen, A.M.; Choi, W.; Yu, W.; Lee, H. The effect of task-oriented activities training on upper-limb function, daily activities, and quality of life in chronic stroke patients: A randomized controlled trial. *Int. J. Environ. Res. Public Health* **2022**, *19*, 14125. [CrossRef] [PubMed]
45. Kokotilo, K.J.; Eng, J.J.; Boyd, L.A. Reorganization of brain function during force production after stroke: A systematic review of the literature. *J. Neurol. Phys. Ther.* **2009**, *33*, 45–54. [CrossRef] [PubMed]
46. Hsieh, C.-L.; Sheu, C.-F.; Hsueh, I.-P.; Wang, C.-H. Trunk control as an early predictor of comprehensive activities of daily living function in stroke patients. *Stroke* **2002**, *33*, 2626–2630. [CrossRef] [PubMed]

Disclaimer/Publisher's Note: The statements, opinions and data contained in all publications are solely those of the individual author(s) and contributor(s) and not of MDPI and/or the editor(s). MDPI and/or the editor(s) disclaim responsibility for any injury to people or property resulting from any ideas, methods, instructions or products referred to in the content.

bioengineering

MDPI

Article

Effects of Action Observation Plus Motor Imagery Administered by Immersive Virtual Reality on Hand Dexterity in Healthy Subjects

Paola Adamo [1,2], Gianluca Longhi [1], Federico Temporiti [1,2], Giorgia Marino [1], Emilia Scalona [3], Maddalena Fabbri-Destro [4], Pietro Avanzini [4] and Roberto Gatti [1,2,*]

1 Physiotherapy Unit, IRCCS Humanitas Research Hospital, Via Manzoni 56, 20089 Rozzano, Milan, Italy
2 Department of Biomedical Sciences, Humanitas University, Via Rita Levi Montalcini 4, 20072 Pieve Emanuele, Milan, Italy
3 Dipartimento di Scienze Medico Chirurgiche, Scienze Radiologiche e Sanità Pubblica (DSMC), Università Degli Studi di Brescia, Viale Europa 11, 25123 Brescia, Brescia, Italy
4 Consiglio Nazionale Delle Ricerche, Istituto di Neuroscienze, Via Volturno, 39-E, 43125 Parma, Parma, Italy
* Correspondence: roberto.gatti@hunimed.eu

Abstract: Action observation and motor imagery (AOMI) are commonly delivered through a laptop screen. Immersive virtual reality (VR) may enhance the observer's embodiment, a factor that may boost AOMI effects. The study aimed to investigate the effects on manual dexterity of AOMI delivered through immersive VR compared to AOMI administered through a laptop. To evaluate whether VR can enhance the effects of AOMI, forty-five young volunteers were enrolled and randomly assigned to the VR-AOMI group, who underwent AOMI through immersive VR, the AOMI group, who underwent AOMI through a laptop screen, or the control group, who observed landscape video clips. All participants underwent a 5-day treatment, consisting of 12 min per day. We investigated between and within-group differences after treatments relative to functional manual dexterity tasks using the Purdue Pegboard Test (PPT). This test included right hand (R), left hand (L), both hands (B), R + L + B, and assembly tasks. Additionally, we analyzed kinematics parameters including total and sub-phase duration, peak and mean velocity, and normalized jerk, during the Nine-Hole Peg Test to examine whether changes in functional scores may also occur through specific kinematic patterns. Participants were assessed at baseline (T0), after the first training session (T1), and at the end of training (T2). A significant time by group interaction and time effects were found for PPT, where both VR-AOMI and AOMI groups improved at the end of training. Larger PPT-L task improvements were found in the VR-AOMI group (*d:* 0.84, CI_{95}: 0.09–1.58) compared to the AOMI group from T0 to T1. Immersive VR used for the delivery of AOMI speeded up hand dexterity improvements.

Keywords: action observation; motor imagery; virtual reality; manual dexterity

Citation: Adamo, P.; Longhi, G.; Temporiti, F.; Marino, G.; Scalona, E.; Fabbri-Destro, M.; Avanzini, P.; Gatti, R. Effects of Action Observation Plus Motor Imagery Administered by Immersive Virtual Reality on Hand Dexterity in Healthy Subjects. *Bioengineering* **2024**, *11*, 398. https://doi.org/10.3390/bioengineering11040398

Academic Editors: Carlo Albino Frigo and Dante Mantini

Received: 30 January 2024
Revised: 3 April 2024
Accepted: 16 April 2024
Published: 19 April 2024

Copyright: © 2024 by the authors. Licensee MDPI, Basel, Switzerland. This article is an open access article distributed under the terms and conditions of the Creative Commons Attribution (CC BY) license (https://creativecommons.org/licenses/by/4.0/).

1. Introduction

Cognitive facilitations including action observation (AO) and motor imagery (MI) have been recently adopted to promote motor learning in athletes or in subjects with motor impairments [1,2].

AO entails the observation of motor tasks through video clips, which are delivered using a laptop screen showing motor contents representing motor acts executed by age and gender-matched subjects [1,3,4]. This stimulation recruits the mirror neuron system (MNS), a frontoparietal network active both during the execution and observation of motor acts. The MNS plays a key role in understanding actions performed by others [5] and is involved in the building of motor memories related to the observed tasks [6]. The neural solicitation driven by AO is dependent on the features of the observed stimuli, such as the observer perspective or the transitive or intransitive nature of the action. For example, a first-person perspective induces higher MNS activation during the observation of upper limb transitive actions, compared to a third-person perspective [7,8]. In addition, the impact of AO on motor learning is larger when the observed actions are accompanied by action-related sounds [9].

MI consists of a cognitive process in which subjects imagine themselves executing motor tasks without performing them [10,11]. MI performed immediately after AO has been reported to further enhance motor learning when compared to the execution of AO alone [11]. The rationale behind this finding lies in the overlap between the neural substrates of AO and MI, both of which share large territories over frontoparietal circuits [10]. In fact, the association of AO and MI (AOMI) has been adopted to enhance the performance of sport-related gestures [12]. Moreover, AOMI-related benefits have been demonstrated in terms of motor re-learning in patients with motor impairments [13] when compared to the administration of AO or MI alone.

When considering treatments integrating action observation (e.g., AOMI), an additional resource is represented by virtual reality (VR), which is a computer-generated simulation of a three-dimensional image or environment that users can interact with in a seemingly real or physical manner. This interaction is facilitated through specialized devices such as helmets with a screen inside or controllers fitted with sensors [14,15]. VR technologies may be "immersive" when the subject is completely immersed in the virtual environment and visual information change according to the movement of the user's head [16]. Immersive VR offers multiple advantages. On one hand, it leads to the illusion of being in a virtual place and the first-person perspective generates the illusion that the virtual body is the observer's own [17]. The observed upper limbs are coherent with the real limbs of the subject, enhancing the embodiment of the observer [18], the sense of agency, the self-attribution of the virtual body [19,20], and the overall engagement during the treatment [21]. On the other hand, neuroimaging studies have proven that action observation presented in 3D evokes a greater sensorimotor response than 2D stimuli [22]. All these advantages induced clinical researchers to adopt immersive VR in the field of motor rehabilitation, demonstrating its efficacy in favoring the upper limb recovery in patients with neurological disorders [23,24]. In this scenario, previous studies have suggested that implicit motor learning, defined as the acquisition of new motor skills without the awareness of the learning process, may be enhanced when subjects are embodied in immersive virtual environments and when they have a first-person perspective [25]. The embodiment with the virtual body might increase AO effects, since the mirror mechanism has been reported as modulated not only by the observed movement, but also by action meaning and environmental, contextual, and emotional factors [26–28].

Few studies have described the link among AO, VR, and MI in terms of motor learning. Choi [29] described higher motor imagery abilities following hand movements observed in immersive VR. Moreover, the administration of AO through immersive VR to patients in a subacute phase after a stroke has shown improvements in upper limb motor function when compared to upper limb motor rehabilitation performed without VR [30]. It is worth noting that the aforementioned cognitive approaches have been frequently associated with exercises aimed at improving hand dexterity [31,32]. Manual dexterity consists of the ability to execute coordinated hand and fingers movements and derives from the integration of hand biomechanics with sensorimotor and cognitive processes [32]. When considering hand function, improvements in manual dexterity after training represent an index of enhanced motor control during hand and fingers movements [33]. However, although literature data suggest the opportunity to enhance the embodiment through the use of immersive VR systems, no studies have investigated the effect of an AOMI training administered through immersive VR on manual dexterity when compared to a conventional AOMI training

Therefore, the aim of this study was to investigate the effects on manual dexterity of AOMI delivered through immersive VR compared to conventional AOMI delivered via a laptop screen in healthy subjects. The study hypothesis was that immersive VR might boost the effects of AOMI on motor learning in healthy subjects.

This modality of administration may increase motor resonance, since AOMI is modulated by contextual and environmental factors associated with observed movement [26,28]. The novelty of the study lies in leveraging the embodiment induced by action observation and motor imagery delivered using immersive VR, alongside the first-person perspective, rather than via a laptop screen to induce motor learning in terms of manual dexterity changes.

To measure the extent and changes in manual dexterity, we identified specific functional tests that were administered both before and after a week of daily training. However, motor learning may manifest as an improvement in a functional test, as well as a higher velocity at which this improvement occurs [34]. For this reason, the functional battery was administered also after the first training session. Additionally, we investigated the movement kinematics during manual dexterity tasks to examine whether changes in functional scores may also occur through specific kinematic patterns.

We provide an outline of the manuscript to guide readers through its content. In the Background section, we highlight the rationale behind why VR may boost AOMI effects on motor learning. Subsequently, in the Methods section, we describe the study design, the sample included in the study, the different characteristics of treatment administered, and a detailed description of the outcome measures chosen to assess manual dexterity. The Results section presents the effects of different treatments, whose explanations, interpretations, and implications are addressed in the Discussion and Conclusions.

2. Materials and Methods

2.1. Participants

The sample size was estimated based on previous reports about the Purdue Pegboard Test ([35], see below). It was estimated that, considering an alpha error of 5%, a minimum of 15 participants would be required in each group to provide 80% power to detect a Cohen's $d = 1.0$ (large effect size) between VR-AOMI and AOMI groups at T2 [36].

Forty-five healthy subjects (15 females, 30 males; mean age 23.4 ± 2.68 years) were enrolled, and their characteristics are reported in Table 1. Inclusion criteria were: (1) age between 20 and 35 years, (2) right-handedness according to the Edinburgh Handedness Inventory [37]. Conversely, the presence of upper limb sensorimotor disorders or recent traumatic injuries, the usual performance of motor activities or sports involving remarkable manual skills (e.g., playing instruments, juggling), a history of epilepsy seizures, and visual impairments that cannot be corrected with lenses were used as exclusion criteria. All participants signed a written informed consent and the study protocol was approved by the Ethical Committee of the Humanitas Clinical and Research Center (approval number: VR-AOT-GR-2019).

Table 1. Characteristics of study participants. Data are shown as mean ± standard deviation.

	VR-AOMI Group (n = 15)	AOMI Group (n = 15)	CTRL Group (n = 15)	*p*-Value
Age (years)	24.06 ± 3.1	23.53 ± 3.24	22.53 ± 1.06	0.32
Weight (kg)	67.43 ± 10.82	66.06 ± 12.52	69.53 ± 8.71	0.67
Height (cm)	176.53 ± 6.76	172.6 ± 11.09	172.73 ± 6.96	0.39
Gender	11M/4F (27%F)	9M/6F (40%F)	10M/5F (33%F)	0.74
KVIQ	36.6 ± 7.22	40.87 ± 6.42	38.27 ± 8.13	0.35

VR-AOMI: action observation and motor imagery through immersive VR; **AOMI**: action observation and motor imagery through laptop; **CTRL**: control; **KVIQ**: kinesthetic and visual imagery questionnaire; **M**: male; **F**: female.

2.2. Intervention

Action stimulus depicted a humanoid avatar playing a pianola with the left hand from a first-person perspective (Figure 1). The action performed by the humanoid avatar replicated the kinematics recorded from a healthy subject [38] by a motion capture system

(Awinda, XSens, Enschede, The Netherlands) incorporating industrial gloves to track hand and finger movements (Manus Prime II Xsens, Enschede, The Netherlands). Given the immersive nature of the stimuli, VR-AOMI participants wearing the headset could explore the surrounding space by moving head and gaze accordingly.

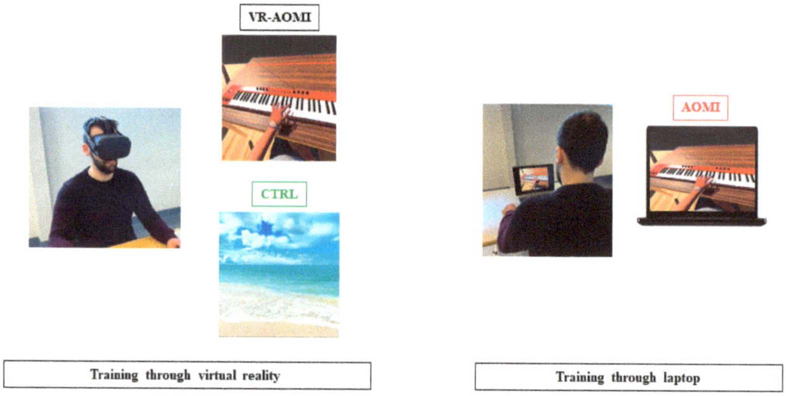

Figure 1. Study participant undergoing the VR-AOMI through the Oculus headset (VR-AOMI), AOMI through a laptop screen (AOMI), and landscape observation through VR (CTRL).

Both VR-AOMI and AOMI participants were asked to carefully observe the action for 3 min, with an audio trace coherent with piano keys pressure. After the observation, they had to imagine themselves performing the action (MI) for one minute, avoiding any active movement. Observation and imagination were repeated three times (12 min in total) for five consecutive days, maintaining constant the daytime of task administration. Finally, participants of the CTRL group observed landscapes (free of any biological motor contents) in immersive VR for the same amount of time as the AOMI groups.

2.3. Study Design, Randomization, and Enrollment

The study has a three-armed, single-blind, randomized, controlled design. Participants were recruited by an independent researcher not involved in the subsequent stages of the study and they were assigned to one of the three experimental groups according to a random computer-generated list. The VR-AOMI group (15 subjects) underwent AOMI through immersive presentation of the action stimuli, the AOMI group (15 subjects) observed the same stimuli via a laptop screen, while the CTRL group (15 subjects) observed landscape videos via the VR visor (Oculus II).

2.4. Functional and Kinematic Assessment

The kinesthetic and visual imagery questionnaire (KVIQ) was administered to participants at baseline to assess their motor imagery abilities [39].

The main outcome of the study pertained to the participants' hand dexterity. For this reason, all subjects underwent a functional and kinematic assessment at baseline (T0), after the first training session (T1—day 1), and at the end of training (T2—day 5). The study timeline is shown in Figure 2.

1° day	2° day	3° day	4° day	5° day
T0 assessment				
1st training (VR-AOMI/AOMI/CTRL)	2nd training (VR-AOMI/AOMI/CTRL)	3rd training (VR-AOMI/AOMI/CTRL)	4th training (VR-AOMI/AOMI/CTRL)	5th training (VR-AOMI/AOMI/CTRL)
T1 assessment				T2 assessment

Figure 2. Study timeline.

The assessment procedures were conducted by a researcher unaware of group allocation at the Motion Analysis Lab of the Humanitas Clinical Institute, Milan, Italy.

The functional assessment consisted of the Purdue Pegboard Test (PPT), including four subtests: PPT-R (right hand), PPT-L (left hand), PPT-B (both hands simultaneously), and an assembly task. Participants had 30 s for each peg-insertion task and 60 s for the assembly task [35].

The kinematic assessment included the Nine-hole peg test (NHPT) [40] and the finger tapping test (FTT) [41]. During the NHPT, participants were seated on a height-adjustable chair with the pegboard positioned on a table in front of them. They were asked to grasp the pegs from a container one by one, place them into a nine-hole board, remove the pegs from the board, and replace them in the container as quickly as possible [40]. Kinematic data were recorded using an optoelectronic system (SMART-DX, BTS, Milan, Italy) equipped with eight infrared cameras and 16 reflective markers, of which three were on the table to define the global reference system and 13 were on anatomical landmarks. The system calibration involved a 10 s static test with four additional markers on the NHPT board.

Markers trajectories were filtered using a fourth-order low-pass Butterworth filter (cut-off 4 Hz). Subsequently, we extracted the total and single-phase times (peg grasp, peg transfer, peg in hole, hand return), normalized jerk, mean and peak velocity during peg transfer and hand return phases [40]. Subjects performed two trials for both the right and the left sides, and the shortest one in terms of total time of execution was considered for the analysis.

Kinematic assessment also included the finger tapping test (FTT), performed with participants seated on a chair, forearms resting on a desk with elbows at 90 degrees flexion. They performed a tapping sequence (thumb, index, middle, ring, and little finger) at maximum speed for 15 s without visual feedback. Fingers were marked distally with smaller reflective markers (6 mm diameter). Errors were recorded for incorrect sequence movements. The total number of errors and the total number of movements considering all the fingers were analyzed [41].

2.5. Statistical Analysis

After verifying the normality assumption through the Shapiro–Wilk test, a parametric pipeline was adopted. Univariate ANOVA and chi-square test were used to investigate between-group differences in terms of demographic characteristics and KVIQ at baseline. A 3 × 3 mixed ANOVA with time as within-subjects and group as between-subjects factors was used to evaluate differences in terms of manual dexterity assessed by functional test and kinematic analysis between groups over time. Post-hoc analyses were Bonferroni-corrected to reduce the false positive ratio. In addition, changes from baseline to post-treatment (deltas) were calculated and the effect size (Cohen's d), with its 95% confidence interval (CI_{95}), was also computed between deltas in the three groups and between each time point in the same group and interpreted as small ($0.2 < d < 0.5$), medium ($0.5 < d < 0.8$), large ($0.8 < d < 1.3$), and very large ($d > 1.3$). Analyses were conducted using SPSS Statistics 29.0 for iOS and the statistical level of significance was set to $\alpha = 0.05$.

3. Results

None of the participants withdrew from the study and no between-group differences were found in terms of baseline characteristics including age, weight, height, gender, and KVIQ score, as shown in Table 1.

3.1. The Effects of AOMI Applied through VR on Manual Dexterity Assessed with the Purdue Pegboard Test

A significant time by group interaction and time effect were found for PPT-R, PPT-L, PPT-R + L + B, and PPT-Assembly tasks, while a time effect was found for PPT-Both.

The graphs in Figure 3 show the PPT scores obtained across the three groups, while numeric values for within- and between-group comparisons of deltas are reported in Tables 2 and 3, respectively.

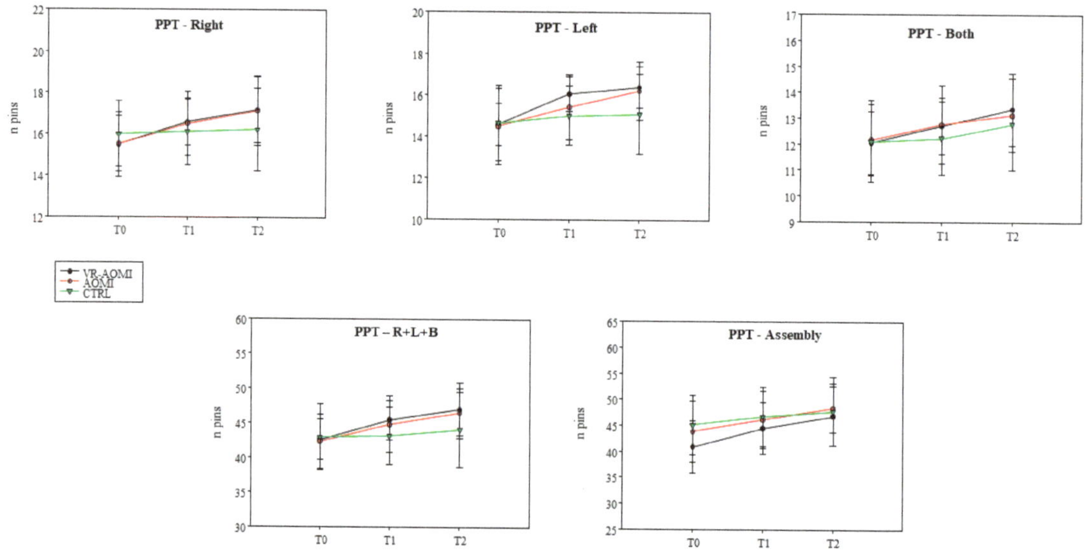

Figure 3. The scores obtained in Purdue Pegboard Test across the three groups are shown. Data are presented as means (dots and triangles) and standard deviation (vertical bars).

Table 2. PPT scores in the VR-AOMI (**A**), AOMI (**B**), and CTRL (**C**) groups at T0, T1, and T2 are expressed as mean ± standard deviation. Within-group comparisons between T1/T0 and T2/T0 are expressed as Cohen's *d* with 95% confidence interval (CI_{95}).

		(A)			
VR-AOMI	**T0**	**T1**	**T2**	***d*** **(CI_{95}) T1–T0**	***d*** **(CI_{95}) T2–T0**
R task	15.49 ± 1.57	16.60 ± 1.14	17.17 ± 1.57	1.03 (0.39, 1.65)	1.51 (0.75, 2.52)
L task	14.58 ± 1.03	16.06 ± 0.85	16.40 ± 0.99	2.37 (1.53, 3.36)	1.84 (0.99, 2.67)
B Task	12.04 ± 1.22	12.71 ± 1.09	13.35 ± 1.38	0.75 (0.16, 1.31)	1.79 (0.95,2.61)
R + L + B task	42.62 ± 2.94	45.42 ± 2.82	46.98 ± 3.80	1.75 (0.92, 2.56)	2.07 (1.15, 2.97)
Assembly task	40.86 ± 5.02	44.55 ± 4.94	46.86 ± 5.70	1.86 (0.99, 2.69)	2.73 (1.60, 3.85)
		(B)			
AOMI	**T0**	**T1**	**T2**	***d*** **(CI_{95}) T1–T0**	***d*** **(CI_{95}) T2–T0**
R task	15.53 ± 1.33	16.51 ± 1.53	17.13 ± 1.67	0.98 (0.34,1.58)	1.51 (0.74, 2.42)
L task	14.49 ± 1.83	15.42 ± 1.55	16.22 ± 1.38	1.35 (0.63, 2.05)	1.70 (0.89, 2.50)
B Task	12.18 ± 1.37	12.77 ± 1.50	13.13 ± 1.41	0.84 (0.24, 1.42)	1.09 (0.43, 1.72)
R + L + B task	42.33 ± 3.94	44.29 ± 3.90	46.46 ± 3.56	1.38 (0.65, 2.08)	1.91 (1.03, 2.76)
Assembly task	43.86 ± 5.91	46.15 ± 5.58	48.40 ± 4.68	0.62 (0.06, 1.17)	1.11 (0.45, 1.75)
		(C)			
CTRL	**T0**	**T1**	**T2**	***d*** **(CI_{95}) T1–T0**	***d*** **(CI_{95}) T2–T0**
R task	16.00 ± 1.58	16.10 ± 1.57	16.22 ± 2.00	0.1 (−0.41, 0.60)	0.22 (−0.30, 0.73)
L task	14.64 ± 1.81	15.02 ± 1.43	15.11 ± 1.92	0.52 (−0.3, 1.05)	0.50 (−0.04, 1.04)
B Task	12.11 ± 1.58	12.24 ± 1.40	12.80 ± 1.77	0.16 (−0.35, 0.67)	1.18 (0.54, 1.84)
R + L + B task	42.95 ± 4.73	43.15 ± 4.18	44.11 ± 5.42	0.09 (−0.42, 0.60)	0.72 (0.14, 1.28)
Assembly task	45.09 ± 5.77	46.78 ± 5.69	47.78 ± 6.53	0.90 (0.28, 1.49)	1.53 (0.76, 2.28)

Abbreviations. VR-AOMI: action observation through immersive virtual reality group; **AOMI**: action observation group; **CTRL**: control group; ***d***: Cohen's *d*; **CI_{95}**: confidence interval; **R**: right; **L**: left; **B**: both.

Table 3. Comparisons of deltas among the VR-AOMI, AOMI, and CTRL groups for the Purdue Pegboard Test. Data were expressed as Cohen's *d* with 95% confidence interval (CI_{95}). Significant results are shown in bold text.

	ΔT1–T0 Cohen's *d* (CI_{95})			**ΔT2–T0 Cohen's *d* (CI_{95})**		
	VR-AOMI/AOMI	**VR-AOMI/CTRL**	**AOMI/CTRL**	**VR-AOMI/AOMI**	**VR-AOMI/CTRL**	**AOMI/CTRL**
R task	0.13 (−0.59–0.84)	**0.90 (0.14–1.65)**	**0.81 (0.06–1.55)**	0.08 (−0.63–0.80)	**1.39 (0.58–2.19)**	**1.35 (0.54–2.13)**
L task	**0.84 (0.09–1.58)**	**1.64 (0.79–2.46)**	**0.79 (0.04–1.52)**	0.09 (−0.63–0.80)	**1.42 (0.60–2.21)**	**1.31 (0.50–2.09)**
B Task	0.08 (−0.63–0.80)	0.62 (−0.12–1.35)	0.60 (−0.14–1.33)	0.44 (−0.29–1.16)	**0.94 (0.18–1.69)**	0.36 (−0.37–1.08)
R + L + B task	0.18 (−0.54–0.90)	**1.38 (0.57–2.17)**	**1.16 (0.38–1.93)**	0.09 (−0.62–0.81)	**1.71 (0.85–2.54)**	**1.57 (0.73–2.38)**
Assembly task	0.48 (−0.26–1.20)	**1.03 (0.26–1.79)**	0.21 (−0.51–0.92)	0.45 (−0.28–1.17)	**1.67 (0.82–2.49)**	0.59 (−0.15–1.32)

Abbreviations. VR-AOMI: action observation performed through immersive virtual reality group; **AOMI**: action observation group; **CTRL**: control group; **CI_{95}**: confidence interval; **R**: right; **L**: left; **B**: both.

Table 2 reports within-group differences expressed as Cohen's d and CI_{95}. Both VR-AOMI and AOMI groups improved from T0 to T1 and further experienced enhanced manual dexterity from T0 to T2 during PPT-R, PPT-L, PPT-Both, PPT-R + L + B, and PPT-Assembly tasks. These results suggest a major effect driven by action observation regardless of its visual display features. The CTRL group improved from T0 to T2 only for PPT-Both, PPT-R + L + B, and PPT-Assembly tasks.

Table 3 reports between-group differences expressed as Cohen's *d* and CI_{95}. Differences between VR-AOMI and AOMI groups in terms of deltas were exclusively found for PPT-L at T1 in favor of VR-AOMI ($p = 0.029$, *d*: 0.84, CI_{95}: 0.09–1.58), indicating that the use of virtual reality speeded up the motor learning process, reaching a significant increase in dexterity already after the first training session (Figure 4). In addition, between-group differences in terms of deltas (T1–T0 and T2–T0) were detected in favor of both AOMI groups compared to the CTRL group in all subtasks of PPT (Table 3).

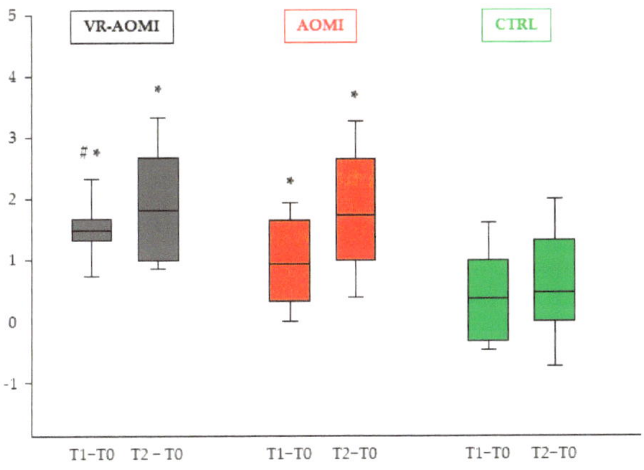

Figure 4. Between-group differences for deltas from T0 to T1 and from T2 to T0 for the PPT-L task. Boxes represent the range between the first and the third quartile, the middle horizontal line is the mean value, and the ends of the vertical line, from top to bottom, are the maximum and minimum values, respectively. Symbols show differences in terms of delta between the VR-AOMI and AOMI groups from T0 to T1 (#) and between the VR-AOMI and AOMI groups with the CTRL group from T0 to T1 and from T0 to T2 (*).

3.2. The Effects of AOMI Applied through VR on Kinematic Assessment during the Nine-Hole Peg Test

Results for the NHPT with the left hand revealed a time effect for removing time ($p = 0.003$), peak return velocity ($p = 0.009$), transfer time ($p = 0.040$), return time ($p = 0.029$), and transfer velocity ($p = 0.042$) (Table S1).

Within-group post-hoc analyses revealed a decrease in removing time in the AOMI group at T2 compared to T0 (*MD:* 0.93, $p = 0.041$, CI_{95}: 0.03, 1.83), while the VR-AOMI group revealed a decrease in transfer time at T2 when compared to T0 (*MD:* 0.38, $p = 0.05$, CI_{95}: 0.76). The CTRL group revealed a decrease in return time at T2 compared to T1 (*MD:* 0.31, $p = 0.013$, CI_{95}: 0.05, 0.57). Both the AOMI (*MD:* −0.11, $p = 0.04$, CI_{95}: −0.19, 0.03) and control groups (*MD:* −0.09, $p = 0.012$, CI_{95}: −0.18, −0.02) achieved higher peak velocity during the return phase at T2 compared to T1.

The NHPT performed with the right hand revealed a time effect for removing time ($p = 0.003$) and a time by group interaction for total test ($p = 0.025$) and peg-in-hole times ($p = 0.043$). Between-group post-hoc analyses revealed a lower peg-in-hole time in the CTRL group compared to the AOMI group at T2 (*MD:* 0.76, $p = 0.036$, CI_{95}: 0.04, 1.48). Finally, both VR-AOMI (*MD:* 0.47, $p = 0.027$, CI_{95}: 0.06, 0.89) and CTRL (*MD:* 0.46, $p = 0.030$, CI_{95}: 0.05, 0.88) groups showed a decrease in terms of removing time from T0 to T2 (Table S2).

3.3. The Effects of AOMI Applied through VR on Kinematic Assessment during the Finger Tapping Test

A time effect was found in total errors in the FTT performed with the left hand ($p = 0.002$). Post-hoc analyses revealed improvement in the AOMI group ($p = 0.027$) at T2 compared to T0. A time effect was found also in total fingers movement ($p < 0.001$). Post-hoc analyses revealed an improvement only in the AOMI group ($p = 0.003$) from T1 to T0, while all the groups showed an improvement ($p < 0.001$) at T2 compared to T0. Only the VR-AOMI group improved ($p = 0.025$) from T1 to T2. No improvements were found during the FTT with the right hand.

4. Discussion

This was the first study aimed at investigating whether immersive VR boosts the effects of AOMI. The main study finding was that AOMI delivered via immersive VR speeded up improvements in manual dexterity in healthy subjects. In fact, although AOMI groups showed similar manual dexterity abilities at the end of training, the VR-AOMI group achieved significantly greater dexterity changes than the AOMI group after the first training session. On the other hand, the AOMI group gradually improved manual dexterity over the five training sessions. Finding the most effective strategies to fasten acquisition of motor skills is fundamental for motor learning relative to both athletes and patients undergoing motor rehabilitation, where one goal is reducing the recovery time of subjects undergoing rehabilitation plannings.

Our study applied this approach to hand dexterity, since the effects of AOMI on the upper limbs are the most investigated in the literature and are described as the most promising in motor rehabilitation [13]. The choice to investigate healthy subjects derived from the desire to avoid any discomfort in patients assigned to the VR-AOMI group, since previous studies have described cybersickness as a potential side effect during the use of immersive virtual reality systems [42]. However, few studies have reported minimal cybersickness using a head-mounted display in healthy subjects and stroke patients during upper limb exercises through fully immersive VR systems [43]. Finally, the choice to associate AO and MI derives from literature data supporting the efficacy of AOMI in promoting the learning of complex motor tasks when compared to the use of AO or MI alone in healthy subjects [44].

The current study demonstrates that virtual reality speeded up motor learning induced by AOMI, as reported by PPT—Left hand and assembly tasks results. Based on these findings, immersive virtual reality may have enhanced embodiment and amplified 3D kinematics details of the motor task, boosting the acquisition of complex motor skills in young healthy subjects [1]. Interestingly, the largest effect in both AOMI groups was found for left 'trained' hand, consistently with previous studies which have demonstrated that motor learning is related to the characteristics of the observed movement [45]. Improvements at the level of the right untrained hand may derive from an interlimb transfer effect from the non-dominant to the dominant hand after a unilateral dexterity training [46].

Nevertheless, a certain heterogeneity in terms of the PPT-Assembly task was found at baseline, where the control group revealed higher mean score than the AOMI groups. However, this heterogeneity may be attributed to higher variability in terms of assembly task performance compared to unimanual tasks during PPT execution in healthy subjects [35]. The opportunity to speed up the effects of AOMI through its association with VR may depend on a greater sense of embodiment, since the motor learning process has been shown to increase when embodiment with the observed movements is higher [15]. In fact, resonance of MNS does not depend only on observed actions, since previous studies have suggested that environment and context factors also modulate corticospinal excitability during AO [26,28]. Particularly, the inferior frontal gyrus and the ventral premotor cortex revealed a significant signal increase when the context suggested the intention associated with hands actions [26]. Familiarity with and expertise of the observed action has also been shown to increase the MNS resonance [47], and the embodiment induced by immersive VR may increase the perception of familiarity with the observed action [48].

Moreover, stereoscopic 3D stimuli may provide a greater number of kinematic details than bidimensional visual stimuli, allowing for a precise assessment and understanding of the observed actions [22,49]. Finally, it is worth noting that the sense of embodiment has been also described as influenced by the person-related perspective, since an egocentric perspective has been demonstrated to trigger the illusion of body ownership and self-attribution of the virtual body [50]. Overall, VR-AO delivered in the first-person-perspective may also be considered as a multisensory stimulation which exploits the simultaneous use of different sensory inputs to enhance motor learning processes [51,52]. In particular, immersive VR systems allow for the integration of realistic scenarios engaging the patient's sensorimotor system and promoting the feeling of being into the virtual environment [53]. In this scenario, the current study findings induce us to consider AO, MI, and VR as mutually beneficial. In fact, immersive VR might enhance motor imagery performance, providing a rich immersive and illusive experience especially in first-person perspective virtual scenarios [29,54].

The current study found improvements after AO treatments in PPT, while differences in the 9HPT and finger tapping test were not detected with respect to AO or CTRL groups. PPT has been previously described as more sensitive than the nine-hole peg test in detecting changes in manual dexterity, especially in healthy subjects [55]. Little improvements in the finger tapping test and bimanual tasks were found in the control group, probably as a result of a learning effect caused by the repetition of the test three times in one week [56].

Although VR seems to speed up improvements in manual dexterity specifically in the left observed hand, changes at the end of the treatment period were the same for immersive and non-immersive AOMI. However, confirmation of this result in clinical practice would have an interesting relevance, mostly in the rehabilitation field, where faster functional recovery is a goal of rehabilitation [57].

5. Conclusions

In conclusion, the results of the study suggest that AOMI delivered through immersive VR speeded up hand dexterity improvements. The study has some limitations. First, a single exercise was delivered during the treatment period and improvements may have been greater with a progression of exercise difficulty within the treatment period. This limitation derived from the fact that exercises recorded in virtual reality were specifically designed for the rehabilitation of post-stroke patients in a subacute phase. Thus, the authors chose the most challenging task for a healthy young subject. Furthermore, visually induced motion sickness was not explored, which has been described as a possible side effect of VR immersion, with symptoms including nausea, disorientation, and oculomotor discomfort [38]. However, no dropouts were reported either in the landscape or the VR-AOMI groups. Finally, the study was unbalanced in terms of gender and age according to the inclusion criteria, as the number of males was twice that of females, and the mean age was lower with respect to the mean age computed from the specified inclusion criteria. However, the proportion of males and females in the VR-AOMI, AOMI, and CTRL groups, as well as the mean age in all three groups, did not show statistically significant differences.

Further studies are needed to confirm the current findings in clinical practice and explore the opportunity to reduce the recovery time in subjects undergoing rehabilitation programs.

Supplementary Materials: The following supporting information can be downloaded at: https://www.mdpi.com/article/10.3390/bioengineering11040398/s1, Table S1: Comparison between VR-AOMI, AOMI and Control groups at T0, T1 and T2 for kinematic indexes during Nine-Hole Peg Test performed with the left hand. Data are shown as mean and standard deviation. Significant results are shown in bold text.; Table S2: Comparison between VR-AOMI, AOMI and Control groups at T0 and T2 for kinematic indexes during Nine-Hole Peg Test (NHPT) performed with right hand. Data are shown as mean and standard deviation. Significant results are shown in bold text.

Author Contributions: Conceptualization, R.G. and P.A. (Paola Adamo); methodology, G.L., G.M. and P.A. (Paola Adamo); software, E.S., M.F.-D. and P.A. (Pietro Avanzini); formal analysis, G.M., P.A. (Paola Adamo) and F.T.; investigation, G.L. and P.A. (Paola Adamo); resources, E.S. and F.T.; data curation, G.L. and P.A. (Paola Adamo); writing—original draft preparation, P.A. (Paola Adamo) and G.L.; writing—review and editing, R.G. and F.T.; visualization, P.A. (Pietro Avanzini), G.M. and M.F.-D.; supervision, P.A. (Pietro Avanzini) and M.F.-D.; project administration, R.G. All authors have read and agreed to the published version of the manuscript.

Funding: This research received no external funding.

Institutional Review Board Statement: The study was conducted in accordance with the Declaration of Helsinki and approved by the Ethical Committee of the Humanitas Clinical and Research Center, Milan, Italy (protocol code: VR-AOT-GR-2019, date of approval: 2 July 2021).

Informed Consent Statement: Informed consent was obtained from all subjects involved in the study. Written informed consent has been obtained from the participants to publish this paper.

Data Availability Statement: The dataset used and analyzed during the current study is available from the corresponding author upon reasonable request.

Conflicts of Interest: The authors declare no conflicts of interest.

Nomenclature

VR	Virtual reality
AOMI	Action observation and motor imagery
CTRL	Control
MNS	Mirror Neuron System
KVIQ	Kinesthetic and Visual Imagery Questionnaire
NHPT	Nine Hole Peg Test
PPT	Purdue Pegboard Test
FTT	Finger Tapping Test
T0	Baseline assessment
T1	Assessment after the first treatment
T2	Assessment at the end of the treatment period

References

1. Rizzolatti, G.; Fabbri-Destro, M.; Nuara, A.; Gatti, R.; Avanzini, P. The role of mirror mechanism in the recovery, maintenance, and acquisition of motor abilities. *Neurosci. Biobehav. Rev.* **2021**, *127*, 404–423. [CrossRef] [PubMed]
2. Simonsmeier, B.A.; Andronie, M.; Buecker, S.; Frank, C. The effects of imagery interventions in sports: A meta-analysis. *Int. Rev. Sport Exerc. Psychol.* **2021**, *14*, 186–207. [CrossRef]

3. Buccino, G. Action observation treatment: A novel tool in neurorehabilitation. *Philos. Trans. R. Soc. Lond. B Biol. Sci.* **2014**, *369*, 20130185. [CrossRef] [PubMed]
4. Buccino, G.; Solodkin, A.; Small, S.L. Functions of the mirror neuron system: Implications for neurorehabilitation. *Cogn. Behav. Neurol.* **2006**, *19*, 55–63. [CrossRef] [PubMed]
5. Rizzolatti, G.; Cattaneo, L.; Fabbri-Destro, M.; Rozzi, S. Cortical mechanisms underlying the organization of goal-directed actions and mirror neuron-based action understanding. *Physiol. Rev.* **2014**, *94*, 655–706. [CrossRef] [PubMed]
6. Stefan, K.; Cohen, L.G.; Duque, J.; Mazzocchio, R.; Celnik, P.; Sawaki, L.; Ungerleider, L.; Classen, J. Formation of a motor memory by action observation. *J. Neurosci.* **2005**, *25*, 9339–9346. [CrossRef] [PubMed]
7. Angelini, M.; Fabbri-Destro, M.; Lopomo, N.F.; Gobbo, M.; Rizzolatti, G.; Avanzini, P. Perspective-dependent reactivity of sensorimotor mu rhythm in alpha and beta ranges during action observation: An EEG study. *Sci. Rep.* **2018**, *8*, 12429. [CrossRef] [PubMed]
8. Gatti, R.; Rocca, M.A.; Fumagalli, S.; Cattrysse, E.; Kerckhofs, E.; Falini, A.; Filippi, M. The effect of action observation/execution on mirror neuron system recruitment: An fMRI study in healthy individuals. *Brain. Imaging Behav.* **2017**, *11*, 565–576. [CrossRef] [PubMed]
9. De Manzano, Ö.; Kuckelkorn, K.L.; Ström, K.; Ullén, F. Action-Perception Coupling and Near Transfer: Listening to Melodies after Piano Practice Triggers Sequence-Specific Representations in the Auditory-Motor Network. *Cereb. Cortex.* **2020**, *30*, 5193–5203. [CrossRef]
10. Hardwick, R.M.; Caspers, S.; Eickhoff, S.B.; Swinnen, S.P. Neural correlates of action: Comparing meta-analyses of imagery, observation, and execution. *Neurosci. Biobehav. Rev.* **2018**, *94*, 31–44. [CrossRef]
11. Bisio, A.; Bassolino, M.; Pozzo, T.; Wenderoth, N. Boosting Action Observation and Motor Imagery to Promote Plasticity and Learning. *Neural Plast.* **2018**, *2018*, 8625861. [CrossRef] [PubMed]
12. Romano-Smith, S.; Wood, G.; Wright, D.J.; Wakefield, C.J. Simultaneous and alternate action observation and motor imagery combinations improve aiming performance. *Psychol. Sport. Exerc.* **2018**, *38*, 100–106. [CrossRef]
13. Chye, S.; Valappil, A.C.; Wright, D.J.; Frank, C.; Shearer, D.A.; Tyler, C.J.; Diss, C.E.; Mian, O.S.; Tillin, N.A.; Bruton, A.M. The effects of combined action observation and motor imagery on corticospinal excitability and movement outcomes: Two meta-analyses. *Neurosci. Biobehav. Rev.* **2022**, *143*, 104911. [CrossRef]
14. Prasertsakul, T.; Kaimuk, P.; Chinjenpradit, W.; Limroongreungrat, W.; Charoensuk, W. The effect of virtual reality-based balance training on motor learning and postural control in healthy adults: A randomized preliminary study. *Biomed. Eng. Online* **2018**, *17*, 124. [CrossRef]
15. Pastel, S.; Petri, K.; Chen, C.H.; Wiegand Cáceres, A.M.; Stirnatis, M.; Nübel, C.; Schlotter, L.; Witte, K. Training in virtual reality enables learning of a complex sports movement. *Virtual Real.* **2023**, *27*, 523–540. [CrossRef]
16. Fusco, A.; Tieri, G. Challenges and Perspectives for Clinical Applications of Immersive and Non-Immersive Virtual Reality. *J. Clin. Med.* **2022**, *11*, 4540. [CrossRef]
17. Slater, M.; Pérez-Marcos, D.; Ehrsson, H.H.; Sanchez-Vives, M.V. Towards a digital body: The virtual arm illusion. *Front. Hum. Neurosci.* **2008**, *2*, 6. [CrossRef]
18. Banakou, D.; Slater, M. Body ownership causes illusory self-attribution of speaking and influences subsequent real speaking. *Proc. Natl. Acad. Sci. USA* **2014**, *111*, 17678–17683. [CrossRef]
19. Kilteni, K.; Normand, J.M.; Sanchez-Vives, M.V.; Slater, M. Extending body space in immersive virtual reality: A very long arm illusion. *PLoS ONE* **2012**, *7*, e40867. [CrossRef] [PubMed]
20. Argelaguet, F.; Hoyet, L.; Trico, M.; Lecuyer, A. The role of interaction in virtual embodiment: Effects of the virtual hand representation. In Proceedings of the 2016 IEEE Virtual Reality (VR), Greenville, SC, USA, 19–23 March 2016; pp. 3–10.
21. Nuara, A.; Fabbri-Destro, M.; Scalona, E.; Lenzi, S.E.; Rizzolatti, G.; Avanzini, P. Telerehabilitation in response to constrained physical distance: An opportunity to rethink neurorehabilitative routines. *J. Neurol.* **2022**, *269*, 627–638. [CrossRef] [PubMed]
22. Jastorff, J.; Abdollahi, R.O.; Fasano, F.; Orban, G.A. Seeing biological actions in 3D: An fMRI study. *Hum. Brain Mapp.* **2016**, *37*, 203–219. [CrossRef] [PubMed]
23. Laver, K.E.; Lange, B.; George, S.; Deutsch, J.E.; Saposnik, G.; Crotty, M. Virtual reality for stroke rehabilitation. *Cochrane Database Syst. Rev.* **2017**, *11*, CD008349. [CrossRef] [PubMed]
24. Burin-Chu, S.; Baillet, H.; Leconte, P.; Lejeune, L.; Thouvarecq, R.; Benguigui, N. Effectiveness of virtual reality interventions of the upper limb in children and young adults with cerebral palsy: A systematic review with meta-analysis. *Clin. Rehabil.* **2024**, *38*, 15–33. [CrossRef] [PubMed]
25. Dejian, L.; Dede, C.; Huang, R.; Richards, J. Implicit Learning through Embodiment in Immersive Virtual Reality. In *Virtual, Augmented, and Mixed Realities in Education*; Springer: Singapore, 2018; pp. 19–33.
26. Iacoboni, M.; Molnar-Szakacs, I.; Gallese, V.; Buccino, G.; Mazziotta, J.C.; Rizzolatti, G. Grasping the intentions of others with one's own mirror neuron system. *PLoS Biol.* **2005**, *3*, e79. [CrossRef]
27. Liuzza, M.T.; Candidi, M.; Sforza, A.L.; Aglioti, S.M. Harm avoiders suppress motor resonance to observed immoral actions. *Soc. Cogn. Affect. Neurosci.* **2015**, *10*, 72–77. [CrossRef] [PubMed]

28. Amoruso, L.; Urgesi, C. Contextual modulation of motor resonance during the observation of everyday actions. *NeuroImage* **2016**, *134*, 74–84. [CrossRef] [PubMed]
29. Choi, J.W.; Kim, B.H.; Huh, S.; Jo, S. Observing Actions Through Immersive Virtual Reality Enhances Motor Imagery Training. *IEEE Tranbs. Neural. Syst. Rehabil. Eng.* **2020**, *28*, 1614–1622. [CrossRef]
30. Mao, H.; Li, Y.; Tang, L.; Chen, Y.; Ni, J.; Liu, L.; Shan, C. Effects of mirror neuron system-based training on rehabilitation of stroke patients. *Brain Behav.* **2020**, *10*, e01729. [CrossRef]
31. Tofani, M.; Santecchia, L.; Conte, A.; Berardi, A.; Galeoto, G.; Sogos, C.; Petrarca, M.; Panuccio, F.; Castelli, E. Effects of Mirror Neurons-Based Rehabilitation Techniques in Hand Injuries: A Systematic Review and Meta-Analysis. *Int. J. Environ. Res. Public Health* **2022**, *19*, 5526. [CrossRef]
32. Agnelli, M.; Libeccio, B.; Frisoni, M.C.; Bolzoni, F.; Temporiti, F.; Gatti, R. Action observation plus motor imagery and somatosensory discrimination training are effective non-motor approaches to improve manual dexterity. *J. Hand Ther.* **2023**, *37*, 94–100. [CrossRef]
33. Chan, T. An investigation of finger and manual dexterity. *Percept. Mot. Skills.* **2000**, *90*, 537–542. [CrossRef]
34. Karni, A.; Meyer, G.; Rey-Hipolito, C.; Jezzard, P.; Adams, M.M.; Turner, R.; Ungerleider, L.G. The acquisition of skilled motor performance: Fast and slow experience-driven changes in primary motor cortex. *Proc. Natl. Acad. Sci. USA* **1998**, *95*, 861–868. [CrossRef]
35. Buddenberg, L.A.; Davis, C. Test-retest reliability of the Purdue Pegboard Test. *Am. J. Occup. Ther.* **2000**, *54*, 555–558. [CrossRef]
36. Lakens, D. Calculating and reporting effect sizes to facilitate cumulative science: A practical primer for *t*-tests and ANOVAs. *Front. Psychol.* **2013**, *4*, 863. [CrossRef]
37. Oldfield, R.C. The assessment and analysis of handedness: The Edinburgh inventory. *Neuropsychologia* **1971**, *9*, 97–113. [CrossRef]
38. Scalona, E.; De Marco, D.; Bazzini, M.C.; Nuara, A.; Zilli, A.; Taglione, E.; Pasqualetti, F.; Della Polla, G.; Lopomo, N.F.; Fabbri-Destro, M.; et al. A Repertoire of Virtual-Reality, Occupational Therapy Exercises for Motor Rehabilitation Based on Action Observation. *Data* **2022**, *7*, 9. [CrossRef]
39. Malouin, F.; Richards, C.L.; Jackson, P.L.; Lafleur, M.F.; Durand, A.; Doyon, J. The Kinesthetic and Visual Imagery Questionnaire (KVIQ) for assessing motor imagery in persons with physical disabilities: A reliability and construct validity study. *J. Neurol. Phys. Ther.* **2007**, *31*, 20–29. [CrossRef]
40. Temporiti, F.; Mandaresu, S.; Calcagno, A. Kinematic evaluation and reliability assessment of the Nine Hole Peg Test for manual dexterity. *J. Hand Ther.* **2023**, *36*, 560–567. [CrossRef]
41. Li, J.; Zhu, H.; Pan, Y.; Wang, H.; Cen, Z.; Yang, D.; Luo, W. Three-Dimensional Pattern Features in Finger Tapping Test for Patients with Parkinson's disease. *Proc. Annu. Int. Conf. IEEE Eng. Med. Biol Soc.* **2020**, *2020*, 3676–3679.
42. Huygelier, H.; Mattheus, E.; Abeele, V.V.; van Ee, R.; Gillebert, C.R. The Use of the Term Virtual Reality in Post-Stroke Rehabilitation: A Scoping Review and Commentary. *Psychol. Belg.* **2021**, *61*, 145–162. [CrossRef]
43. Lee, S.H.; Jung, H.Y.; Yun, S.J.; Oh, B.M.; Seo, H.G. Upper Extremity Rehabilitation Using Fully Immersive Virtual Reality Games with a Head Mount Display: A Feasibility Study. *PMR* **2020**, *12*, 257–262. [CrossRef]
44. Romano Smith, S.; Wood, G.; Coyles, G.; Roberts, J.W.; Wakefield, C.J. The effect of action observation and motor imagery combinations on upper limb kinematics and EMG during dart-throwing. *Scand. J. Med. Sci. Sports* **2019**, *29*, 1917–1929. [CrossRef]
45. Gatti, R.; Tettamanti, A.; Gough, P.M.; Riboldi, E.; Marinoni, L.; Buccino, G. Action observation versus motor imagery in learning a complex motor task: A short review of literature and a kinematics study. *Neurosci. Lett.* **2013**, *540*, 37–42. [CrossRef]
46. Pereira, E.A.; Raja, K.; Gangavalli, R. Effect of training on interlimb transfer of dexterity skills in healthy adults. *Am. J. Phys. Med. Rehabil.* **2011**, *90*, 25–34. [CrossRef]
47. Calvo-Merino, B.; Grèzes, J.; Glaser, D.E.; Passingham, R.E.; Haggard, P. Seeing or Doing? Influence of Visual and Motor Familiarity in Action Observation. *Curr. Biol.* **2006**, *16*, 1905–1910. [CrossRef]
48. Turhan, B.; Gümüş, Z.H. A Brave New World: Virtual Reality and Augmented Reality in Systems Biology. *Front. Bioinform.* **2022**, *2*, 873478. [CrossRef]
49. Ferri, S.; Pauwels, K.; Rizzolatti, G.; Orban, G.A. Stereoscopically Observing Manipulative Actions. *Cereb. Cortex.* **2016**, *26*, 3591–3610. [CrossRef]
50. Petkova, V.I.; Khoshnevis, M.; Ehrsson, H.H. The perspective matters! Multisensory integration in ego-centric reference frames determines full-body ownership. *Front. Psychol.* **2011**, *2*, 35. [CrossRef]
51. Šlosar, L.; Peskar, M.; Pišot, R.; Marusic, U. Environmental enrichment through virtual reality as multisensory stimulation to mitigate the negative effects of prolonged bed rest. *Front. Aging Neurosci.* **2023**, *15*, 1169683. [CrossRef]
52. Tinga, A.M.; Visser-Meily, J.M.; van der Smagt, M.J.; Van der Stigchel, S.; van Ee, R.; Nijboer, T.C. Multisensory Stimulation to Improve Low- and Higher-Level Sensory Deficits after Stroke: A Systematic Review. *Neuropsychol. Rev.* **2016**, *26*, 73–91. [CrossRef]
53. Brugada-Ramentol, V.; Bozorgzadeh, A.; Jalali, H. Enhance VR: A multisensory approach to cognitive training and monitoring. *Front. Digit. Health* **2022**, *4*, 916052. [CrossRef] [PubMed]
54. Waltemate, T.; Gall, D.; Roth, D.; Botsch, M.; Latoschik, M.E. The impact of avatar personalization and immersion on virtual body ownership, presence, and emotional response. *IEEE Trans. Vis. Comput. Graph.* **2018**, *24*, 1643–1652. [CrossRef] [PubMed]
55. Proud, E.L.; Miller, K.J.; Bilney, B.; Morris, M.E.; McGinley, J.L. Construct validity of the 9-Hole Peg Test and Purdue Pegboard Test in people with mild to moderately severe Parkinson's disease. *Physiotherapy* **2020**, *107*, 202–208. [CrossRef] [PubMed]

56. Tseng, Y.C.; Chang, K.Y.; Liu, P.L.; Chang, C.C. Applying the purdue pegboard to evaluate precision assembly performance. In Proceedings of the IEEE International Conference on Industrial Engineering and Engineering Management (IEEM) 2017, Singapore, 10–13 December 2017; pp. 1179–1183.
57. Cortes, J.C.; Goldsmith, J.; Harran, M.D.; Xu, J.; Kim, N.; Schambra, H.M.; Luft, A.R.; Celnik, P.; Krakauer, J.W.; Kitago, T. A short and distinct time window for recovery of arm motor control early after stroke revealed with a global measure of trajectory kinematics. *Neurorehabilit. Neural Repair* **2017**, *31*, 552–560. [CrossRef] [PubMed]

Disclaimer/Publisher's Note: The statements, opinions and data contained in all publications are solely those of the individual author(s) and contributor(s) and not of MDPI and/or the editor(s). MDPI and/or the editor(s) disclaim responsibility for any injury to people or property resulting from any ideas, methods, instructions or products referred to in the content.

bioengineering

MDPI

Article

Balance Evaluation Based on Walking Experiments with Exoskeleton Interference

Liping Wang [1,2], Xin Li [1,2], Yiying Peng [1,2], Jianda Han [1,2,3] and Juanjuan Zhang [1,2,3,*]

1 Tianjin Key Laboratory of Intelligent Robotics, Institute of Robotics and Automatic Information System, Nankai University, Tianjin 300350, China; lipingwang@mail.nankai.edu.cn (L.W.); 2120190365@mail.nankai.edu.cn (X.L.); yiyingpeng@mail.nankai.edu.cn (Y.P.); hanjianda@nankai.edu.cn (J.H.)
2 College of Artificial Intelligence, Nankai University, Tianjin 300350, China
3 School of Materials Science and Engineering, Smart Sensing Interdisciplinary Science Center, Nankai University, Tianjin 300350, China
* Correspondence: juanjuanzhang@nankai.edu.cn

Abstract: The impairment of walking balance function seriously affects human health and will lead to a significantly increased risk of falling. It is important to assess and improve the walking balance of humans. However, existing evaluation methods for human walking balance are relatively subjective, and the selected metrics lack effectiveness and comprehensiveness. We present a method to construct a comprehensive evaluation index of human walking balance. We used it to generate personal and general indexes. We first pre-selected some preliminary metrics of walking balance based on theoretical analysis. Seven healthy subjects walked with exoskeleton interference on a treadmill at 1.25 m/s while their ground reaction force information and kinematic data were recorded. One subject with Charcot–Marie–Tooth walked at multiple speeds without the exoskeleton while the same data were collected. Then, we picked a number of effective evaluation metrics based on statistical analysis. We finally constructed the Walking Balance Index (WBI) by combining multiple metrics using principal component analysis. The WBI can distinguish walking balance among different subjects and gait conditions, which verifies the effectiveness of our method in evaluating human walking balance. This method can be used to evaluate and further improve the walking balance of humans in subsequent simulations and experiments.

Keywords: walking balance; statistical analysis; principal component analysis; ankle exoskeleton

Citation: Wang, L.; Li, X.; Peng, Y.; Han, J.; Zhang, J. Balance Evaluation Based on Walking Experiments with Exoskeleton Interference. *Bioengineering* **2024**, *11*, 386. https://doi.org/10.3390/bioengineering11040386

Academic Editor: Carlo Albino Frigo

Received: 17 March 2024
Revised: 8 April 2024
Accepted: 11 April 2024
Published: 16 April 2024

Copyright: © 2024 by the authors. Licensee MDPI, Basel, Switzerland. This article is an open access article distributed under the terms and conditions of the Creative Commons Attribution (CC BY) license (https://creativecommons.org/licenses/by/4.0/).

1. Introduction

Walking balance has always been a concern in the field of medical rehabilitation. Impaired walking balance function seriously affects human health, especially for patients and the elderly with mobility difficulties. The human body can use fast and powerful muscle contraction and relaxation to move joints quickly to maintain posture and balance. The strength of muscles is critical for providing the power to regulate balance [1]. Decreased muscle strength, slowed muscle response, and limited range of motion of joints all result in the ineffective implementation of balance strategies, which affects human locomotion ability [2], disturbs the stability of posture [3], and eventually leads to frequent falls [4]. Falling can easily lead to a variety of health problems, such as fracture, joint deformity, dislocation, and soft tissue injury [5]. Therefore, the walking balance of humans, especially those with limited mobility, is a matter of concern. It is crucial to provide assistance for them and improve their walking balance. We first need to evaluate the state of human walking balance and aim to optimize it in future works.

The evaluation methods for balance are usually grouped into clinical tests, scaling methods, and quantitative measurements by instruments. Clinical tests, such as the Romberg Test [6,7], are easy to apply but inaccurate and subjective. These methods are only suitable for the preliminary screening of patients with suspected balance problems [8].

Scaling methods include the Berg Balance Scale [9], Tinetti Test [10], and so on. Compared to clinical tests, scaling methods are quantified by levels and are relatively more accurate. However, the scaling methods still have a certain subjectivity [11,12]. The determination of grading boundaries depends on the personal judgment of evaluators [13,14]. Quantitative measurement methods, such as the pressure plate method [15], have the advantage of being able to accurately measure the state of balance. Existing quantitative measurement methods select metrics only based on the intuition and experience of professionals and cover fewer metrics.

Researchers have used a lot of metrics to evaluate balance. The commonly selected metrics are the human body's center of mass (COM) [4,16,17] and the trajectory of the center of pressure (COP) [18–20]. When the human is standing still, the body's stability can be evaluated by the position of the COM [16]. Whether an additional step is required to avoid a fall can also be obtained by analyzing the position and the velocity of the COM [21]. In static balance, measuring the swing trajectory of the COP with a force plate has become a standard for evaluating static balance ability [18,22]. Additionally, it has shown that the distance and velocity between the anterior and lateral displacements of the COP contribute to explaining the walking balance [19,20]. Acceleration information from human body can also be used to assess human walking balance [23,24]. The Lyapunov index of the COP's speed and the trunk acceleration of the elderly who are prone to falling are significantly higher than those of normal people [24]. Experts put forward the stability criterion of the Zero Moment Point (ZMP), which provides an effective solution for the balanced judgment of human walking [25]. The relative position between the centroidal moment pivot (CMP) and ZMP can supplement information about the rotational dynamics of the human body. In this study, we considered the COP and ZMP as the same points [26] so we replaced the position of the ZMP with the position of the COP. In addition, the margin of stability (MOS) contains information on position and velocity which can describe changes in human motion [21,27]. However, a single metric is not effective or comprehensive as it does not consider or reflect all factors influencing balance during human locomotion. It is necessary to combine multiple metrics and establish a comprehensive quantitative index to evaluate the walking balance of different individuals under different walking conditions.

The purpose of this study was to present a method to construct an evaluation index of human walking balance that (1) comprehensively selects more preliminary metrics related to walking balance in theory, (2) reduces human factors through being driven by pure data, and (3) can quantify and evaluate human walking balance in a way that can distinguish between different individuals under different walking conditions. We collected gait data from eight subjects, seven healthy subjects and one Charcot–Marie–Tooth (CMT) patient. The main clinical symptoms for patients with CMT, like muscle weakness and atrophy, seriously affect their walking balance. Then, we used our method to generate personal and general indexes. We verified the effectiveness of this method in evaluating human walking balance and analyzed the main influencing metrics of balance during human walking. This study is expected to provide guidance for future research on walking balance evaluation, explore the potential factors that influence human walking balance, and further improve the walking balance function for the required population in the future.

2. Methods

We pre-selected a list of preliminary metrics for human walking balance through theoretical analysis. Then, we recruited eight subjects, seven healthy subjects and one subject with CMT. We conducted walking experiments with exoskeleton interference to collect data for the selected metrics from the seven healthy subjects. We also collected the same data from the CMT subject at multiple speeds. We further picked subsets of these metrics that significantly deviated between different gait conditions through statistical analysis. We eventually constructed a comprehensive quantitative evaluation index of walking balance by combining multiple metrics using principal component analysis (PCA).

We computed and analyzed the results of our method to verify its effectiveness in evaluating human walking balance.

2.1. Experimental Platform

We used a gait experimental platform to collect data on the preliminary metrics of human walking balance (Figure 1). The experimental platform consisted of a force-measuring treadmill, an optical motion capture system, and an ankle exoskeleton system. The ankle exoskeleton system was driven by an off-board actuation system with a Bowden cable and controlled by a real-time control (DS1202, dSPACE, Paderborn, GmbH) system.

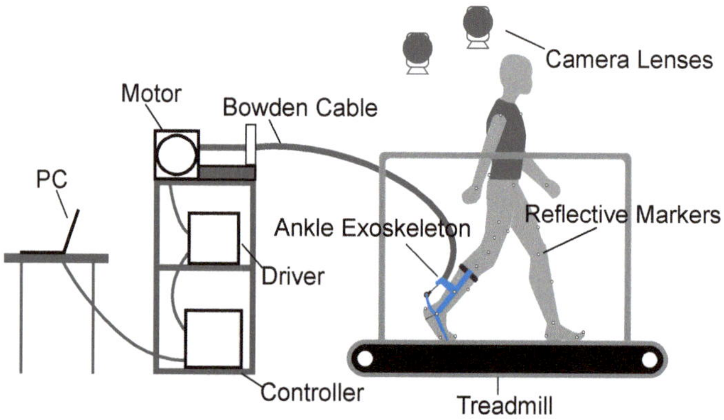

Figure 1. Gait experimental platform. The gait experimental platform included a control system, an actuation system, transmissions, a motion capture system, and an ankle exoskeleton.

2.1.1. Treadmill

We used the gait analysis treadmill (Bertec, Columbus, OH, USA) in this study. The treadmill can measure forces and moments in three directions of an orthogonal coordinate system. Some related balance metrics, such as the positions of the COP, can be calculated by measuring the real-time ground reaction forces and moments.

2.1.2. Motion Capture System

We used an optical motion capture system (Oqus 700+, Qualisys, Sweden) with ten high-speed camera lenses at a sampling rate of 100 Hz to accurately catch the motion of subjects with reflective markers. The basic motion data of subjects, such as the velocities and accelerations of limbs, can be calculated through analysis software.

2.1.3. Ankle Exoskeleton System

We used the ankle exoskeleton to apply ankle interference torques and to obtain relatively unbalanced gait data. The ankle exoskeleton is mainly composed of three components: a calf frame, a foot frame, and a forefoot frame. The interference torque is transmitted through a Bowden cable and applied at the ankle joint. The concrete structure of the ankle exoskeleton can be found in the previous article [28].

The ankle exoskeleton uses several sensors to measure human–robot interactions. A magnetic encoder (PQY 18, ACCNT, Dongguan, China) was fixed on the lateral shaft of the exoskeleton to measure ankle joint angles. A load cell (DYMH-106, DAYSENSOR, Bengbu, China) was seated at the ankle lever to measure Bowden cable force. A footswitch (KW12, Risym, Shenzhen, China) was installed on the heel of the shoe to detect heel strike during gait cycles.

2.1.4. Control and Actuation System

The control and actuation system were mounted on a shelf next to the treadmill. We used the real-time control system (DS1202, dSPACE, Paderborn, Germany) to sample data from sensors at 5000 Hz and drive the actuation system at 500 Hz. The actuation system was composed of an AC servo motor, a 5:1 planetary gear, and a motor driver (BSM90N-175AA, GBSM90-MRP120-5, and MF180-04AN-16A, ABB, Zurich, Switzerland).

2.2. Pre-Selection of Preliminary Metrics

In accordance with previous studies, we pre-selected the following 14 balance metrics related to human walking balance:

Group A, the positions and velocities of the COP:
(1) COP_x, the position of the COP on the sagittal axis;
(2) V_{COP_x}, the velocity of the COP on the sagittal axis;
(3) COP_y, the position of the COP on the coronal axis;
(4) V_{COP_y}, the velocity of the COP on the coronal axis.
Group B, the accelerations of the COM:
(5) A_{COM_x}, the acceleration of the COM on the sagittal axis;
(6) A_{COM_y}, the acceleration of the COM on the coronal axis;
(7) A_{COM_z}, the acceleration of the COM on the vertical axis;
(8) A_{COM}, the resultant acceleration of the COM.
Group C, the positions of the COM:
(9) COM_x, the position of the COM on the sagittal axis;
(10) COM_y, the position of the COM on the coronal axis;
(11) COM_z, the position of the COM on the vertical axis.
Group D, other metrics:
(12) COP_CMP, the distance between the CMP and COP;
(13) MOS, the MOS of the trunk;
(14) A_{ANG}, the angular acceleration of the trunk.

The real-time positions of the COP can be calculated by the ground reaction forces and moments as follows:

$$COP_x = \frac{-h \cdot F_{ground,x} - M_{ground,y}}{F_{ground,z}} \tag{1}$$

$$COP_y = \frac{h \cdot F_{ground,y} - M_{ground,x}}{F_{ground,z}} \tag{2}$$

where $F_{ground,x}$, $F_{ground,y}$, and $F_{ground,z}$ are the components of the ground reaction forces along the three coordinate axes, respectively. $M_{ground,x}$ and $M_{ground,y}$ are the ground reaction moments in the sagittal and coronal plane, and h is the height difference between the belt and the x–y plane of the treadmill coordinate system.

The velocities of the COP can be obtained by:

$$V_{COP_x} = \frac{d}{dt}COP_x, V_{COP_y} = \frac{d}{dt}COP_y \tag{3}$$

The position of the human body's COM can be expressed by:

$$COM = \sum_{i=1}^{n} \frac{m_i \cdot com_i}{M} \tag{4}$$

where com_i and m_i, respectively, represent the centroid position and mass of the ith segment of the human body, M represents the total mass of the human body, and n is the number of human body segments.

The accelerations of the COM can be obtained by:

$$A_{COM_x} = \frac{d^2}{dt^2}COM_x, A_{COM_y} = \frac{d^2}{dt^2}COM_y, A_{COM_z} = \frac{d^2}{dt^2}COM_z, \quad (5)$$

$$A_{COM} = \sqrt{A_{COM_x}^2 + A_{COM_y}^2 + A_{COM_z}^2} \quad (6)$$

The position of the CMP can be calculated from ground reaction forces and the position of the COM as follows:

$$CMP_x = COM_x - \frac{F_{ground,x}}{F_{ground,z}} \cdot COM_Z \quad (7)$$

$$CMP_y = COM_y - \frac{F_{ground,y}}{F_{ground,z}} \cdot COM_Z \quad (8)$$

The MOS proposed by Terry et al. [27] was adopted in this study. The MOS can be expressed as:

$$MOS = \left|(COP + \frac{V_{COP}}{\omega_0}) - (COM + \frac{V_{COM}}{\omega_0})\right| \quad (9)$$

$$\omega_0 = \sqrt{\frac{g}{l}} \quad (10)$$

where g is the acceleration of gravity, and l is the height of the COM.

We also selected three time domain characteristics as the gait features for each metric above: root mean square (RMS), variance (Var), and range (Range). The RMS has a non-negative feature which is often used to characterize the effective value of the data. Var characterizes the degree of dispersion of the data distribution. Range reflects the range of the data, emphasizing the extreme values. The three selected time domain characteristics show different aspects of the collected data. Therefore, 14 balance metrics were selected, and three time domain characteristics were selected for each balance metric. A total of 42 preliminary metrics were selected in this study.

2.3. Experimental Protocol

Seven healthy subjects (five males and two females; age = 23.3 ± 1.0 years; body mass = 67.9 ± 12.4 kg; height = 175.4 ± 6.5 cm) participated in this study. We also recruited one male subject with CMT (age = 14 years; body mass = 35.3 kg; height = 160 cm) to verify the feasibility of our method for the patient (Table 1). The study was conducted according to the Helsinki Declaration and approved by the ethical committee of Nankai University (reference number: NKUIRB2021054). All subjects provided written informed consent before completing the protocol. All methods used in this study were performed in accordance with the relevant guidelines and regulations.

We designed and conducted walking experiments with exoskeleton interference based on the experimental platform to collect the preliminary metric data for the healthy subjects in relatively balanced and unbalanced gait conditions. The human ankle plays a vital role in balance, propulsion, and locomotion in a wide range of environments encountered. Two important mechanisms for balance control are the stepping strategy and the lateral ankle strategy. The ankle strategy is faster than the stepping strategy [29]. The timing, magnitude, and speed of torque produced by the ankle joint all affect walking balance [30–32]. Therefore, we used the ankle exoskeleton to apply ankle interference torque, affecting walking balance.

Table 1. Subjects of experiments.

Subject	Height (cm)	Weight (kg)	Peak Torque (N·m)
1	172	53	22
2	170	58	22
3	173	65	22
4	188	90	40
5	175	72	17
6	170	75	15
7	180	62	20
8	160	35.3	-

To create external interference for subjects during their walking, we first defined four interference torque profiles, sinusoidal interference torque (SIN), constant interference torque (CON), random interference torque (RAN), and mixed interference torque (MIX). The first three profiles are shown in Figure 2. SIN consisted of two positive continuous sinusoids which started at heel strike and ended before toe off. The SIN had two peak values that were given at the ankle joint mainly during the mid-stance and terminal stance phase. The CON was similar to a step signal whose maximum torque remained the same. We used two cubic splines to form the rise and fall phase of the CON. The RAN was generated by random combinations of sine and cosine signals. Its curves changed each stride, and peak values were applied to the ankle joint at any time during the stance phase. The MIX was a combination of the first three interference torque profiles whose order was SIN–CON–RAN–SIN–CON–RAN, and the duration of each application of interference torque was the same. We used two parameters to limit the generation of torque profiles: peak torque and applied time. Peak torque refers to the maximum allowable torque that the subjects could bear. The peak torque of each subject is shown in Table 1. Applied time refers to the time when the interference torque was applied to subjects from the first heel strike to the ground. For safety reasons, we only applied ankle interference in the supporting phase (0–60% of a gait cycle) during walking. The interference torque was tracked using a combination of proportional–derivative control and an iterative learning algorithm [33,34]. The effectiveness of this torque control method has been demonstrated in the previously published literature. The torque controller can be found in previous articles [35,36].

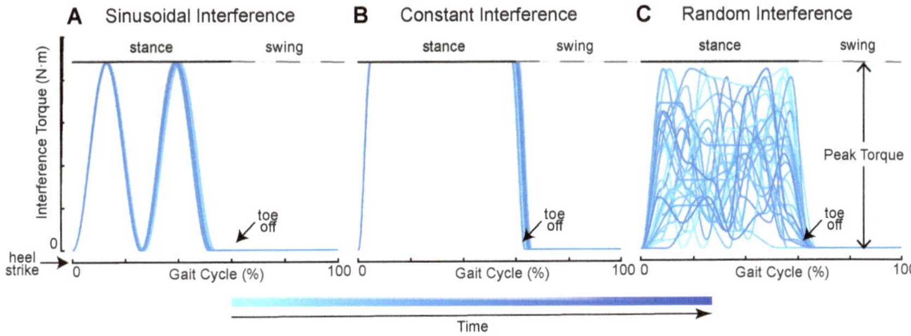

Figure 2. Three ankle interference torque profiles. These torques were applied only in the support phase (0–60% of a gait cycle) for safety. (**A**) Sinusoidal interference torque. It consisted of two continuous sinusoids. (**B**) Constant interference torque. Its maximum torque remained the same. (**C**) Random interference torque. The torque profile was generated randomly.

The walking experiments with exoskeleton interference for healthy subjects were divided into two parts: zero torque (ZT), where subjects walked with the exoskeleton but no torque applied, and ankle interference torque, where subjects walked with the exoskeleton applying the interference torque. The seven healthy subjects walked on the treadmill at 1.25 m/s and wore the right ankle exoskeleton. The first healthy subject walked under a 30 s ZT condition followed by a 6 min MIX condition with a 2 min break in between. The other six healthy subjects first performed under a ZT condition for 30 s and then a SIN condition for 2 min, a CON condition for 2 min, a RAN condition for 2 min, and, finally, a MIX condition for 6 min with a 2 min rest between each condition. The CMT subject walked without the ankle exoskeleton, which was called normal walking (NW), on the treadmill at multiple speeds, which he could accept. We collected his gait data for 30 s at 0.45 m/s, 0.55 m/s, and 0.65 m/s. After a period of rest, a second round of experiments was carried out in reverse order, and the gait data were collected again. For each subject, 38 reflective markers (six on each foot, eight on each calf and thigh, three on the pelvis, one on the chest, and three on each arm) were affixed to the whole body during all experiments. We collected the motion capture data and treadmill data during all the experiments. The collected experimental data were used in the construction and validation of the comprehensive index for human walking balance.

2.4. *Construction of Walking Balance Index*

We processed the experimental data from the eight subjects and obtained the data for the selected metrics. We picked subsets of these metrics using statistical analysis and constructed the index by combining multiple metrics using PCA. The data processing was conducted using OpenSim 4.1 [37,38] and MATLAB (MathWorks, Natick, MA, USA). The statistical analysis and PCA were conducted using MATLAB.

2.4.1. Data Processing

We synchronized and collected the three-dimensional ground reaction forces and moments (1000 Hz) data and motion capture (100 Hz) data of the eight subjects using the real-time controller. We pre-processed the experimental data with low-pass filters and scaled the musculoskeletal model with the height and weight of the subjects. We performed inverse kinematics using the Rajagopal2015 model in OpenSim [39]. We then calculated the 14 aforementioned balance metrics using the positions of marks and the masses of the human body segments. We segmented the metric data according to the gait cycles. For each metric in each gait cycle, we calculated all three time domain characteristics and summarized them into three data sets. We finally obtained the data sets of the 42 selected preliminary metrics under each gait condition.

2.4.2. Statistical Analysis

Some of these metrics may be ineffective and even interfere with the evaluation of walking balance. It was necessary to further distinguish the effective metrics with evaluation ability from the preliminary metrics using statistical analysis.

From the data processing, we obtained data sets for the 42 selected preliminary metrics under each gait condition. For each subject, we used the significance tests to test all the preliminary metrics data with exoskeleton interference (SIN, CON, RAN, and MIX) and the preliminary metrics data without exoskeleton interference (ZT) in pairs.

We first used the Lilliefors test to test the normality of these data and the Bartlett test to test their homogeneity of variance. According to the results of normality and variance homogeneity tests, we chose the t-test for the data conforming to the normality and variance homogeneity tests, the corrected t-test for the data conforming to the normality but not homogeneity of variance, and the rank-sum test for the other data. All the above were two-sided tests. Considering the sample size (42 metrics per stride), the significance level was set at $\alpha = 0.01$. We finally picked subsets of the preliminary metrics that significantly deviated between different gait conditions.

2.4.3. Principal Component Analysis

There are always some correlations between the effective metrics we picked above so these needed to be further extracted.

The principle of PCA is to recombine original variables into a new set of unrelated comprehensive variables. This method can transform high-dimensional problems into low-dimensional problems to reduce the correlations. The new variables formed by a linear combination of the initial variables are the principal components. These principal components are independent of each other and largely resistant to man-made interference. The principal component coefficients and eigenvalues are also generated only based on data. Therefore, we used PCA to reduce the correlations among these effective metrics and constructed the final index.

We first standardized and decentralized the data of the effective metrics picked above. For subjects with different heights and weights, the values of distance metrics were different in the relatively consistent gait conditions. The units and orders of magnitude of these metrics were also different. Therefore, all the distance metrics were scaled according to the height and weight of each subject, and all the metrics needed to be dimensionless. We used the Kaiser–Meyer–Olkin test to test whether these effective metrics were correlated and could be analyzed by PCA.

We then calculated the covariance matrix of the metric data and calculated the principal component coefficients and eigenvalues of the covariance matrix through PCA. The principal component coefficients were used to compose principal components, and the eigenvalues were used to calculate the weights of the principal components. We used the principal components whose cumulative contribution rate exceeded 85% to replace the original metrics and constructed the comprehensive quantitative index, which we called the Walking Balance Index (WBI). The WBI can be expressed as follows:

$$y_i = \sum_{j=1}^{m} w_{ij} x_j \tag{11}$$

$$WBI = \sum_{i=1}^{n} \frac{\lambda_i}{\sqrt{\sum_{j=1}^{n} \lambda_i^2}} y_i \tag{12}$$

where x_j and w_{ij} are the jth effective metric and the coefficient of x_j in principal component i, and m is the number of effective metrics after statistical analysis. y_i and λ_i are the ith principal component and the eigenvalue of y_i, and n is the number of principal components that we used.

2.5. Validation

We collected gait data from the seven healthy subjects under five gait conditions: ZT, SIN, CON, RAN, and MIX. We also collected gait data from the CMT subject at three speeds. Each combination of these data with and without exoskeleton interference can be used to construct the index. All experimental data can be used as validation data and to calculate the value of the WBI.

We mainly used the gait data of the seven healthy subjects obtained under ZT and MIX conditions to construct the personal indexes and compute their values. We also used the gait data of one healthy subject obtained under ZT and MIX conditions to construct the general index, and we used the data of six healthy subjects obtained under multiple gait conditions as validation data to compute the values of the general index. The data of the CMT subject were used to compute the value of the WBI constructed from subject1's gait data obtained under ZT and MIX conditions. In addition, we also used the data of the CMT subject and subject1 to construct the index and compared the potential correlations and differences in the balance metrics between the CMT patient and healthy individuals.

Meanwhile, we obtained four gait combinations by combining the condition without interference (ZT) and four conditions with interference (SIN, CON, RAN, MIX) in pairs for the healthy subjects. We constructed and calculated the WBI using all combinations so as to further explore the possible disadvantages of this method, the main factors affecting the walking balance of humans, and the consistencies and differences of strategies for adjusting walking balance for different individuals.

3. Results

Figure 3 shows the results of the preliminary metrics for the CMT patient (subject8) under the NW condition and healthy subject2 under ZT and MIX conditions. Compared to under the ZT condition, there were significant differences in these metrics under the NW condition for subject8, but there were fewer differences in them under the MIX conditions for subject2. The fluctuation range of the CMT patient was much larger than that of the healthy subject.

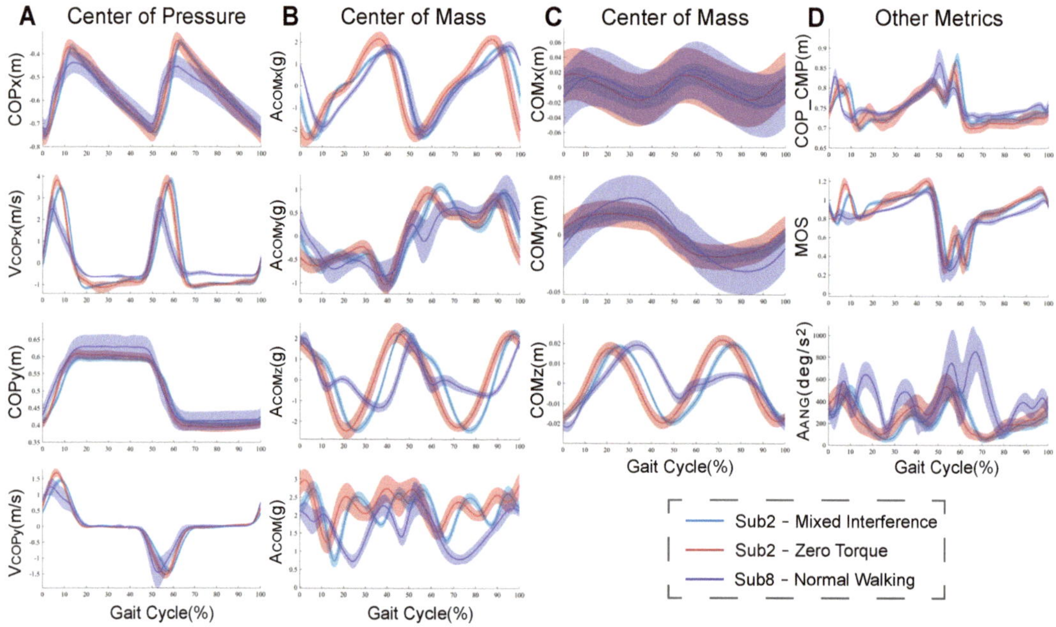

Figure 3. The changes in the preliminary metrics. (**A**) The changes in the positions and velocities of the COP. (**B**) The changes in the accelerations of the COM in three directions and the resultant acceleration. (**C**) The changes in the positions of the COM. (**D**) The changes in the relative position between the COP and CMP, MOS, and trunk angular acceleration. The blue curves show the changes for subject2 under the ZT condition. The red curves show the changes for subject2 under the MIX condition. The purple curves show the changes for subject8 with CMT under the NW condition. The shadow areas show the fluctuation range of different gait cycles.

The results of the personal indexes for the seven healthy subjects under ZT and MIX conditions are shown in Figure 4. Compared to under the ZT condition, the values of their personal indexes under MIX conditions are bigger and have a wider range of fluctuations. The results of the constructed indexes under two gait combinations for subject2 are shown in Figure 5. The values of the WBI under the two combinations are too similar to be distinguished (Figure 5A). The means and standard deviations of the indexes under the ZT condition are still smaller than those under CON and RAN conditions (Figure 5B). The general index constructed from one of the healthy subjects under ZT and MIX conditions was used to calculate the means and standard deviations for the six healthy subjects under multiple gait conditions (Figure 6). The mean of the general index for each subject under the ZT condition was lower than that under other conditions.

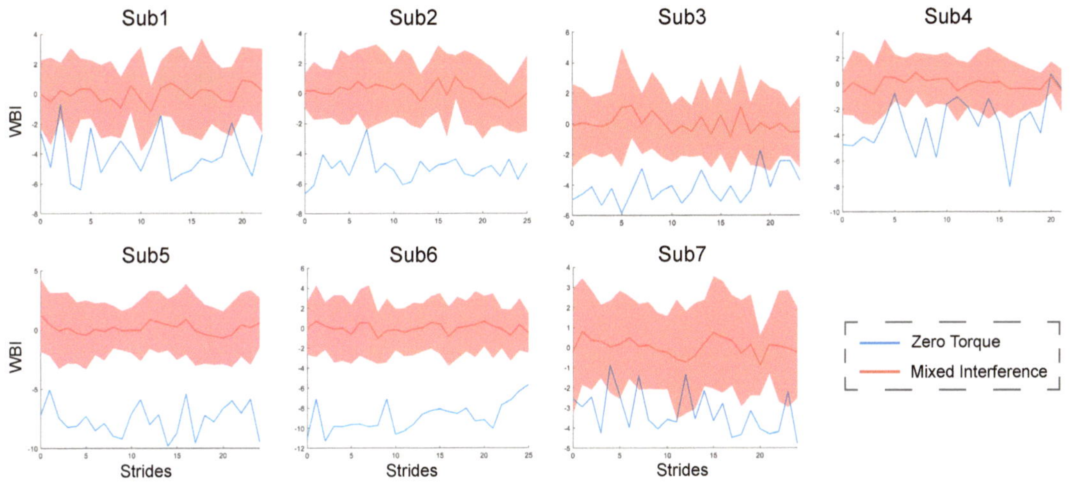

Figure 4. The values of personal WBI for seven healthy subjects. The blue curves are the WBI of each stride under the ZT condition. The red curves and areas are the mean and fluctuation range of the WBI under the MIX condition.

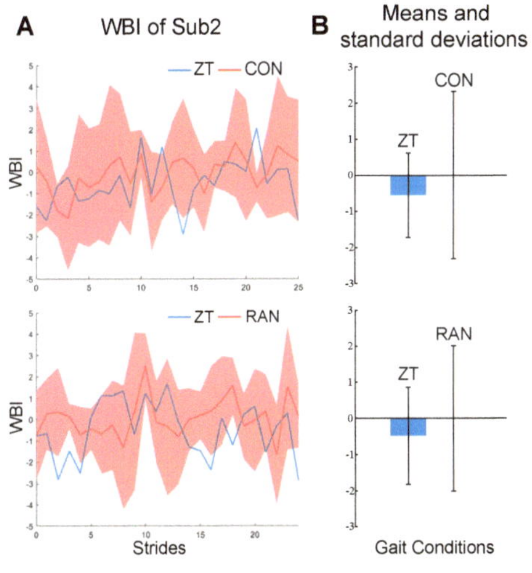

Figure 5. The results of WBI for subject2. (**A**) The WBI under ZT and CON conditions (above). The WBI under ZT and RAN conditions (below). (**B**) Bars and whiskers are means and standard deviations of the WBI for subject2 under ZT and CON conditions (above) and ZT and RAN conditions (below).

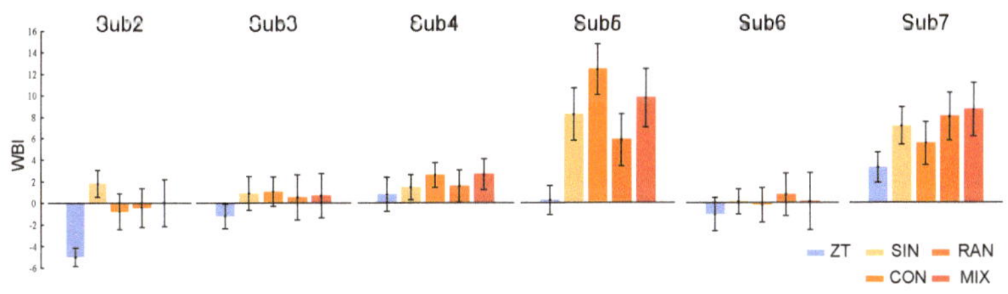

Figure 6. The values of the WBI for the six healthy subjects using the general index constructed from one subject's data obtained under ZT and MIX conditions. Bars and whiskers are the means and standard deviations of the WBI for six healthy subjects under different gait conditions.

Table 2 shows the results of the statistical analysis for subject1 under ZT and MIX conditions. A total of 25 effective metrics were picked from 42 preliminary metrics. They showed significant differences under the two different gait conditions. There was no significant difference in the information for COP_x, COP_CMP, or A_{COM_x} ($p > 0.01$, Table 2). There were significant differences in the characteristics from the three time domains, COP_y, V_{COP}, A_{COM_y}, A_{COM_z}, and COM_z ($p < 0.01$, Table 2). Other balance metrics showed significant differences in partial time domain characteristics. Table 3 shows the number of times that the metrics showed significant differences under four gait combinations for the six healthy subjects.

Table 2. Statistical analysis for subject1.

Time Domain Features	*p*-Value						
	Preliminary Metrics						
	COP_x	COP_y	V_{COP_x}	V_{COP_y}	MOS	A_{ANG}	COP_CMP
Var	0.664	0.000 *	0.002 *	0.000 *	0.000 *	0.002 *	0.081
RMS	0.727	0.000 *	0.006 *	0.000 *	0.041	0.079	0.039
Range	0.144	0.000 *	0.001 *	0.000 *	0.000 *	0.000 *	0.012
	COM_x	COM_y	COM_z	A_{COM_x}	A_{COM_y}	A_{COM_z}	A_{COM}
Var	0.756	0.020	0.000 *	0.107	0.000 *	0.000 *	0.070
RMS	0.001 *	0.011	0.000 *	0.105	0.000 *	0.000 *	0.000 *
Range	0.988	0.004 *	0.000 *	0.058	0.002 *	0.000 *	0.075

* statistically significant difference between ZT and MIX conditions.

Table 3. The number of times that the metrics showed significant differences under four gait combinations* for six subjects.

Time Domain Features	Preliminary Metrics				
	COP_x	COP_y	V_{COP_x}	V_{COP_y}	COP_CMP
Var	4/4/3/2/4/4 **	4/4/4/4/0/4	4/3/2/4/4/2	4/4/4/4/0/4	3/4/2/2/4/1
RMS	0/0/0/0/0/0	0/0/1/1/3/0	4/3/2/4/4/2	4/4/4/4/0/4	0/1/0/1/4/0
Range	4/4/2/2/4/4	4/4/4/4/1/4	4/4/1/4/4/2	4/4/4/4/0/4	2/4/0/2/4/2
	COM_x	COM_y	COM_z	MOS	A_{ANG}
Var	4/4/1/3/4/1	0/4/1/1/0/4	2/4/4/4/4/4	4/4/4/3/2/2	4/3/2/3/4/4
RMS	4/1/1/2/1/1	3/4/1/0/2/0	2/4/4/4/4/4	4/4/0/4/3/0	3/4/2/3/4/4
Range	4/3/1/3/4/2	1/4/1/1/0/4	2/4/4/4/4/4	4/4/4/4/4/2	3/4/2/3/4/4
	A_{COM_x}	A_{COM_y}	A_{COM_z}	A_{COM}	
Var	4/0/1/3/3/3	0/4/4/4/4/0	3/0/1/4/4/4	1/2/1/4/4/4	
RMS	4/0/1/3/3/3	0/4/4/4/4/0	3/0/1/4/4/4	3/0/0/4/4/4	
Range	4/1/2/4/4/3	0/4/4/4/0/2	2/0/0/4/4/4	2/3/1/4/4/2	

* All six subjects completed testing under the zero torque gait condition and four interference torque gait conditions. Four gait combinations were obtained by combining the zero torque condition without interference and four conditions with interference. ** Each element of the table represents the number of times that the metric showed significant differences under the four combinations for subject2/subject3/subject4/subject5/subject6/subject7.

The results of the PCA for subject1 are shown in Table 4. Table 5 shows the coefficient matrix of principal components for subject1. The contributions of the first three principal components were more than 10% (Table 4). The V_{COP}, COP_y, COM_z, A_{COM_z}, and MOS had larger coefficients of principal components compared to the other metrics (Tables 4 and 5).

Table 4. Principal component analysis for subject1.

Principal Components	Eigenvalue	Cumulative Contribution (%)	Cumulative Contribution (%)
1	7.42	**33.74**	33.74
2	5.36	**24.36**	58.1
3	3.62	**16.43**	74.53
4	1.79	8.14	82.67
5	1.48	6.75	89.42

Bold font represents the principal components which contributed more than 10%.

Table 5. The coefficient matrix of principal components for subject1.

Time Domain Features	Effective Balance Metrics	Principal Components				
		1	2	3	4	5
Var	COP_y	**0.330**	−0.075	−0.039	−0.178	−0.157
	COM_z	0.073	0.257	**0.331**	−0.025	0.110
	A_{COMy}	0.278	−0.002	−0.182	0.121	**0.419**
	A_{COMz}	0.169	0.265	0.286	0.000	−0.040
	V_{COPx}	−0.102	**0.380**	−0.150	−0.073	0.002
	V_{COPy}	**0.325**	−0.010	−0.128	−0.145	−0.212
	MOS	-0.047	**0.313**	−0.316	−0.140	−0.088
	A_{ANG}	0.122	0.092	−0.118	**0.588**	−0.278
RMS	A_{COMy}	0.278	−0.005	−0.179	0.121	**0.425**
	A_{COMz}	0.170	0.265	0.285	−0.001	−0.044
	A_{COM}	0.203	0.266	0.156	0.081	0.078
	V_{COPx}	−0.102	**0.381**	−0.149	−0.072	0.003
	V_{COPy}	**0.327**	−0.010	−0.123	−0.147	−0.210
Range	COP_y	**0.324**	−0.079	−0.004	−0.184	−0.161
	COM_y	0.245	−0.078	−0.022	−0.170	0.046
	COM_z	0.073	0.238	0.281	0.033	0.174
	A_{COMy}	0.250	0.020	−0.176	0.198	**0.454**
	A_{COMz}	0.146	0.168	**0.379**	0.004	−0.095
	V_{COPx}	−0.110	**0.346**	−0.248	−0.075	0.005
	V_{COPy}	**0.318**	0.002	−0.157	−0.158	−0.204
	MOS	−0.059	**0.317**	−0.311	−0.110	−0.045
	A_{ANG}	0.104	0.059	−0.096	**0.606**	−0.326

Bold font represents that the coefficient of the metric in this principal component is larger than others.

Figure 7 shows the results of the WBI for the CMT patient (subject8) under the NW condition with respect to subject1 under ZT and MIX conditions. The index in Figure 7A was constructed from subject8's NW condition and subject1's MIX condition. The index in Figure 7B was constructed from subject1's ZT and MIX conditions. In both cases, the mean and fluctuation range of the CMT subject's WBI was much larger than that of the healthy subject.

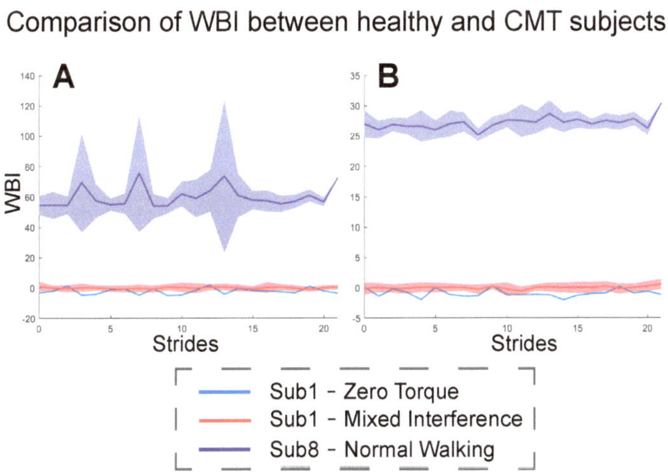

Figure 7. WBI of healthy subject1 under ZT and MIX conditions and subject8 with CMT under the NW condition. (**A**) The results of the WBI constructed from subject8's NW condition and subject1's MIX condition. (**B**) The results of the WBI constructed from subject1's ZT and MIX conditions. The blue curves show subject1's WBI for each stride under the ZT condition. The red curves and areas show the mean and fluctuation range of subject1's WBI under the MIX condition. The purple curves and areas show the mean and fluctuation range of subject8's WBI under the NW condition.

4. Discussion

This study presents a method to construct a comprehensive evaluation index of human walking balance, and we used this method to generate personal and general indexes. We designed walking experiments with exoskeleton interference to collect data from seven healthy subjects and collected data from one CMT subject at multiple speeds. We used statistical analysis and PCA to construct the Walking Balance Index. We verified the feasibility of this method in evaluating walking balance for healthy individuals and one CMT patient.

Healthy subjects were not significantly affected by the ankle interference torque (Figure 3). This may be because the applied torque was small (for safety). Due to the adjustment mechanism of human ankles, the effect of interference torque on one ankle joint is not obvious, and the impact on other parts of the human body is not great either. The fluctuation range of these metrics increased under the MIX condition with respect to the ZT condition, indicating that interference torque can affect the amplitude of human motion. Especially the metrics related to COM were affected more obviously. When human lower limbs are disturbed, the human will unconsciously adjust the trunk's posture to keep the COM within the support surface. Studies have shown that insufficient trunk muscle strength affects the balance ability of humans and increases the risk of falling [4]. The CMT subject's metrics fluctuated more widely with respect to those of the healthy subject with exoskeleton interference, indicating that the state of walking balance for the CMT subject is worse than that for the healthy subject. The walking strategies of patients are different from that of healthy individuals. The ankle movements of CMT patients are limited due to the muscle imbalance between multiple muscle groups. They can only use the hip joint and the stride adjustment mechanism frequently to maintain balance. This results in a large increase in body swing for CMT patients [40]. The stepping strategy and the lateral ankle strategy of balance control both need the contribution of the ankle joint. For healthy subjects, the ankle interference torque disturbs the timing, magnitude, and speed of biological torque produced by the ankle joint. For the subject with CMT, the main clinical symptoms, like muscle weakness and atrophy, cause them to have low strength and slow response in their lower limb muscles, which cannot provide the appropriate ankle torque to maintain balance during walking. Although the decrease in walking balance was found in changes in multiple metrics, the decrease in or loss of ankle joint function may be one of the main factors affecting walking balance.

The personal index for each healthy subject could effectively distinguish the state of walking balance under different gait conditions, indicating that the method we proposed is feasible in evaluating human walking balance (Figure 4). A smaller value of the WBI indicates that the subject is more balanced, while a larger value of the WBI indicates that the subject is more unbalanced. The results of the WBI for subject2, shown in Figure 5A, are difficult to distinguish, probably because the effect of a small amount of ankle interference torque is not obvious for healthy subjects who have better balance ability. The means of the WBI in Figure 5B still indicate that our method is effective in evaluating the walking balance, even for a subject with good balance ability.

The general index constructed from the same metrics and weights could still distinguish the state of walking balance for the six subjects (Figure 6). This suggests that our proposed method may be generalized to different individuals and gait conditions. The metrics and weights that were used to construct the general index have certain universality. However, the values of the WBI shown in Figure 6 are different from those shown in Figure 4 because the metrics and weights used in the construction of the WBI are different. This may lead to the loss of some information related to walking balance, but the relative change trends of the index between different gait conditions did not transform. This indicates that the relative rather than the absolute changes of the index are more worthy of consideration.

Not all the preliminary metrics have the ability to evaluate walking balance in our experiments. Some of them showed no significant difference between ZT and MIX conditions (Tables 2 and 3). Both COP_y and V_{COP_y} showed strong significant differences in the three time domain characteristics, suggesting that they can well reveal the adjustment strategies for maintaining balance during walking. Relevant studies have shown that the position of the COP in the coronal plane has a certain predictive ability for falling and walking stability [24]. The adjustment strategies of walking balance for different individuals show consistencies and differences. The total number of times there were significant differences in V_{COP}, COP_x, COM_z, A_{ANG}, and MOS for the six subjects was relatively higher (Table 3), indicating that the six subjects made changes in these metrics to maintain their balance. These metrics are more likely to be used as general metrics to construct the general index and evaluate the walking balance of different individuals. Five subjects showed the most frequent significant differences in V_{COP_y}, but subject6 did not. This result indicates that different subjects have different adjustment strategies to combat ankle interference and maintain walking balance. Certain metrics of the same subject showed significant differences under each gait combination, such as subject2's COM_x, subject3's COM_y, and subject4's COM_z. This may be because the same subject adopted the same walking strategies under different gait conditions, and they did not change their walking strategies because of different external interferences. It is also possibly because the interference was only applied to the ankle joint.

More than 85% of the data information can be represented by five principal components, indicating a great correlation between these metrics (Table 4). From Table 5, it can be seen that the changes in the COP in the direction of the coronal and sagittal axes are closely related to the walking balance of subject1. In addition, we used PCA and analyzed the principal components that contributed more than 10% under ZT and MIX conditions for the other healthy subjects. The V_{COP} and COP_y of most of the healthy subjects played an important role in their walking balance, but other metrics were different according to individual differences. This suggests that different individuals have different walking strategies to keep balance. The COM_z of subject4 showed significant differences under all gait combinations (Table 3) but only accounted for one item in principal component 4. It was not the most important influencing factor of the walking balance for subject4, which illustrated the importance of PCA in this work. Some balance metrics may be common across different individuals while others not. Therefore, the index calculated by general metrics and weights may have a different effect in evaluating the walking balance of different individuals. If the general metrics and weights used to construct the WBI are not significantly influenced by changes in the walking balance of one subject, it will not lead to an obvious effect on the evaluation of the specific subject's walking balance. However, the relative change trend of the WBI value is consistent among different walking conditions.

To compare with the existing evaluation methods, the Berg Balance Scale and Tinetti Test were applied after the main experiments for the seven healthy subjects. Results showed that the two scores remained at the maximum for all the healthy subjects even when wearing the unilateral ankle exoskeleton with and without the interference torques. However, subjects said their walking balance was influenced under the interference conditions. We observed that some subjects even changed their walking pattern to prevent themselves from falling. This may be because subjects still had certain ability to maintain walking balance but decreased the state of walking balance when they were walking with interference torques. Such situations cannot be reflected by the scores of Berg Balance Scale or Tinetti Test, but the WBI can reflect the changing of walking balance from the variation trend of its values.

The index constructed by our proposed method could also be used to evaluate the walking balance of the CMT subject, and it was universal between the CMT patient and the healthy subject1 (Figure 7). The state of walking balance for this patient was bad and varied considerably between gait cycles. This result directly reflects the pathological gaits of the patient. The results of the WBI in Figure 7A are different from those in Figure 7B because the metrics and weights used to construct the index were different. To further explore the differences in walking balance between the CMT subject and a healthy subject, we compared the results of the PCA in the two cases (Figure 7A,B). We found that some metrics greatly affected the values of the WBI for the healthy subject and the CMT subject, such as COM_z, MOS, V_{COP_x}, and A_{ANG}. In the principal components whose total contribution rate was more than 85%, the CMT subject contained more metrics, and some of them did not even appear in the principal components of the healthy subjects, such as *COP_CMP*. This result indicates that the CMT patient adopted more balance adjustment strategies while walking than the healthy people. Meanwhile, we found that the healthy subject was more inclined to change in the coronal direction, while the CMT subject was more inclined to change in the sagittal direction.

Our findings suggest that a single metric may not be enough to evaluate walking balance. Each balance metric has the ability to assess static and dynamic balance, but it will be changed among different individuals and gait conditions. When human walking balance is influenced, the performance is multi-aspect and personal to different individuals. The variation trend of different metrics reveals the different strategies humans use to maintain walking balance. Personalized customization of the WBI may be more effective during rehabilitation processes. Our work provides a possible method to consider multiple metrics comprehensively in the field of walking balance evaluation. Our proposed method is expected to provide evidence and guidance for walking balance evaluation in future clinical treatment and rehabilitation. The WBI can be used to identify whether the walking balance of human changes on different days and under different gait conditions. The rate of the WBI may be used to predict the changing trend of walking balance and the possibility of falling. By combining Human-in-the-Loop optimization, the WBI can be considered as an objective function of exoskeleton assistance to improve the walking balance of people in need.

There are some possible disadvantages and limitations in this study. The WBI was calculated offline, which means that we could not evaluate the walking balance synchronously when subjects were walking. Since we used the motion capture system and treadmill to collect metrics data, the method is limited to use in the laboratory environment. Our method needs at least two sets of data under relatively balanced and unbalanced gait conditions to construct the WBI. In our experiments, we only applied unilateral ankle interference to the healthy subjects but did not apply multiple external interferences to other parts of the human body. Similar external interference could be added to knee, hip joints, or multiple joints. Our subjects were young people and included only one CMT patient so it is not certain whether this method can be used to evaluate the walking balance of all age groups and different patients. Our proposed method only considers kinematic metrics without revealing the intrinsic physiological mechanism of maintaining walking balance. For future works, we plan to incorporate more physiological and biomechanical metrics into the construction of the Walking Balance Index. Meanwhile, we will recruit more subjects and apply multiple external interferences in future experiments. We hope to apply this method to more people and gait conditions.

5. Conclusions

We presented a method to construct a comprehensive evaluation index of human walking balance. We used this method to generate personal and general indexes for different individuals and evaluate the state of their walking balance. We designed walking experiments with exoskeleton interference to collect data from seven healthy subjects and collected data from one CMT subject at multiple speeds. We used statistical analysis to pick the effective metrics from the preliminary metrics and used PCA to construct the index by combining multiple metrics. We not only demonstrated the feasibility and universality of this method in evaluating walking balance for healthy people but also in one patient. The results show consistencies and differences in walking balance adjustment strategies for different individuals under different gait conditions. This study is expected to provide evidence for walking balance evaluation and guidance for future clinical treatment and rehabilitation, improving the walking balance of people in need and allowing further consideration of the WBI as an optimization target of exoskeleton assistance for them in the future.

Author Contributions: Conceptualization, L.W. and X.L.; methodology, L.W. and X.L.; software, L.W. and X.L.; validation, L.W. and J.Z.; formal analysis, L.W.; investigation, L.W.; resources, L.W.; data curation, L.W.; writing—original draft preparation, L.W.; writing—review and editing, L.W., J.Z., J.H. and Y.P.; visualization, L.W.; supervision, L.W. and J.Z.; project administration, L.W. and J.Z.; funding acquisition, J.Z. and J.H. All authors have read and agreed to the published version of the manuscript.

Funding: This work was supported in part by National Natural Science Foundation of China, grant number 62073179.

Institutional Review Board Statement: The study was conducted in accordance with the Declaration of Helsinki and approved by the ethical committee of Nankai University (reference number: NKUIRB2021054).

Informed Consent Statement: Informed consent was obtained from all subjects involved in the study. Written informed consent was obtained from the patient to publish this paper.

Data Availability Statement: The raw data supporting the conclusions of this article will be made available by the authors on request.

Conflicts of Interest: The authors declare no conflicts of interest.

References

1. Jadelis, K.; Miller, M.E.; Ettinger, W.H. Jr.; Messier, S.P. Strength, balance, and the modifying effects of obesity and knee pain: Results from the Observational Arthritis Study in Seniors (oasis). *J. Am. Geriatr. Soc.* **2001**, *49*, 884–891. [CrossRef] [PubMed]
2. Dreher, T.; Wolf, S.; Heitzmann, D.; Fremd, C.; Klotz, M.; Wenz, W. Tibialis posterior tendon transfer corrects the foot drop component of cavovarus foot deformity in Charcot-Marie-Tooth disease. *J. Bone Jt. Surg. Am.* **2014**, *96*, 456–462. [CrossRef] [PubMed]
3. van der Linden, M.H.; van der Linden, S.C.; Hendricks, H.T.; van Engelen, B.G.; Geurts, A.C. Postural instability in Charcot-Marie-Tooth type 1A patients is strongly associated with reduced somatosensation. *Gait. Posture* **2010**, *31*, 483–488. [CrossRef] [PubMed]
4. Granacher, U.; Gollhofer, A.; Hortobágyi, T.; Kressig, R.W.; Muehlbauer, T. The importance of trunk muscle strength for balance, functional performance, and fall prevention in seniors: A systematic review. *Sport. Med.* **2013**, *43*, 627–641. [CrossRef] [PubMed]
5. Kannus, P.; Parkkari, J.; Koskinen, S.; Niemi, S.; Palvanen, M.; Järvinen, M.; Vuori, I. Fall-induced injuries and deaths among older adults. *JAMA* **1999**, *281*, 1895–1899. [CrossRef]
6. Reicke, N. The Romberg head-shake test within the scope of equilibrium diagnosis. *HNO* **1992**, *40*, 195–201. [PubMed]
7. Agrawal, Y.; Carey, J.P.; Hoffman, H.J.; Sklare, D.A.; Schubert, M.C. The modified Romberg Balance Test: Normative data in US adults. *Otol. Neurotol.* **2011**, *32*, 1309–1311. [CrossRef]
8. Wei, W.; Mcelroy, C.; Dey, S. Using Sensors and Deep Learning to Enable On-Demand Balance Evaluation for Effective Physical Therapy. *IEEE Access* **2020**, *8*, 99889–99899. [CrossRef]
9. Blum, L.; Korner-Bitensky, N. Usefulness of the Berg Balance Scale in stroke rehabilitation: A systematic review. *Phys. Ther.* **2008**, *88*, 559–566. [CrossRef]
10. Tinetti, M.E.; Williams, T.F.; Mayewski, R. Fall risk index for elderly patients based on number of chronic disabilities. *Am. J. Med.* **1986**, *80*, 429–434. [CrossRef]
11. Craig, J.J.; Bruetsch, A.P.; Lynch, S.G.; Horak, F.B.; Huisinga, J.M. Instrumented balance and walking assessments in persons with multiple sclerosis show strong test-retest reliability. *J. Neuroeng. Rehabil.* **2017**, *14*, 43. [CrossRef] [PubMed]
12. Ming, Z.; Hong, W.; Linlin, Z.; Qingyun, M.; Ling, S.; Feng, Z. The Study and Application of Algorithm for Indexes of Man's Balance Capacity. In Proceedings of the 2018 10th International Conference on Measuring Technology and Mechatronics Automation (ICMTMA), Changsha, China, 10–11 February 2018; pp. 476–480. [CrossRef]
13. Nilsagård, Y.; Lundholm, C.; Denison, E.; Gunnarsson, L.G. Predicting accidental falls in people with multiple sclerosis—A longitudinal study. *Clin. Rehabil.* **2009**, *23*, 259–269. [CrossRef] [PubMed]
14. Cameron, M.H.; Thielman, E.; Mazumder, R.; Bourdette, D. Predicting falls in people with multiple sclerosis: Fall history is as accurate as more complex measures. *Mult. Scler. Int.* **2013**, *2013*, 496325. [CrossRef] [PubMed]
15. Terekhov, Y. Stabilometry as a diagnostic tool in clinical medicine. *Can. Med. Assoc. J.* **1976**, *115*, 631. [PubMed]
16. Kagawa, T.; Suzuki, R. Balance map analysis for visualization and quantification of balance in human walking. *IEEE Trans. Neural Syst. Rehabil. Eng.* **2021**, *29*, 2153–2163. [CrossRef] [PubMed]
17. Beck, O.N.; Shepherd, M.K.; Rastogi, R.; Martino, G.; Ting, L.H.; Sawicki, G.S. Exoskeletons need to react faster than physiological responses to improve standing balance. *Sci. Robot.* **2023**, *8*, eadf1080. [CrossRef]
18. Huurnink, A.; Fransz, D.P.; Kingma, I.; van Dieën, J.H. Comparison of a laboratory grade force platform with a Nintendo Wii Balance Board on measurement of postural control in single-leg stance balance tasks. *J. Biomech.* **2013**, *46*, 1392–1395. [CrossRef] [PubMed]

19. Biswas, A.; Lemaire, E.D.; Kofman, J. Dynamic gait stability index based on plantar pressures and fuzzy logic. *J. Biomech.* **2008**, *41*, 1574–1581. [CrossRef] [PubMed]
20. Piirtola, M.; Era, P. Force platform measurements as predictors of falls among older people–a review. *Gerontology* **2006**, *52*, 1–16. [CrossRef]
21. Hof, A.; Gazendam, M.; Sinke, W. The condition for dynamic stability. *J. Biomech.* **2005**, *38*, 1–8. [CrossRef]
22. Brech, G.C.; Plapler, P.G.; de Souza Meirelles, E.; Marcolino, F.M.D.; Greve, J.M.D. Evaluation of the association between osteoporosis and postural balance in postmenopausal women. *Gait Posture* **2013**, *38*, 321–325. [CrossRef] [PubMed]
23. Howcroft, J.; Kofman, J.; Lemaire, E.D.; McIlroy, W.E. Analysis of dual-task elderly gait in fallers and non-fallers using wearable sensors. *J. Biomech.* **2016**, *49*, 992–1001. [CrossRef] [PubMed]
24. Choi, H.S.; Baek, Y.S. Effects of the degree of freedom and assistance characteristics of powered ankle-foot orthoses on gait stability. *PLoS ONE* **2020**, *15*, e0242000. [CrossRef] [PubMed]
25. Vukobratović, M.; Borovac, B. Zero-moment point—Thirty five years of its life. *Int. J. Humanoid Robot.* **2004**, *1*, 157–173. [CrossRef]
26. Sardain, P.; Bessonnet, G. Forces acting on a biped robot. Center of pressure-zero moment point. *IEEE Trans. Syst. Man Cybern.-Part A Syst. Hum.* **2004**, *34*, 630–637. [CrossRef]
27. Terry, K.; Stanley, C.; Damiano, D. A new perspective on the walking margin of stability. *J. Appl. Biomech.* **2014**, *30*, 737–741. [CrossRef] [PubMed]
28. Chen, J.; Han, J.; Zhang, J. Design and Evaluation of a Mobile Ankle Exoskeleton With Switchable Actuation Configurations. *IEEE/ASME Trans. Mechatron.* **2022**, *27*, 1846–1853. [CrossRef]
29. Reimann, H.; Fettrow, T.D.; Thompson, E.D.; Agada, P.; McFadyen, B.J.; Jeka, J.J. Complementary mechanisms for upright balance during walking. *PLoS ONE* **2017**, *12*, e0172215. [CrossRef] [PubMed]
30. Simoneau, M.; Corbeil, P. The effect of time to peak ankle torque on balance stability boundary: Experimental validation of a biomechanical model. *Exp. Brain Res.* **2005**, *165*, 217–228. [CrossRef]
31. Sieri, T.; Beretta., G. Fall risk assessment in very old males and females living in nursing homes. *Disabil. Rehabil.* **2004**, *26*, 718–723. [CrossRef]
32. Robinovitch, S.N.; Heller, B.; Lui, A.; Cortez, J. Effect of strength and speed of torque development on balance recovery with the ankle strategy. *J. Neurophysiol.* **2002**, *88*, 613–620. [CrossRef] [PubMed]
33. Zhang, J.; Cheah, C.C.; Collins, S.H. Experimental comparison of torque control methods on an ankle exoskeleton during human walking. In Proceedings of the 2015 IEEE International Conference on Robotics and Automation (ICRA), Seattle, WA, USA, 26–30 May 2015; pp. 5584–5589. [CrossRef]
34. Zhang, J.; Cheah, C.C.; Collins, S.H. Chapter 5—Torque Control in Legged Locomotion. In *Bioinspired Legged Locomotion*; Elsevier: Amsterdam, The Netherlands, 2017; pp. 347–400. [CrossRef]
35. Wang, W.; Chen, J.; Ji, Y.; Jin, W.; Liu, J.; Zhang, J. Evaluation of lower leg muscle activities during human walking assisted by an ankle exoskeleton. *IEEE Trans. Ind. Inform.* **2020**, *16*, 7168–7176. [CrossRef]
36. Han, H.; Wang, W.; Zhang, F.; Li, X.; Chen, J.; Han, J.; Zhang, J. Selection of muscle-activity-based cost function in human-in-the-loop optimization of multi-gait ankle exoskeleton assistance. *IEEE Trans. Neural Syst. Rehabil. Eng.* **2021**, *29*, 944–952. [CrossRef] [PubMed]
37. Delp, S.L.; Anderson, F.C.; Arnold, A.S.; Loan, P.; Habib, A.; John, C.T.; Guendelman, E.; Thelen, D.G. OpenSim: Open-Source Software to Create and Analyze Dynamic Simulations of Movement. *IEEE Trans. Biomed. Eng.* **2007**, *54*, 1940–1950. [CrossRef] [PubMed]
38. Seth, A.; Hicks, J.L.; Uchida, T.K.; Habib, A.; Dembia, C.L.; Dunne, J.J.; Ong, C.F.; DeMers, M.S.; Rajagopal, A.; Millard, M.; et al. OpenSim: Simulating musculoskeletal dynamics and neuromuscular control to study human and animal movement. *PLoS Comput. Biol.* **2018**, *14*, e1006223. [CrossRef] [PubMed]
39. Rajagopal, A.; Dembia, C.L.; DeMers, M.S.; Delp, D.D.; Hicks, J.L.; Delp, S.L. Full-body musculoskeletal model for muscle-driven simulation of human gait. *IEEE Trans. Biomed. Eng.* **2016**, *63*, 2068–2079. [CrossRef]
40. Lynn, S.G.; Sinaki, M.; Westerlind, K.C. Balance characteristics of persons with osteoporosis. *Arch. Phys. Med. Rehabil.* **1997**, *78*, 273–277. [CrossRef]

Disclaimer/Publisher's Note: The statements, opinions and data contained in all publications are solely those of the individual author(s) and contributor(s) and not of MDPI and/or the editor(s). MDPI and/or the editor(s) disclaim responsibility for any injury to people or property resulting from any ideas, methods, instructions or products referred to in the content.

bioengineering

MDPI

Article

Direct Current Stimulation over the Primary Motor Cortex, Cerebellum, and Spinal Cord to Modulate Balance Performance: A Randomized Placebo-Controlled Trial

Jitka Veldema [1,*], Teni Steingräber [1], Leon von Grönheim [1], Jana Wienecke [2], Rieke Regel [1], Thomas Schack [1] and Christoph Schütz [1]

1 Faculty of Psychology and Sports Science, Bielefeld University, 33615 Bielefeld, Germany; teni.unciyan@uni-bielefeld.de (T.S.); leonvongroenheim@gmx.de (L.v.G.); rieke.regel@uni-bielefeld.de (R.R.); thomas.schack@uni-bielefeld.de (T.S.); christoph.schuetz@uni-bielefeld.de (C.S.)

2 Department of Exercise and Health, Paderborn University, 33098 Paderborn, Germany; wienecke@sportmed.uni-paderborn.de

* Correspondence: jitka.veldema@uni-bielefeld.de; Tel.: +49-(0)151-44-64-83-71; Fax: +49-521-106-6432

Abstract: Objectives: Existing applications of non-invasive brain stimulation in the modulation of balance ability are focused on the primary motor cortex (M1). It is conceivable that other brain and spinal cord areas may be comparable or more promising targets in this regard. This study compares transcranial direct current stimulation (tDCS) over (i) the M1, (ii) the cerebellum, and (iii) trans-spinal direct current stimulation (tsDCS) in the modulation of balance ability. Methods: Forty-two sports students were randomized in this placebo-controlled study. Twenty minutes of anodal 1.5 mA t/tsDCS over (i) the M1, (ii) the cerebellum, and (iii) the spinal cord, as well as (iv) sham tDCS were applied to each subject. The Y Balance Test, Single Leg Landing Test, and Single Leg Squat Test were performed prior to and after each intervention. Results: The Y Balance Test showed significant improvement after real stimulation of each region compared to sham stimulation. While tsDCS supported the balance ability of both legs, M1 and cerebellar tDCS supported right leg stand only. No significant differences were found in the Single Leg Landing Test and the Single Leg Squat Test. Conclusions: Our data encourage the application of DCS over the cerebellum and spinal cord (in addition to the M1 region) in supporting balance control. Future research should investigate and compare the effects of different stimulation protocols (anodal or cathodal direct current stimulation (DCS), alternating current stimulation (ACS), high-definition DCS/ACS, closed-loop ACS) over these regions in healthy people and examine the potential of these approaches in the neurorehabilitation.

Keywords: tDCS; tsDCS; balance; postural control; primary motor cortex; cerebellum; spinal cord; healthy people

Citation: Veldema, J.; Steingräber, T.; von Grönheim, L.; Wienecke, J.; Regel, R.; Schack, T.; Schütz, C. Direct Current Stimulation over the Primary Motor Cortex, Cerebellum, and Spinal Cord to Modulate Balance Performance: A Randomized Placebo-Controlled Trial. *Bioengineering* **2024**, *11*, 353. https://doi.org/10.3390/bioengineering11040353

Academic Editor: Carlo Albino Frigo

Received: 5 March 2024
Revised: 25 March 2024
Accepted: 26 March 2024
Published: 4 April 2024

Copyright: © 2024 by the authors. Licensee MDPI, Basel, Switzerland. This article is an open access article distributed under the terms and conditions of the Creative Commons Attribution (CC BY) license (https://creativecommons.org/licenses/by/4.0/).

1. Introduction

Non-invasive DCS is a powerful tool modulating neural processing and can be successfully used for research and therapies. DCS consists of the application of a low-intensity direct current that flows between two or more electrodes. Present data indicate that a single session of tDCS can induce neurophysiological changes up to 120 min beyond the stimulation period [1–3], and its persistence increases linearly with the duration and the intensity of current applied [1,2]. A simplified theory distinguishes between anodal tDCS (with anode placed over the region of interest and cathode over another cranial or extracranial region) and cathodal tDCS (with reverse electrode positioning). Anodal tDCS should induce depolarization of neurons and increase corticospinal excitability. In contrast, cathodal stimulation should lead to a hyperpolarization of neurons and decrease corticospinal excitability [4–6]. Indeed, the real data show a large variability outside of this theoretical scope [7]. A key factor that determines the tDCS-induced effects is electrode positioning. A current systematic review indicates that tDCS applied over different regions modulates different aspects of walking in healthy people. While application over the primary motor cortex (M1) and cerebellum improved speed, synchronization, and variability during simple walking, dorsolateral prefrontal cortex (DLPFC) stimulation improved gait parameters under dual-task conditions [8]. However, another systematic review points to the fact that diverse interactions exist between tDCS specifications (M1/cerebellum, unilateral/bilateral/central, single/multiple sessions) and motor task interactions (uni/bimanual, greater/less difficulty) [9]. This makes it difficult to draw clear conclusions. In addition, the reference electrode positioning may significantly impact the tDCS-induced effects. A simulation study (based on a numerical body model) compared six different cathode positions (right temporal lobe, right supraorbital region, right deltoid, left deltoid, under the chin, and right buccinator muscle) during anodal tDCS over the left M1 [10]. The results indicate that extracephalic electrodes may be more effective in the modulation of the spinal cord and similar or less effective in the modulation of the brainstem, than cephalic electrodes [10]. Another modeling study shows that a multipolar tDCS, with two anodes (over the right and the left M1) and one cathode (either over the spinal cord or over the right deltoid) may be effective in the modulation of deep brain structures, such as the thalamus, midbrain, and brainstem [11]. Numerous authors suggest that tsDCS, with one electrode over the spinal cord and the other electrode over another extracephalic region (such as spinal cord, deltoid muscle, iliac crest, etc.) are promising alternatives to conventional cranial applications [12]. Our study extends the knowledge on this field and investigates (in a direct comparison) the effects of three different electrode placements in modulating balance ability.

Balance and postural control are complex sensorimotor functions controlled by integrated brain and spinal networks [13–15]. Their neural background is still not fully understood. A recent systematic review with a meta-analysis emphasized the key role of the brainstem, cerebellum, basal ganglia, thalamus, and several cortical regions based on (functional) magnetic resonance imaging ((f) MRI) and positron emission tomography (PET) data [13]. Similarly, another systematic review indicated the key role of the cerebellum and brainstem, followed by the basal ganglia, thalamus, hippocampus, inferior parietal cortex, and frontal lobe regions, using MRI investigations [14]. Additionally, the spinal cord seems to play a crucial role in balance and postural control, as indicated by electrophysiological studies [15]. It has been repeatedly demonstrated that balance training leads (in addition to an improved balance ability) to spinal adaptations in the form of a suppressed Hoffmann reflex (H-reflex) [15,16].

Although the available data indicate that several cortical and subcortical brain regions, the cerebellum, and the spinal cord are crucially involved during motor control [13–15], the present applications of DCS focus mainly on M1 [17–19]. The evidence for the remaining central and peripheral nervous system is insufficient, similar to studies that directly compare the stimulation over different areas [17–19]. Therefore, the question arises whether other regions may be comparable or even more promising for DCS applications. Our study investigates and compares the effectiveness of t/tsDCS over the M1, cerebellum, and spinal cord [20–22].

2. Methods

2.1. Study Design

This was a randomized placebo-controlled crossover study. Three single sessions of real t/tsDCS (over the (i) M1, (ii) cerebellum, and (iii) spinal cord) and one session of sham tDCS were applied to each participant in a randomized order (PC-generated) with a washout period of at least 48 h in between. Balance ability was evaluated immediately before and after each intervention. The study was conducted according to the standards established by the Declaration of Helsinki, approved by the Ethics Committee of Bielefeld University (2022-043), and entered in the German Clinical Trial Register on 28 September 2023 (DRKS00032749).

2.2. Participants

The inclusion criteria were as follows: (1) age between 18 and 25 years, (2) no contraindications for tDCS (checked by safety screening questionnaire [23]), and (3) no relevant neurological, psychiatric, or orthopedic disorders. All subjects provided their written informed consent prior to participation. A G*power analysis (effect size = 0.25, α error probability $p < 0.05$, Power = 0.95) revealed that a sample size of at least 40 participants is needed to detect statistically significant effects using ANOVA with four interventions and two timepoints.

2.3. Intervention

Each subject completed four separate 20 min interventional sessions: (1) 1.5 mA tDCS over the M1, (2) 1.5 mA tDCS over the cerebellum, (3) 1.5 mA tsDCS over the spinal cord, and (4) sham tDCS (stimulator turned off after 5 s) over M1. A DC-stimulator PLUS (NeuroConn Gmbh, Ilmenau, Germany) and two saline-soaked sponge electrodes (5 cm × 7 cm) were used. For M1 stimulation, the anode was placed over the Cz, and the cathode was placed over the right supraorbital area (Fp2). For cerebellar stimulation, the anode was placed over the O2, and the cathode was placed over the right buccinator muscle. The electrodes positioning for M1 and cerebellar tDCS is in line with previous studies [24,25]. For spinal stimulation, the anode was placed over the spinal cord at the Th8 level, and the cathode was placed over L2. A simulation study indicated that this electrode placement is superior (in comparison to deltoid, umbilicus, and iliac crest cathode placements) regarding the electric field generated in lumbar and sacral spinal segments [26]. The international 10/20 EEG system [27] and palpation method [28,29] were used to determine electrode positioning during M1, cerebellar and spinal t/tsDCS. Figure 1 shows the electrodes' placements used in this study.

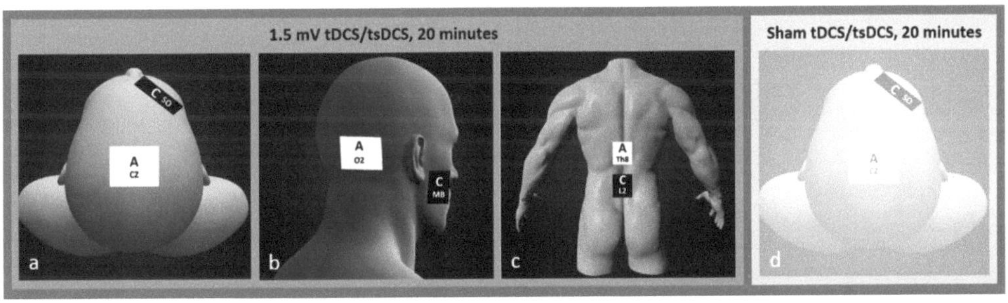

Figure 1. Electrodes positioning used for (**a**) M1 tDCS, (**b**) cerebellar tDCS, (**c**) spinal tDCS and (**d**) sham tDCS.

2.4. Assessments

Three different assessments (Y Balance Test, the Single Leg Landing Test, and the Single Leg Squat Balance Test) were used to evaluate balance ability. The right and the left leg were tested in a randomized order during each test. The investigators were blinded to intervention allocation.

The Y Balance Test was performed using a test kit (FMS, Chatham, VA, USA). The maximal reach of the free lower leg in the (a) anterior, (b) posterolateral, and (c) posteromedial directions was determined during a one leg stance on the opposite leg [30]. A better balance ability was associated with a greater reach distance. Five trials were performed for each leg and direction. The mean value of the five trials was used for analysis.

During the Single Leg Landing Test, participants were instructed to perform a forward jump (50% of their body height), land on a single limb, and achieve a stable position as quickly as possible [31–33]. The center of gravity (COG) in the anterior–posterior and medial–lateral directions and the time taken to regain balance were recorded using a force plate (AMTI, Watertown, MA, USA). A smaller COG area and a faster time to stabilize indicated better balance. Five trials were performed for each leg. The mean value was used for the analysis.

During the Single Leg Squat Test, probands performed five consecutive single-leg squats (10% of their body height) [31,34]. The center of gravity (COG) in the anterior–posterior and medial–lateral directions was recorded using the force plate described above. The smaller the COG area was, the better the balance. Two trials were performed for each leg. The mean values were used for the analysis.

2.5. Analysis

The SPSS software package, version 27 (International Business Machines Corporation Systems, IBM, Ehningen, BW, Germany), was used to analyze the data collected during this study. The independent sample *t*-tests evaluated pre-interventional comparability. Repeated-measure ANOVAs with the factors "intervention" and "time" compared the pre–post changes across interventions. Mauchly's sphericity tests and Greenhouse–Geisser corrections were applied. Due to multiple comparisons, a *p*-value of ≤ 0.01 was considered statistically significant. The outliers (mean ± 3 SD) were excluded from the analysis. The researcher performing statistical analysis was not blind to intervention allocation.

3. Results

Overall, 42 participants were randomized (age 25.1 ± 3.2 years, 19 females, 23 males, 36 right-footed, and 6 left-footed). The foot preferred to kick the ball was considered to be dominant [35]. All participants tolerated the interventions well without severe adverse events. Four participants reported less severe side effects, such as a burning sensation and nausea (one participant after M1 stimulation) and a metallic taste in the mouth (three participants after cerebellar stimulation). The pre-interventional data did not differ significantly across interventions. Table 1 summarizes the data on balance collected during the experiment. The outliers (4% of values) were removed. The ANOVAs detected significant time*intervention interactions on the Y Balance Test, but not on the Single Leg Landing Test and the Single Leg Squat Test. The effects were observed more frequently for the left leg than for the right leg. For the left leg, a significant improvement of balance ability (in comparison to the sham tDCS) was detected after M1 ($F_{1,40}$ = 8.999; p = 0.005) ($F_{1,36}$ = 18.624; p < 0.001), cerebellar ($F_{1,40}$ = 8.796; p = 0.005) ($F_{1,36}$ = 16.291; $p \leq$ 0.001) and spinal ($F_{1,39}$ = 13.55; $p \leq$ 0.001) ($F_{1,34}$ = 8.799; p = 0.005) application for the posterior-lateral and posterior-medial directions, respectively. For the right leg, only tsDCS induced significantly greater effects than the sham tDCS for both the posterior-lateral ($F_{1,39}$ = 11.53; p = 0.002) and posterior-medial ($F_{1,39}$ = 7.943; p = 0.008) directions. No significant effects were observed for the anterior direction. The intervention-induced effects did not significantly differ across real t/tsDCS interventions. Figures 2 and 3 illustrate the intervention-induced changes.

Table 1. Balance performance (means and SD) at both time-points (pre, post).

				Sham tDCS	M1 tDCS	Cerebellar tDCS	Spinal tDCS
Y Balance Test	Right leg	Anterior direction (cm)	pre	57.29 ± 7.41	56.10 ± 5.57	57.54 ± 7.45	56.42 ± 5.31
			post	57.53 ± 7.80	56.43 ± 5.80	58.36 ± 7.76	56.91 ± 5.54
		Posterolateral direction (cm)	pre	105.21 ± 12.07	103.72 ± 10.54	104.45 ± 12.76	104.00 ± 11.09
			post	106.26 ± 12.47	106.48 ± 12.31	107.14 ± 13.35	107.06 ± 12.47 **
		Posteromedial direction (cm)	pre	101.19 ± 12.76	99.18 ± 13.42	102.17 ± 13.99	99.47 ± 11.23
			post	103.03 ± 13.05	103.48 ± 14.42	105.53 ± 14.03	103.45 ± 11.70 **

Table 1. *Cont.*

				Sham tDCS	M1 tDCS	Cerebellar tDCS	Spinal tDCS
	Left leg	Anterior direction (cm)	pre	57.76 ± 7.02	56.96 ± 5.77	56.98 ± 5.87	57.02 ± 4.85
			post	57.75 ± 7.08	56.92 ± 5.57	57.77 ± 5.92	57.60 ± 5.27
		Posterolateral direction (cm)	pre	103.78 ± 11.17	102.50 ± 10.41	103.58 ±± 12.24	102.54 ± 10.14
			post	104.68 ± 11.12	106.54 ± 12.50 **	106.56 ± 12.74 **	106.06 ± 11.24 ***
		Posteromedial direction (cm)	pre	101.58 ± 11.15	99.88 ± 13.66	101.48 ± 14.02	100.30 ± 11.15
			post	102.99 ± 11.12	103.9 ± 14.11 ***	104.98 ± 14.27 ***	103.75 ± 10.65 **
Single Leg Landing Test	Right leg	Center of gravity area (mm^2)	pre	5163 ± 1486	5363 ± 1470	5363 ± 1678	5406 ± 1690
			post	5619 ± 1975	5530 ± 1665	5752 ± 1910	5938 ± 1764
		Time to stabilization (ms)	pre	1.196 ± 0.180	1.211 ± 0.185	1.141 ± 0.154	1.191 ± 0.154
			post	1.219 ± 0.188	1.243 ± 0.206	1.203 ± 0.147	1.240 ± 0.160
	Left leg	Center of gravity area (mm^2)	pre	5437 ± 1224	5257 ± 1017	5017 ± 1173	5305 ± 1282
			post	5997 ± 1580	5154 ± 1281	5574 ± 1546	6133 ± 1501
		Time to stabilization (ms)	pre	1.232 ± 0.153	1.273 ± 0.177	1.227 ± 0.187	1.264 ± 0.170
			post	1.274 ± 0.196	1.329 ± 0.192	1.285 ± 0.210	1.326 ± 0.161
Single Leg Squat Test	Right leg	Center of gravity area (mm^2)	pre	3020 ± 1246	2967 ± 1312	3006 ± 988	3232 ± 1057
			post	2950 ± 1071	3147 ± 1089	3085 ± 1201	3043 ± 1073
	Left leg	Center of gravity area (mm^2)	pre	3241 ± 1274	3223 ± 1159	3179 ± 987	3298 ± 973
			post	3087 ± 1299	3223 ± 1302	3329 ± 817	3093 ± 898

Notes: intervention-induced changes in comparison to sham ** = $p \leq 0.01$; *** = $p \leq 0.001$.

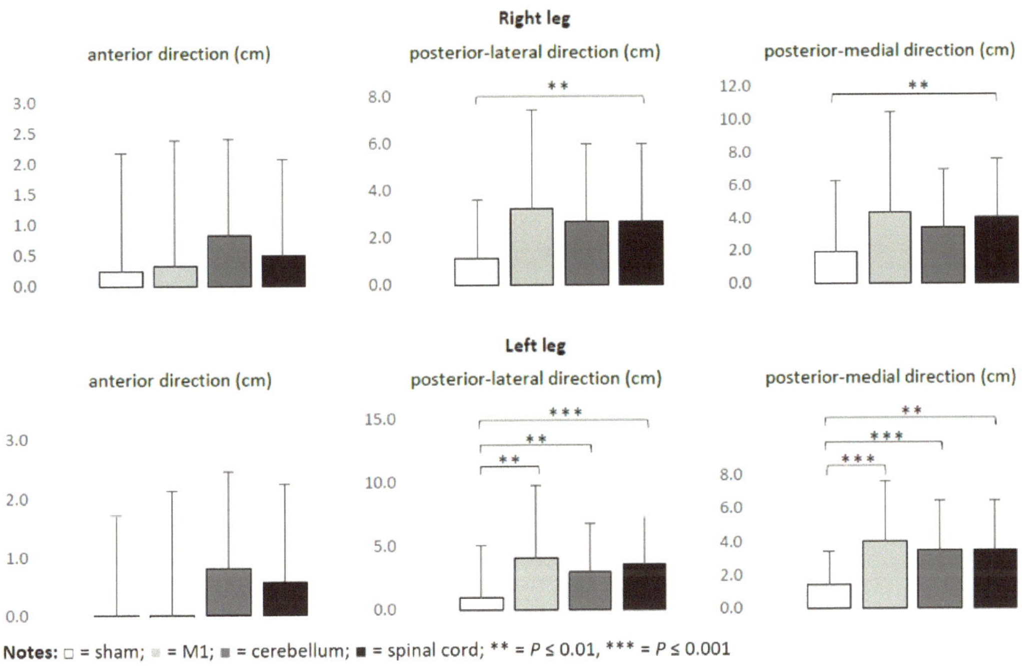

Figure 2. Intervention-induced changes (means and SD) in the Y Balance Test in relation to baseline.

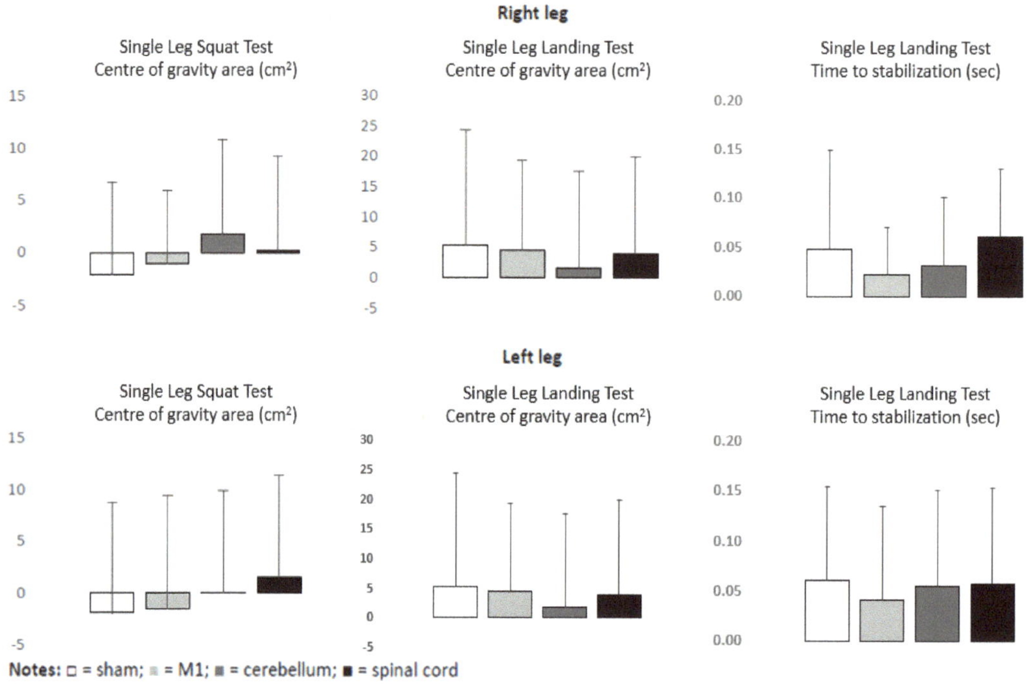

Figure 3. Intervention-induced changes (means and SD) in Single Leg Squat Test and Single Leg Landing Test in relation to baseline. Notes: □ = sham; ■ = M1; ■ = cerebellum; ■ = spinal.

4. Discussion

The aim of this study is to investigate and compare the effects of 1.5 mA t/tsDCS applied over the M1, cerebellum, and spinal cord on balance and postural control. The data show that (1) stimulation of each region significantly improved balance and postural control during the Y Balance Test but not during the Single Leg Landing Test and the Single Leg Squat Test, and (2) spinal stimulation improved the balance ability of both legs, while M1 and cerebellar stimulation improved the right leg stand only.

4.1. Stimulated Area Specific Modulation

Although several neuroimaging data indicate that several cortical and subcortical regions, the cerebellum, the brainstem, and the spinal cord, are crucially involved during balance and postural control [13–15], the majority of existing studies have applied tDCS over the M1 [17–19]. The previous evidence for the remaining regions was insufficient. Direct comparisons of different regions regarding t/tsDCS-induced effects on balance and postural control were almost non-existent [17–19]. We have demonstrated that the cerebellum and spinal cord are promising targets for the application of t/tsDCS in supporting balance control, in addition to M1. Thus, our results provide an important contribution to this field. Accordingly, a review suggests that the core systems of the automatic process of postural control are mostly achieved by the brainstem and spinal cord, while the forebrain structures and cerebellum act on the brainstem–spinal cord systems so that the cognitive processes of postural control can be achieved [36]. A model developed in the 1990s indicated that so-called central pattern generators (CPGs) could play a crucial role in gait and posture control [37–39]. CPGs are located in the lower thoracic and lumbar regions of the vertebrate spinal cord and drive rhythmic and stereotyped motor behavior such as walking or swimming without input from higher brain areas [37–39]. It is assumed that spinal reflex networks are crucially involved in these self-organizing neural circuits [40,41]. This finding is supported by studies that detected the suppression of H-reflexes after balance training, in parallel to balance and postural control improvement [15,16]. Besides this, it is cogitable that the orientation of neurons within the spinal cord (highly orientated axons extending along the craniocaudal axis) [42] leads to more consistent tsDCS-induced effects in comparison to cerebral tDCS application (inconsistent axons extending within the gyral banks) [43,44].

4.2. Leg-Specific Modulation

Our data show a greater improvement in the balance ability for standing on the left leg than on the right leg. This is true for M1 and cerebral tDCS, but not for tsDCS. This can be caused by electrode positioning in relation to the sagittal body plane in our study. The electrodes were placed symmetrically during tsDCS (anode over Th8 and cathode over L2). In contrast, a stronger right-hemispheric modulation was expected from M1 tDCS (anode over Cz and cathode over right supraorbital area) [45] and cerebellar tDCS (anode over O2 and cathode over right buccinator muscle) [46]. Indeed, the effects detected in our study are not consistent with the theory that the cerebrum controls the contralateral hemi body and the cerebellum controls the ipsilateral hemi body [47,48]. This theory (among others) is based on fMRI investigations that show that the active movement of a single lower limb is associated with increased neural activation of the primary sensorimotor cortex, supplementary motor area, cingulate motor area, secondary somatosensory cortex, and basal ganglia of the contralateral hemisphere, but with increased neural activation within the ipsilateral anterior lobe of the cerebellum [47].

A growing number of studies have demonstrated hemispheric asymmetries of motor control [47,49,50]. FMRI data show that brain activation during a movement of the non-dominant limb is more bilateral than during the same movement performed with the dominant extremity [47]. A TMS study demonstrated that the voluntary movement of a hand resulted in an increase in MEP amplitude in the non-task hand. This increase was more pronounced during left hand movements than during left hand tasks [50]. Accordingly, lesion studies indicate that the non-affected hemisphere can compensate for damage to the non-dominant hemisphere rather than for damage to the dominant hemisphere [49,51]. Hand motor recovery after left hemispheric stroke is two to three times slower than that after a right hemispheric incident [49,51]. Thus, one may assume that non invasive brain stimulation (NIBS) over the dominant hemisphere has the potential to modulate neural processing and/or motor control within the whole body, while the targeting of the non-dominant hemisphere modulates the non-dominant hemi body only. Unfortunately, there exists insufficient evidence in this regard. Existing NIBS research strongly focuses on the dominant hemisphere (and leg) and neglects the non-dominant hemisphere (and extremities). Future research should address this gap.

4.3. Balance Task-Specific Modulation

Our data demonstrate that the choice of assessment significantly influences the effects. While numerous significant t/tsDCS-induced improvements were found on the Y Balance Test, no effects were detected on the Single Leg Landing Test and the Single Leg Squat. Accordingly, numerous studies have shown little consistency in balance performance when using different assessments [52,53]. Balance performance during (i) single leg landing, (ii) stance on unstable platform, and (iii) forward falls correlated only weakly in young healthy people [53]. Similarly, performance during (i) the bipedal stance, (ii) stance on unstable platform, and (iii) the Functional Reach Test were not correlated in children aged 7–10 years [52]. It can be assumed that different mechanisms are responsible for balance control. An interesting perspective on this topic offers the Balance Evaluation Systems Test [54]. This test battery differentiates between six balance control systems (biomechanical constraints, stability limits/verticality, anticipatory postural adjustments, postural responses, sensory orientation, and stability in gait). This assessment was developed to identify the underlying cause for poor functional balance in several cohorts [54]. Participants with balance deficiencies in one category do not necessarily show deficits in other categories [54]. Future studies should closely evaluate the relationships between balance performance measured by differential assessments and the neural processing within brain and spinal cord networks. Besides this, the relationships between deep and superficial sensitivity and balance and/or gait performance should be investigated. The existing data show that balance and postural control are complex neural processes, and their neural background has not been fully understood to date.

5. Strengths and Limitations

This is the first placebo-controlled study that compared the effects of t/tsDCS over the different areas of balance ability in healthy participants. Our results provide additional insights into the neural background of balance and postural control and support the development of innovative therapy strategies in several cohorts. A weakness of our experiments is the limited number of participants; missing neuro-navigation to determine the exact location of optimal tDCS application; and differential electrode positioning in relation to the sagittal body plane: (1) both electrodes over the midline during tsDCS, (2) anode over the midline and cathode over the right supraorbital area during M1 tDCS, and (3) anode over the right cerebellum and cathode over the right buccinator muscle during cerebellar stimulation.

6. Conclusions

Our study indicates that both cerebellum and spinal cord are promising areas for the application of NIBS in supporting balance ability in young healthy people. While tsDCS supported balance in both legs, M1 and cerebellar tDCS supported only right leg balance. It is an open question whether, and to what extent, our finding can be transferred to people of different ages and disabled populations. Future research should fill the gap of evidence on this field and investigate the effects of tDCS applications over different brain and spinal cord regions in different cohorts. The incorporation of neuro-navigation to precisely target stimulation areas can improve the specificity and efficacy of the results. The tsDCS should be more closely examined in the framework of motor rehabilitation. The motor disabilities caused by several neurological diseases are associated not only with changes in neural processing within the brain but also within the spinal networks [55–58]. E.g., an increased spinal reflex is observed in stroke victims compared to healthy people and its normalization correlates with a successful gait recovery [55,56]. A desynchronization of spinal reflex loop oscillations seems to be one of the main sources of tremor in Parkinson's disease [57]. An increasing number of dystonia cases (traditionally considered a disorder of the basal ganglia, brainstem, and cerebellum) are reported in patients with spinal cord pathology [58]. tsDCS can be a promising alternative to "traditional" cerebral applications in these cohorts [55–58].

Author Contributions: J.V., C.S. and T.S. (Thomas Schack) conceptualized and designed the study. T.S. (Teni Steingräber), L.v.G., J.W. and R.R. performed the acquisition of the data. J.V., C.S. and T.S. (Teni Steingräber) analyzed and interpreted the data. J.V. wrote the first version of the manuscript. J.V., C.S., T.S. (Thomas Schack), T.S. (Teni Steingräber), L.v.G., J.W. and R.R. reviewed the final version of the manuscript. All authors have read and agreed to the published version of the manuscript.

Funding: This research received no specific grant from any funding agency in the public, commercial, or non-profit sectors. We acknowledge the financial support of the Open Access Publication Fund of Bielefeld University for the article processing charge.

Institutional Review Board Statement: The study was approved by the Ethics Committee of Bielefeld University (EC no. 2022-043 and date of approval 28 March 2022). All participants gave their written informed consent.

Informed Consent Statement: Informed consent was obtained from all subjects involved in the study.

Data Availability Statement: The datasets generated during and/or analyzed during the current study are available from the corresponding author on reasonable request.

Conflicts of Interest: The authors declare that there are no conflicts of interest.

References

1. Nitsche, M.A.; Paulus, W. Excitability changes induced in the human motor cortex by weak transcranial direct current stimulation. *J. Physiol.* **2000**, *527 Pt 3*, 633–639. [CrossRef] [PubMed]
2. Woods, A.J.; Antal, A.; Bikson, M.; Boggio, P.S.; Brunoni, A.R.; Celnik, P.; Cohen, L.G.; Fregni, F.; Herrmann, C.S.; Kappenman, E.S.; et al. A technical guide to tDCS, and related non-invasive brain stimulation tools. *Clin. Neurophysiol.* **2016**, *127*, 1031–1048. [CrossRef] [PubMed]
3. Kuo, H.I.; Bikson, M.; Datta, A.; Minhas, P.; Paulus, W.; Kuo, M.F.; Nitsche, M.A. Comparing cortical plasticity induced by conventional and high-definition 4 × 1 ring tDCS: A neurophysiological study. *Brain Stimul.* **2013**, *6*, 644–648. [CrossRef] [PubMed]
4. Giordano, J.; Bikson, M.; Kappenman, E.S.; Clark, V.P.; Coslett, H.B.; Hamblin, M.R.; Hamilton, R.; Jankord, R.; Kozumbo, W.J.; McKinley, R.A.; et al. Mechanisms and Effects of Transcranial Direct Current Stimulation. *Dose Response* **2017**, *15*, 1559325816685467. [CrossRef] [PubMed]
5. Herrmann, C.S.; Rach, S.; Neuling, T.; Strüber, D. Transcranial alternating current stimulation: A review of the underlying mechanisms and modulation of cognitive processes. *Front. Hum. Neurosci.* **2013**, *7*, 279. [CrossRef] [PubMed]
6. Nasseri, P.; Nitsche, M.A.; Ekhtiari, H. A framework for categorizing electrode montages in transcranial direct current stimulation. *Front. Hum. Neurosci.* **2015**, *9*, 54. [CrossRef] [PubMed]
7. Wiethoff, S.; Hamada, M.; Rothwell, J.C. Variability in response to transcranial direct current stimulation of the motor cortex. *Brain Stimul.* **2014**, *7*, 468–475. [CrossRef]
8. Halakoo, S.; Ehsani, F.; Hosnian, M.; Kheirkhahan, A.; Samaei, A.; Emadi, A. The comparative effects of anodal and cathodal trans-cranial direct current stimulation on balance and posture: A systematic review of literature and meta-analysis. *J. Clin. Neurosci.* **2023**, *107*, 68–76. [CrossRef] [PubMed]
9. Guimarães, A.N.; Porto, A.B.; Marcori, A.J.; Lage, G.M.; Altimari, L.R.; Alves Okazaki, V.H. Motor learning and tDCS: A systematic review on the dependency of the stimulation effect on motor task characteristics or tDCS assembly specifications. *Neuropsychologia* **2023**, *179*, 108463. [CrossRef] [PubMed]
10. Im, C.H.; Park, J.H.; Shim, M.; Chang, W.H.; Kim, Y.H. Evaluation of local electric fields generated by transcranial direct current stimulation with an extracephalic reference electrode based on realistic 3D body modeling. *Phys. Med. Biol.* **2012**, *57*, 2137–2150. [CrossRef]
11. Guidetti, M.; Maria Bianchi, A.; Parazzini, M.; Maiorana, N.; Bonato, M.; Ferrara, R.; Libelli, G.; Montemagno, K.; Ferrucci, R.; Priori, A.; et al. Monopolar tDCS might affect brainstem reflexes: A computational and neurophysiological study. *Clin. Neurophysiol.* **2023**, *155*, 44–54. [CrossRef] [PubMed]
12. Guidetti, M.; Giannoni-Luza, S.; Bocci, T.; Pacheco-Barrios, K.; Bianchi, A.M.; Parazzini, M.; Ionta, S.; Ferrucci, R.; Maiorana, N.V.; Verde, F.; et al. Modeling Electric Fields in Transcutaneous Spinal Direct Current Stimulation: A Clinical Perspective. *Biomedicines* **2023**, *11*, 1283. [CrossRef] [PubMed]
13. Dijkstra, B.W.; Bekkers, E.M.J.; Gilat, M.; de Rond, V.; Hardwick, R.M.; Nieuwboer, A. Functional neuroimaging of human postural control: A systematic review with meta-analysis. *Neurosci. Biobehav. Rev.* **2020**, *115*, 351–362. [CrossRef] [PubMed]
14. Surgent, O.J.; Dadalko, O.I.; Pickett, K.A.; Travers, B.G. Balance and the brain: A review of structural brain correlates of postural balance and balance training in humans. *Gait Posture* **2019**, *71*, 245–252. [CrossRef] [PubMed]
15. Taube, W.; Gruber, M.; Gollhofer, A. Spinal and supraspinal adaptations associated with balance training and their functional relevance. *Acta Physiol.* **2008**, *193*, 101–116. [CrossRef] [PubMed]
16. Chen, Y.S.; Zhou, S. Soleus H-reflex and its relation to static postural control. *Gait Posture* **2011**, *33*, 169–178. [CrossRef] [PubMed]
17. Baharlouei, H.; Ali Salehinejad, M.; Talimkhani, A.; Nitsche, M.A. The Effect of Non-invasive Brain Stimulation on Gait in Healthy Young and Older Adults: A Systematic Review of the Literature. *Neuroscience* **2023**, *516*, 125–140. [CrossRef]
18. Guo, Z.; Bao, D.; Manor, B.; Zhou, J. The Effects of Transcranial Direct Current Stimulation (tDCS) on Balance Control in Older Adults: A Systematic Review and Meta-Analysis. *Front. Aging Neurosci.* **2020**, *12*, 275. [CrossRef]
19. de Moura, M.C.D.S.; Hazime, F.A.; Marotti Aparicio, L.V.; Grecco, L.A.C.; Brunoni, A.R.; Hasue, R.H. Effects of transcranial direct current stimulation (tDCS) on balance improvement: A systematic review and meta-analysis. *Somatosens. Mot. Res.* **2019**, *36*, 122–135. [CrossRef]
20. Priori, A.; Ciocca, M.; Parazzini, M.; Vergari, M.; Ferrucci, R. Transcranial cerebellar direct current stimulation and transcutaneous spinal cord direct current stimulation as innovative tools for neuroscientists. *J. Physiol.* **2014**, *592*, 3345–3369. [CrossRef]
21. Manto, M.; Argyropoulos, G.P.D.; Bocci, T.; Celnik, P.A.; Corben, L.A.; Guidetti, M.; Koch, G.; Priori, A.; Rothwell, J.C.; Sadnicka, A.; et al. Consensus Paper: Novel Directions and Next Steps of Non-invasive Brain Stimulation of the Cerebellum in Health and Disease. *Cerebellum* **2022**, *21*, 1092–1122. [CrossRef] [PubMed]

22. Guidetti, M.; Ferrucci, R.; Vergari, M.; Aglieco, G.; Naci, A.; Versace, S.; Pacheco-Barrios, K.; Giannoni-Luza, S.; Barbieri, S.; Priori, A.; et al. Effects of Transcutaneous Spinal Direct Current Stimulation (tsDCS) in Patients With Chronic Pain: A Clinical and Neurophysiological Study. *Front. Neurol.* **2021**, *12*, 695910. [CrossRef] [PubMed]
23. Keel, J.C.; Smith, M.J.; Wassermann, E.M. A safety screening questionnaire for transcranial magnetic stimulation. *Clin. Neurophysiol.* **2001**, *112*, 720. [CrossRef] [PubMed]
24. Kaminski, E.; Hoff, M.; Rjosk, V.; Steele, C.J.; Gundlach, C.; Sehm, B.; Villringer, A.; Ragert, P. Anodal Transcranial Direct Current Stimulation Does Not Facilitate Dynamic Balance Task Learning in Healthy Old Adults. *Front. Hum. Neurosci.* **2017**, *11*, 16. [CrossRef]
25. Veldema, J.; Gharabaghi, A. Non-invasive brain stimulation for improving gait, balance, and lower limbs motor function in stroke. *J. Neuroeng. Rehabil.* **2022**, *19*, 84. [CrossRef] [PubMed]
26. Fernandes, S.R.; Salvador, R.; Wenger, C.; de Carvalho, M.; Miranda, P.C. Transcutaneous spinal direct current stimulation of the lumbar and sacral spinal cord: A modelling study. *J. Neural Eng.* **2018**, *15*, 036008. [CrossRef] [PubMed]
27. Herwig, U.; Satrapi, P.; Schönfeldt-Lecuona, C. Using the international 10-20 EEG system for positioning of transcranial magnetic stimulation. *Brain Topogr.* **2003**, *16*, 95–99. [CrossRef] [PubMed]
28. Chakraverty, R.; Pynsent, P.; Isaacs, K. Which spinal levels are identified by palpation of the iliac crests and the posterior superior iliac spines? *J. Anat.* **2007**, *210*, 232–236. [CrossRef] [PubMed]
29. Cooperstein, R.; Haneline, M.T. Spinous process palpation using the scapular tip as a landmark vs a radiographic criterion standard. *J. Chiropr. Med.* **2007**, *6*, 87–93. [CrossRef]
30. Plisky, P.; Schwartkopf-Phifer, K.; Huebner, B.; Garner, M.B.; Bullock, G. Systematic Review and Meta-Analysis of the Y-Balance Test Lower Quarter: Reliability, Discriminant Validity, and Predictive Validity. *Int. J. Sports Phys. Ther.* **2021**, *16*, 1190–1209. [CrossRef]
31. Lynall, R.C.; Campbell, K.R.; Mauntel, T.C.; Blackburn, J.T.; Mihalik, J.P. Single-Legged Hop and Single-Legged Squat Balance Performance in Recreational Athletes With a History of Concussion. *J. Athl. Train.* **2020**, *55*, 488–493. [CrossRef] [PubMed]
32. Byrne, A.; Lodge, C.; Wallace, J. Test-Retest Reliability of Single-Leg Time to Stabilization Following a Drop-Landing Task in Healthy Individuals. *J. Sport Rehabil.* **2021**, *30*, 1242–1245. [CrossRef] [PubMed]
33. Troester, J.C.; Jasmin, J.G.; Duffield, R. Reliability of Single-Leg Balance and Landing Tests in Rugby Union; Prospect of Using Postural Control to Monitor Fatigue. *J. Sports Sci. Med.* **2018**, *17*, 174–180.
34. Mengarelli, A.; Verdini, F.; Cardarelli, S.; Di Nardo, F.; Burattini, L.; Fioretti, S. Balance assessment during squatting exercise: A comparison between laboratory grade force plate and a commercial, low-cost device. *J. Biomech.* **2018**, *71*, 264–270. [CrossRef] [PubMed]
35. Schneiders, A.G.; Sullivan, S.J.; O'Malley, K.J.; Clarke, S.V.; Knappstein, S.A.; Taylor, L.J. A valid and reliable clinical determination of footedness. *PM&R* **2010**, *2*, 835–841.
36. Takakusaki, K.; Takahashi, M.; Obara, K.; Chiba, R. Neural substrates involved in the control of posture. *Adv. Robot.* **2017**, *31*, 2–23. [CrossRef]
37. Dutta, S.; Parihar, A.; Khanna, A.; Gomez, J.; Chakraborty, W.; Jerry, M.; Grisafe, B.; Raychowdhury, A.; Datta, S. Programmable coupled oscillators for synchronized locomotion. *Nat. Commun.* **2019**, *10*, 3299. [CrossRef] [PubMed]
38. Dzeladini, F.; van den Kieboom, J.; Ijspeert, A. The contribution of a central pattern generator in a reflex-based neuromuscular model. *Front. Hum. Neurosci.* **2014**, *8*, 371. [CrossRef] [PubMed]
39. Hooper, S.L. Central pattern generators. *Curr. Biol.* **2000**, *10*, R176. [CrossRef]
40. Geyer, H.; Herr, H. A muscle-reflex model that encodes principles of legged mechanics produces human walking dynamics and muscle activities. *IEEE Trans. Neural Syst. Rehabil. Eng.* **2010**, *18*, 263–273. [CrossRef]
41. Ramadan, R.; Geyer, H.; Jeka, J.; Schöner, G.; Reimann, H. A neuromuscular model of human locomotion combines spinal reflex circuits with voluntary movements. *Sci. Rep.* **2022**, *12*, 8189. [CrossRef]
42. Koser, D.E.; Moeendarbary, E.; Hanne, J.; Kuerten, S.; Franze, K. CNS cell distribution and axon orientation determine local spinal cord mechanical properties. *Biophys. J.* **2015**, *108*, 2137–2147. [CrossRef] [PubMed]
43. Lee, J.; Bestmann, S.; Evans, C. A future of current flow modelling for transcranial electrical stimulation? *Curr. Behav. Neurosci. Rep.* **2021**, *8*, 150–159. [CrossRef]
44. Liu, A.; Vöröslakos, M.; Kronberg, G.; Henin, S.; Krause, M.R.; Huang, Y.; Opitz, A.; Mehta, A.; Pack, C.C.; Krekelberg, B.; et al. Immediate neurophysiological effects of transcranial electrical stimulation. *Nat Commun.* **2018**, *9*, 5092. [CrossRef]
45. DaSilva, A.F.; Truong, D.Q.; DosSantos, M.F.; Toback, R.L.; Datta, A.; Bikson, M. State-of-art neuroanatomical target analysis of high-definition and conventional tDCS montages used for migraine and pain control. *Front. Neuroanat.* **2015**, *9*, 89. [CrossRef]
46. Rice, L.C.; D'Mello, A.M.; Stoodley, C.J. Differential Behavioral and Neural Effects of Regional Cerebellar tDCS. *Neuroscience* **2021**, *462*, 288–302. [CrossRef] [PubMed]
47. Kapreli, E.; Athanasopoulos, S.; Papathanasiou, M.; Van Hecke, P.; Strimpakos, N.; Gouliamos, A.; Peeters, R.; Sunaert, S. Lateralization of brain activity during lower limb joints movement. An fMRI study. *Neuroimage* **2006**, *32*, 1709–1721. [CrossRef] [PubMed]
48. Palesi, F.; De Rinaldis, A.; Castellazzi, G.; Calamante, F.; Muhlert, N.; Chard, D.; Tournier, J.D.; Magenes, G.; D'Angelo, E.; Gandini Wheeler-Kingshott, C.A.M. Contralateral cortico-ponto-cerebellar pathways reconstruction in humans in vivo: Implications for reciprocal cerebro-cerebellar structural connectivity in motor and non-motor areas. *Sci. Rep.* **2017**, *7*, 12841. [CrossRef] [PubMed]

49. Lüdemann-Podubecká, J.; Bösl, K.; Theilig, S.; Wiederer, R.; Nowak, D.A. The Effectiveness of 1 Hz rTMS Over the Primary Motor Area of the Unaffected Hemisphere to Improve Hand Function After Stroke Depends on Hemispheric Dominance. *Brain Stimul.* **2015**, *8*, 823–830. [CrossRef]
50. Ziemann, U.; Hallett, M. Hemispheric asymmetry of ipsilateral motor cortex activation during unimanual motor tasks: Further evidence for motor dominance. *Clin. Neurophysiol.* **2001**, *112*, 107–113. [CrossRef]
51. Kimura, D. Acquisition of a motor skill after left-hemisphere damage. *Brain* **1977**, *100*, 527–542. [CrossRef] [PubMed]
52. Muehlbauer, T.; Besemer, C.; Wehrle, A.; Gollhofer, A.; Granacher, U. Relationship between strength, balance and mobility in children aged 7-10 years. *Gait Posture* **2013**, *37*, 108–112. [CrossRef] [PubMed]
53. Ringhof, S.; Stein, T. Biomechanical assessment of dynamic balance: Specificity of different balance tests. *Hum. Mov. Sci.* **2018**, *58*, 140–147. [CrossRef] [PubMed]
54. Horak, F.B.; Wrisley, D.M.; Frank, J. The Balance Evaluation Systems Test (BESTest) to differentiate balance deficits. *Phys. Ther.* **2009**, *89*, 484–4988. [CrossRef] [PubMed]
55. Cho, S.H.; Lee, J.H. Comparison of the Amplitudes of the H-reflex of Post-stroke Hemiplegia Patients and Normal Adults during Walking. *J. Phys. Ther. Sci.* **2013**, *25*, 729–732. [CrossRef] [PubMed]
56. Kawaishi, Y.; Matsumoto, N.; Nishiwaki, T.; Hirano, T. Postactivation depression of soleus H-reflex increase with recovery of lower extremities motor functions in patients with subacute stroke. *J. Phys. Ther. Sci.* **2017**, *29*, 1539–1542. [CrossRef] [PubMed]
57. Anastasopoulos, D. Tremor in Parkinson's Disease May Arise from Interactions of Central Rhythms with Spinal Reflex Loop Oscillations. *J. Park. Dis.* **2020**, *10*, 383–392. [CrossRef] [PubMed]
58. Sarin, S.; Lawal, T.; Abboud, H. Spinal dystonia and other spinal movement disorders. *Dystonia* **2023**, *2*, 11303. [CrossRef]

Disclaimer/Publisher's Note: The statements, opinions and data contained in all publications are solely those of the individual author(s) and contributor(s) and not of MDPI and/or the editor(s). MDPI and/or the editor(s) disclaim responsibility for any injury to people or property resulting from any ideas, methods, instructions or products referred to in the content.

bioengineering

MDPI

Article

Post-Stroke Functional Changes: In-Depth Analysis of Clinical Tests and Motor-Cognitive Dual-Tasking Using Wearable Sensors

Masoud Abdollahi [1], Ehsan Rashedi [1,*], Pranav Madhav Kuber [1], Sonia Jahangiri [1], Behnam Kazempour [1], Mary Dombovy [2] and Nasibeh Azadeh-Fard [1]

[1] Department of Industrial and Systems Engineering, Rochester Institute of Technology, Rochester, NY 14623, USA; ma8489@rit.edu (M.A.); pmk2015@rit.edu (P.M.K.); sj1374@rit.edu (S.J.); bk9656@g.rit.edu (B.K.); nafeie@rit.edu (N.A.-F.)

[2] Department of Rehabilitation and Neurology, Unity Hospital, Rochester, NY 14626, USA; mary.dombovy@rochesterregional.org

* Correspondence: exreie@rit.edu; Tel.: +1-585-475-7260

Abstract: Clinical tests like Timed Up and Go (TUG) facilitate the assessment of post-stroke mobility, but they lack detailed measures. In this study, 21 stroke survivors and 20 control participants underwent TUG, sit-to-stand (STS), and the 10 Meter Walk Test (10MWT). Tests incorporated single tasks (STs) and motor-cognitive dual-task (DTs) involving reverse counting from 200 in decrements of 10. Eight wearable motion sensors were placed on feet, shanks, thighs, sacrum, and sternum to record kinematic data. These data were analyzed to investigate the effects of stroke and DT conditions on the extracted features across segmented portions of the tests. The findings showed that stroke survivors (SS) took 23% longer for total TUG ($p < 0.001$), with 31% longer turn time ($p = 0.035$). TUG time increased by 20% ($p < 0.001$) from STs to DTs. In DTs, turning time increased by 31% ($p = 0.005$). Specifically, SS showed 20% lower trunk angular velocity in sit-to-stand ($p = 0.003$), 21% longer 10-Meter Walk time ($p = 0.010$), and 18% slower gait speed ($p = 0.012$). As expected, turning was especially challenging and worsened with divided attention. The outcomes of our study demonstrate the benefits of instrumented clinical tests and DTs in effectively identifying motor deficits post-stroke across sitting, standing, walking, and turning activities, thereby indicating that quantitative motion analysis can optimize rehabilitation procedures.

Keywords: stroke; neurological disorders; movement analysis; kinematics; rehabilitation

Citation: Abdollahi, M.; Rashedi, E.; Kuber, P.M.; Jahangiri, S.; Kazempour, B.; Dombovy, M.; Azadeh-Fard, N. Post-Stroke Functional Changes: In-Depth Analysis of Clinical Tests and Motor-Cognitive Dual-Tasking Using Wearable Sensors. *Bioengineering* **2024**, *11*, 349. https://doi.org/10.3390/bioengineering11040349

Academic Editors: Franz Konstantin Fuss and Carlo Albino Frigo

Received: 4 March 2024
Revised: 26 March 2024
Accepted: 28 March 2024
Published: 2 April 2024

Copyright: © 2024 by the authors. Licensee MDPI, Basel, Switzerland. This article is an open access article distributed under the terms and conditions of the Creative Commons Attribution (CC BY) license (https://creativecommons.org/licenses/by/4.0/).

1. Introduction

Stroke remains a challenge for healthcare professionals in the 21st century, ranking among the foremost causes of disability in the adult population with ~795,000 new cases each year [1–3]. More importantly, two-thirds of surviving patients, amounting to ~7 million individuals in the United States alone, struggle with compromised mobility. Even after completing a rehabilitation program, only 40% of stroke survivors (SS) are able to completely reclaim their lost functional capabilities [4]. These losses include diminished sensory functions, as well as a decline in both motor and cognitive capabilities. Notably, post-stroke hemiparesis, a condition characterized by weakness on one side of the body, is common. This condition significantly impedes the ability to perform daily tasks, even the most fundamental tasks, such as getting in and out of bed independently. Such activities are collectively referred to as 'Activities of Daily Living' (ADLs). The impact of post-stroke changes in body movement, including aspects such as increased risk of fall, asymmetry, and reduced speed while performing ADLs, has been investigated in the literature [5–7]. These alterations remain a topic of interest in the stroke research community because they can increase the risk of injury to SS who are elderly (~70 years on average), making even the most trivial injuries fatal. Beyond physical movement, stroke also affects an individual's cognitive performance, affecting sensory function and coordination, as noted by Dennis et al. [8]. The interconnectedness of neuro-muscular systems often compromises the ability to perform tasks that demand both motor and cognitive functions [9]. Effective monitoring and assessment of these functions can not only help in maintaining proper care but can also serve as a vital tool for ensuring effective rehabilitation of such patients.

Screening of deficiencies in the physiological functions of SS is widely conducted using clinical tests [10]. Among these tests, Timed Up and Go (TUG), sit-to-stand (STS), and the 10 Meter Walk Test (10MWT) are some of the most prevalent tools employed to evaluate affected motor capabilities [11–13]. The TUG test includes a timed sequence of actions, which includes (a) the process of rising from a seated position in a chair, (b) walking 10 feet, turning, and (c) returning to the initial seated position. Similarly, the STS test involves the evaluation of a patient's ability to transition from a seated position to a standing position and vice versa. In contrast, the 10MWT assesses walking speed and endurance over a designated distance, providing insights into gait patterns and mobility. However, it is noteworthy that conventional applications of tests like TUG, STS, and the 10MWT include measurements of general parameters (such as overall time) and tend to lack the precision required for the evaluation of specific factors, such as a patient's speed/time for specific body movements and the intricate movement patterns associated with the task. For instance, a TUG time of ~13 s was found to best discriminate between the performance of SS and healthy individuals (HI) [11]. An in-depth examination of the specific motions and patterns exhibited during these tests can offer several insights into the deficiencies caused by stroke. This level of scrutiny not only enhances the detection of affected functions but also holds the potential to refine and optimize targeted rehabilitation strategies.

Cognitive demands, especially those arising during the execution of secondary tasks, have been found to cause irregular body movements. Clinicians have increasingly recognized this aspect, with ~79% of physiotherapists incorporating assessments of cognitive contributions [14]. Researchers have also explored a dual-task (DT) paradigm to account for the intricate interactions between motor and cognitive functions in their evaluations. For instance, Plummer-D'Amato et al. [15] conducted an assessment of the impact of concurrent cognitive tasks, including speech, visuospatial, and working-memory exercises, on gait parameters and observed a decline in the parameters during DT performance. In another study, deterioration in gait parameters was associated with an increased risk of fall [16]. Similarly, a study that evaluated prefrontal cortex activation during DT walking found that the activation may prioritize physical demands in SS but might prioritize cognitive demands in HI [17]. The findings were also aligned with those from a study that suggested the prioritization of balance over cognitive tasks [18]. Overall, studies show potential benefits of incorporating DTs with tests that involve motor activities (like walking/turning) because of their similarity to real-world tasks [19–22].

Recent strides in sensor technologies have played a pivotal role in enabling detailed performance assessment through motion analysis. Patterson et al. demonstrated the efficacy of motion analysis in assessing post-stroke changes in gait parameters, and Bae et al. showcased the utility of wearable sensors in evaluating conditions like Parkinson's disease [23,24]. Motion analysis can be helpful in advancing the precision and effectiveness of targeted rehabilitation protocols. Recent studies explored the effects of motor-cognitive DTs [25,26] and their effects during traditional clinical tests such as TUG and the 10MWT [20,27–33]. Researchers have also implemented motion analysis to detect the effects of stroke on movement patterns during daily activities like walking in the presence of DTs [34–36]. Traditionally, movement patterns have been studied using optoelectronic camera-based motion-capture systems, which are laboratory-restrictive despite being highly accurate. To facilitate clinical evaluations, more accessible approaches are being developed. Although video-based pose detection methods can be highly accessible, their capabilities in accurately detecting diverse sets of postures remain a matter of debate. Meanwhile, sensors based on Inertial Measurement Unit (IMU) technology are highly accessible, are sufficiently accurate, and allow field evaluations. Wearable sensors are also present in most modern electronic devices (smartphones, smartwatches, etc.) and can be efficiently used to objectively detect movement variations in clinical populations. By recording movement variations using wearable sensors, it may be possible to detect deficits in specific body functions, enabling more targeted rehabilitation approaches.

Our study is the first to evaluate motor-cognitive DT performance in instrumented clinical tests using multiple wearable motion sensors. The efficiency in detecting affected functions in patients using clinical tests alone, motion analysis alone, or the use of motion analysis in combination with clinical tests remains a point of debate in the stroke community. For instance, one study found that TUG and the step test may allow better prediction of falls than gait speed/static balance measures [37]. The novelty of this study lies in the level of depth in analyzing the impacts of DTs due to the inclusion of wearable sensors. Building on our earlier studies, wherein we employed wearable Inertial Measurement Unit (IMU) sensors to detect movement changes in SS during on-spot turning [38], this study aims to extend our understanding by infusing TUG, STS, and 10MWT clinical tests with detailed motion analysis. This study encompasses multiple objectives: (a) to assess movement patterns between SS and HI during instrumented clinical tests by utilizing multiple wearable sensors; (b) to determine the differences across specific activities within clinical tests and identify the most effective activities to detect motor changes within the common clinical tests; and (c) to assess the impacts of using a DT paradigm on movement patterns in SS. We hypothesize that combining motion analysis (using wearable sensors) with common clinical tests may be more beneficial in identifying specific activities that are more challenging for SS. Furthermore, we expected that inducing additional cognitive demands while performing motor tasks can stress neuro-musculoskeletal systems impacting the individual's physical performance during tests. The overarching objective of this study is to contribute to the development of effective rehabilitation programs, ultimately enhancing the quality of life and overall well-being of post-stroke patients.

2. Materials and Methods

2.1. Participants

We recruited 41 participants, including 21 SS/stroke and 20 HI/control. The sample size was determined based on a preliminary power analysis using G*Power software (version 3.1). The results of our analysis demonstrated that for effect size, significance level, and powers of 0.8, 0.05, and 0.8, we needed a sample size of 21 subjects in each group. There were no statistically significant differences in anthropometric measurements (mean (SD)) between the two groups (p-values for age, height, weight, and BMI were >0.05), as summarized in Table 1. Prior to the commencement of the study, all participants provided their informed consent by signing a written consent form. This study was conducted in accordance with the approval granted by the Institutional Review Board of both the Rochester Institute of Technology and Rochester Regional Health. Inclusion criteria for the stroke group required participants to be capable of walking independently for distances exceeding 10 m, to have experienced a stroke at least six months prior to their participation, and to be free from any severe medical conditions that might significantly impact their physical performance. Likewise, the healthy control group's inclusion criteria excluded individuals with neurological diseases or musculoskeletal issues that might affect functional performance during the experiment.

2.2. Study Approach

The body movement of the participants was recorded as they performed clinical tests in two separate conditions, with and without cognitive loading. We selected clinical tests based on the activities involved: sitting/standing, walking, and turning, consisting of TUG, STS, and the 10MWT. All three tests are widely implemented in the stroke research community for evaluating the motor, locomotor, and balance performance of post-stroke patients [7,12,39]. The participants completed five different tests (TUG, STS, the Balance Test with open/closed eyes, and the 10MWT) in a random order. Owing to the scope of this study, we excluded outcomes from the Balance Test. Additionally, for each test, participants either started with a DT or single-task (ST), and these selections were made randomly. To induce cognitive loading during DT, prior studies suggested that subtraction-based tasks rely more on working memory and can impose a higher cognitive load than verbal fluency [40]. Among the methods used to impose cognitive loading, such as reverse counting, reciting alternate alphabets, and naming items (e.g., animals), reverse counting in steps of 3 and 7 is the most common [41–43]. However, during pilot studies, our observations showed that post-stroke patients were unable to perform reverse counting in steps of 3 and 7 throughout the tasks because of their challenging nature. Thus, we used a simplified version that involved counting backward from 200 in decrements of 10 while following the prescribed path. If participants reached zero during any tests, they were instructed to restart the counting from 200. Importantly, participants were explicitly guided to ensure that their performance remained independent of the numerical counting task.

Table 1. Demographic data for study participants.

Parameters	Stroke (N = 21)	Control (N = 20)	*p*-Value (*t*-Test)
Gender	11 males 10 females	8 males 12 females	-
Age (year)	66 (10)	60 (8)	0.053
Height (cm)	173.8 (8.4)	172.6 (9.6)	0.669
Weight (kg)	86.3 (14.7)	81.8 (18.2)	0.39
BMI (kg/s^2)	28.5 (4)	27.2 (3.9)	0.288

2.2.1. Procedural Details of Clinical Assessments

The initial assessment, the TUG test, entailed a sequence of five timed locomotion tasks, which are visually depicted in Figure 1: (1) initiating from a seated position and transitioning to a standing posture; (2) advancing to a designated cone positioned approximately 3 m (equivalent to roughly 10 feet) from the chair's starting point; (3) executing a turn around the cone; (4) returning to the chair; and (5) performing a turn and reseating on the chair. Each participant completed the TUG test twice, once while concurrently engaging in a cognitive task and once without such DT demands. The selection of the TUG test for this experiment was motivated by its widespread use in clinical settings. Subsequently, the second assessment involved the STS test, during which participants performed five trials of sitting and standing from a chair. Lastly, the third test, the 10MWT, required participants to walk in a straight line for 10 m. Like in the TUG test, participants were asked to count aloud during both the STS and 10MWT assessments.

2.2.2. Experimental Design and Independent Factors

A 2 × 2 experimental design was selected for this study, with independent factors of group (stroke, control) and cognitive load (ST, DT). As participants performed each of the three tests (TUG, STS, 10MWT), their movement was recorded using eight IMUs manufactured by Movella (Xsens, Enschede, The Netherlands) placed on each of the feet, shanks, and thighs and on the sternum and sacrum of the participant in the pre-defined 'lower-body and trunk' configuration (Figure 1). Motion data were collected from each sensor simultaneously at a sampling frequency of 100 Hz using the MVN Analyze® software package (Xsens, Enschede, The Netherlands) as the participants performed each of the tests.

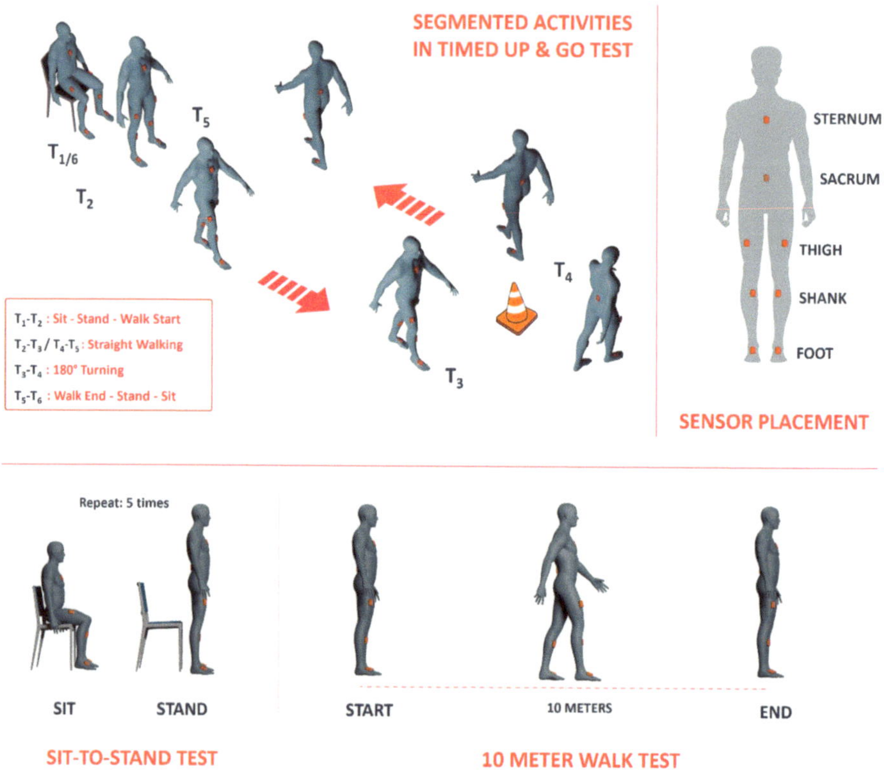

Figure 1. Illustration depicting (top left) segmentation of tasks while performing the Timed Up and Go (TUG) test, (bottom left) the sit-to-stand (STS) test, (bottom right) the 10 Meter Walk Test (10MWT), and (top-right) the placement locations of inertial sensors on the body.

2.3. *Signal Processing and Feature Extraction*

We employed a custom-tailored MATLAB (MathWorks, Natick, MA, USA) code to analyze the kinematic data. The data, encompassing segment angles, angular velocity, and linear acceleration, underwent initial preprocessing and extraction using the MVN Analyze® software package (Xsens, Enschede, The Netherlands). To optimize data quality, a 4th-order low-pass Butterworth filter with a 5 Hz cut-off frequency was applied to the imported data in MATLAB (version R2022b). For each test, a custom MATLAB code was designed for automatically detecting events in each test, segmentation of the tests for each participant, and feature extraction. Following the code's execution for each trial, all segmentation processes were confirmed visually by plotting graphs using markers for detected events. In cases where discrepancies arose between the code-generated segmentation and the defined strategy, manual adjustments were made to the segmentation. As the tests encompassed various tasks, each test was subjected to a distinct signal processing, segmentation, and event-detection approach, which will be elaborated on in the subsequent sections.

2.3.1. TUG Test

To conduct a thorough analysis of the TUG test, it was imperative to break the test down into distinct sections. As previously mentioned, we divided the test into five specific sections: (1) standing up from the chair, (2) walking toward the cone, (3) turning around the cone, (4) walking toward the chair, and (5) turning and sitting on the chair. To identify the start and end points of these sections, we defined six key events to be detected in each trial, as illustrated in Figure 1:

T1: Start of the task by the initiation of standing up from the chair

T2: Start of walking toward the cone (also indicating the end of the standing-up phase)

T3: Start of the turn around the cone (also indicating the end of walking toward the cone)

T4: Start of walking toward the chair (also indicating the turning around the cone)

T5: Start of turning and sitting on the chair (also indicating the end of walking toward the chair)

T6: Conclusion of the test by sitting back down in the chair

To identify the T1 and T6 events, we analyzed three signals: thorax resultant angular velocity, right thigh resultant angular velocity, and left thigh resultant angular velocity. The commencement of the task corresponded to the initial movement of any of these body segments. When we observed a significant rise in the signal of any of these segments, we marked that point as T1. Conversely, to determine T6 (the end of the task), we monitored all three signals, expecting them to reach stable, near-zero levels as participants completed the task.

To detect the T2, T3, T4, and T5 events, we used resultant linear acceleration from both the left and right feet, along with the thorax resultant angular velocity. These events denoted the start and end points of the walking sections (toward the cone and chair), so we aimed to identify them based on the periodic patterns present in the walking signals. The thorax resultant angular velocity signal was used to enhance the accuracy of segmentation during walking during the turning portion, as the commencement and completion of the turn were not clearly discernible from the foot acceleration signals alone. Specifically, we selected the end of a stride (T3) as the point at which a spike in thorax angular velocity occurred, with T4 representing the point when the thorax angular velocity dropped to a minimum after reaching a peak during the turn around the cone. Figure 2 provides a sample of the segmented signals for a participant in the control group, illustrating the effectiveness of this approach. Overall, ten features were calculated, including the total time to complete TUG; the time to perform sit-to-stand, walk toward the cone, turn around the cone, walk toward the chair, and turn and sit; total steps toward the cone and chair; and cadence toward the cone and chair.

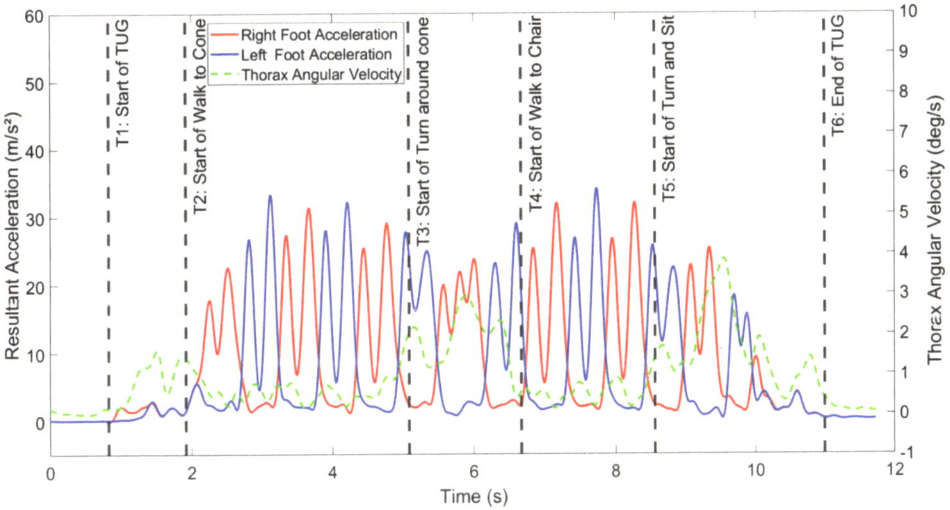

Figure 2. Illustration of the raw signals and segmented sections of the TUG task for a sample participant from the control group.

2.3.2. STS Test

A sit-to-stand test was implemented to evaluate patients' ability to transition between sitting and standing positions, and vice versa. As this task primarily involves flexion and extension movements in various body segments, we focused our analysis on the angular velocity of the sensors, rather than linear acceleration. Each test consisted of five repetitions of the sit–stand–sit activity. To segment the signals and measure the duration of each phase, we utilized the resultant angular velocity data from sensors placed on the thighs (upper legs). Within each repetition or cycle of sit–stand–sit, we observed two distinct convex shapes in the angular velocity data. Each of these convex shapes represented either the sitting or standing phases, and we considered the time between two sequences of convex shapes as an event during the segmentation process. For each sit-to-stand trial, we calculated a total of eight features, which included the total time taken for the sit–stand–sit activity, the mean duration of the sit–stand–sit phase, the mean duration of the sit–stand phase, the mean duration of the stand–sit phase, and the root mean square of angular velocity for the thorax, pelvic region, and right/left thighs during the test.

2.3.3. 10MWT

The 10MWT involved a 10 m walk and a comprehensive gait analysis relying on kinematic data. To analyze motion, especially in walking patterns, a crucial step was identifying heel-strike (HS) and toe-off (TO) events. Precise event timing was determined using resultant linear acceleration data collected from sensors on the participants' shanks, a method consistent with our prior studies [44,45]. The linear acceleration signal displayed two distinct peaks within each stride, one corresponding to TO and the other to HS. These events were identified using segmentation techniques on the sensor data from each shank. It is noteworthy that the gait event could be detected using other sensor data, such as from sensors on the feet or sacrum [46,47]. Subsequently, seven parameters were calculated to quantify various kinematic variables and factors related to walking. These parameters encompassed total walk time, the number of steps during the test, cadence, mean swing total, mean single support portion in strides, gait speed, and stride duration.

2.4. Statistical Analysis

To analyze the data, we conducted a repeated-measures analysis of variance (ANOVA). This involved creating models that accounted for the effects of each of the 25 different measures (shown in Table 2) obtained from the tests (TUG: 10 measures, STS: 8 measures, 10MWT: 7 measures). The key independent variables were group (consisting of stroke and control) and cognitive load (encompassing single task (ST) and DT, as well as their interaction). Before performing the analysis, we confirmed that the prerequisites for repeated-measures ANOVA were satisfied, which included verifying that the continuous dependent variable followed a normal distribution (implementing the Shapiro–Wilk method) and ensuring there were no outliers in the repeated measurements. We also calculated effect sizes (partial eta-squared: η^2) for each analysis, categorizing them as small if η^2 was less than 0.06, medium if it fell between 0.06 and 0.14, and large if it exceeded 0.14 [48]. In all our analyses, we maintained a significance level of 0.05, and we conducted these statistical computations using the JMP software (version 16.2) from the SAS Institute in North Carolina, USA.

Table 2. Main and interaction effects of the group (stroke and control) and cognitive load (single-task (ST) and dual-task (DT)) on the measures from TUG: Timed Up and Go, STS: sit-to-stand, and 10MWT: 10 Meter Walk Test. (Note: The three columns headed Group, CL, and Group × CL show the *p*-values for the null hypotheses of no significant effect of these factors on the dependent factors. The partial effect size (η^2) of each relevant effect analysis is reported in the column to its right. CL: cognitive load).

Measures	Group	η^2	CL	η^2	Group × CL	η^2
TUG Test						
Total time (s)	**<0.001**	0.244 (L)	**<0.001**	0.085 (M)	0.124	0.004 (S)
Sit-to-stand (s)	**0.036**	0.079 (M)	0.056	0.023 (S)	0.542	0.002 (S)
Walk toward cone (s)	**0.003**	0.166 (L)	**<0.001**	0.059 (S)	0.396	0.003 (S)
Turn around the cone (s)	**0.035**	0.069 (M)	**0.005**	0.068 (M)	0.116	0.020 (S)
Walk toward chair (s)	**0.006**	0.152 (L)	**<0.001**	0.067 (M)	0.140	0.007 (S)
Turn and sit (s)	**0.001**	0.206 (L)	**0.008**	0.033 (S)	0.835	<0.001 (S)
Steps toward cone	**0.002**	0.202 (L)	0.113	0.005 (S)	0.371	0.001 (S)
Steps toward chair	**0.005**	0.170 (L)	**0.022**	0.018 (S)	0.440	0.001 (S)
Cadence toward cone (step/min)	0.916	0.001 (S)	**<0.001**	0.107 (L)	0.592	<0.001 (S)
Cadence toward chair (step/min)	0.434	0.014 (S)	**0.002**	0.058 (S)	0.405	0.005 (S)
STS (×5) Test						
Total time (s)	0.289	0.019 (S)	0.150	0.002 (S)	0.215	0.008 (S)
Mean STS (s)	0.289	0.019 (S)	0.150	0.002 (S)	0.215	0.008 (S)
Mean STS up (s)	0.166	0.038 (S)	0.353	0.001 (S)	0.288	0.005 (S)
Mean STS down (s)	0.481	0.006 (S)	0.061	0.002 (S)	0.199	0.009 (S)
RMS of ang vel thorax (deg/s)	**0.003**	0.187 (L)	0.390	<0.001 (S)	0.464	0.001 (S)
RMS of ang vel pelvis (deg/s)	0.182	0.037 (S)	0.285	0.001 (S)	0.143	0.005 (S)
RMS of ang vel right thigh (deg/s)	0.569	0.004 (S)	**0.026**	0.004 (S)	0.204	0.005 (S)
RMS of ang vel left thigh (deg/s)	0.391	0.013 (S)	**0.032**	0.004 (S)	0.261	0.005 (S)
10MWT						
Walk time (s)	**0.010**	0.135 (M)	**<0.001**	0.064 (M)	0.257	0.005 (S)
Steps	**0.019**	0.124 (M)	**<0.001**	0.029 (S)	0.504	<0.001 (S)
Cadence (step/min)	0.184	0.036 (S)	**<0.001**	0.070 (M)	0.152	0.007 (S)
Mean swing total (%)	0.133	0.050 (S)	**<0.001**	0.037 (S)	0.076	0.007 (S)
Single support (%)	0.781	0.001 (S)	**0.044**	0.025 (S)	0.178	0.008 (S)
Gait speed (m/s)	**0.012**	0.124 (M)	**<0.001**	0.079 (M)	0.528	0.001 (S)
Stride duration (s)	0.111	0.050 (S)	**<0.001**	0.067 (M)	0.113	0.010 (S)

3. Results

The statistical analysis presented in Table 2 revealed significant main effects for both group and cognitive load on various key outcome measures in the TUG, STS, and 10MWT assessments. Although parameters such as STS time and RMS angular velocity were primarily influenced by group but not by cognitive load, others, like cadence and swing time, were significantly affected by cognitive load but not by group membership. Findings show an absence of significant interactions between group and cognitive load. Effect sizes for group-related main effects were substantial, particularly in multiple TUG parameters, including total time, walking duration, and turning sections. In contrast, the main effects of cognitive load generally had minor to moderate effects on the measured outcomes. The introduction of a cognitive DT further exacerbated participants' physical limitations during the clinical assessment tests.

Table 3 displays the mean (SD) values of key outcome measures for both single-task (ST) and dual-task (DT) conditions within each group. The analysis highlighted specific activities that posed significantly greater challenges for stroke survivors. In particular, they exhibited a 23% increase in the total time needed to complete the TUG test, with notable variations in the TUG subsections: 20% longer for walking to the cone, 15% for turning around the cone, and 29% for turning and sitting back down, with the most substantial difference observed in the turn and sit subsection. In the STS test, stroke survivors exhibited a 20% reduction in trunk angular velocity, and in the 10MWT, they recorded a 21% longer walk time and a reduced gait speed when compared with the control group.

Table 3. Mean (SD) values of the key outcome measures of the three tests (TUG: Timed Up and Go, STS: sit-to-stand, and 10MWT: 10 Meter Walk Test) for each group in single-task (ST) and dual-task (DT) conditions. Statistically significant values are denoted by: *. Ranges are defined as follows: p-value > 0.05: -, p-value ≤ 0.05: *, p-value ≤ 0.01: **, p-value ≤ 0.001: ***).

Measures	Control (N = 20)		Stroke (N = 21)		Group	CL
	Control ST	Control DT	Stroke ST	Stroke DT	*(Control ST vs. Stroke ST)*	*(Control DT vs. Stroke DT)*
TUG Test						
Time total (s)	13.34 (2.34)	14.8 (1.91)	16.15 (2.97)	18.44 (3.17)	***	***
Sit-to-stand (s)	1.37 (0.5)	1.58 (0.38)	1.84 (0.82)	1.94 (0.75)	*	-
Walk toward cone (s)	3.95 (0.8)	4.34 (0.78)	4.68 (1.05)	5.27 (1.07)	**	***
Turn around the cone (s)	1.75 (0.48)	1.88 (0.46)	1.87 (0.46)	2.3 (0.63)	*	**
Walk toward chair (s)	3.33 (0.72)	3.72 (0.75)	3.96 (1)	4.71 (1.32)	**	***
Turn and sit (s)	2.94 (0.57)	3.28 (0.6)	3.79 (0.92)	4.22 (1.2)	**	**
Steps toward cone	6.05 (1)	6.15 (1.04)	7.19 (1.54)	7.57 (1.54)	**	-
Steps toward chair	4.8 (0.83)	5.15 (1.04)	6.05 (1.56)	6.71 (2.55)	**	*
Cadence toward cone (step/min)	93.21 (12.11)	85.43 (7.61)	92.78 (9.97)	87.47 (14.44)	-	***
Cadence toward chair (step/min)	88.08 (12.77)	83.69 (10.61)	92.21 (11.17)	85.19 (15.83)	-	**
STS (×5) Test						
Time total (s)	17.81 (3.23)	19.06 (4.25)	20.04 (6.59)	20.6 (6.83)	-	-
Mean STS (s)	3.56 (0.65)	3.81 (0.85)	4.01 (1.32)	4.12 (1.37)	-	-
Mean STS up (s)	1.75 (0.32)	1.85 (0.42)	2.01 (0.65)	2.06 (0.63)	-	-
Mean STS down (s)	1.81 (0.35)	1.96 (0.45)	2 (0.75)	2.06 (0.79)	-	-
RMS of ang vel thorax (deg/s)	1.04 (0.23)	1.01 (0.23)	0.82 (0.21)	0.81 (0.18)	**	-
RMS of ang vel pelvis (deg/s)	0.91 (0.26)	0.86 (0.23)	0.78 (0.28)	0.78 (0.25)	-	-
RMS of ang vel right thigh (deg/s)	1.03 (0.21)	0.97 (0.21)	0.97 (0.24)	0.95 (0.21)	-	*
RMS of ang vel left thigh (deg/s)	1.03 (0.21)	0.98 (0.21)	0.96 (0.23)	0.93 (0.2)	-	*
10MWT						
Walk time (s)	10.98 (2.12)	12.18 (2.36)	12.94 (3.41)	15.1 (3.81)	**	***
Steps	17.7 (2.3)	18.8 (2.63)	20.14 (4.04)	21.57 (4.42)	*	***
Cadence (step/min)	98.05 (8.87)	93.83 (9.1)	95.58 (13.23)	87.5 (12.81)	-	***
Mean swing total (%)	0.38 (0.03)	0.37 (0.02)	0.37 (0.03)	0.35 (0.04)	-	***
Single support (%)	0.6 (0.06)	0.6 (0.04)	0.61 (0.04)	0.58 (0.06)	-	*
Gait speed (m/s)	0.94 (0.19)	0.85 (0.15)	0.82 (0.2)	0.7 (0.16)	*	***
Stride duration (s)	1.21 (0.12)	1.27 (0.13)	1.26 (0.19)	1.38 (0.21)	-	***

The results are further illustrated in Figures 3 and 4, which compare key outcome measures between stroke survivors and healthy individuals in ST and DT conditions. Figure 3 shows that stroke survivors took significantly longer to complete the total TUG test and its subsections, especially for turning activities, compared with healthy individuals. DTs increased times across both groups. Figure 4 shows that stroke survivors had more steps toward the chair in the TUG test, reduced swing portion in the 10MWT, lower trunk velocity during sit-to-stand transitions, and slower gait speed in the 10MWT compared with controls. Implementing DTs worsened these metrics for both groups. The figures demonstrate deficits in stroke survivors across sitting, standing, walking, and turning movements, which were further impacted by cognitive demands during the clinical assessment tests.

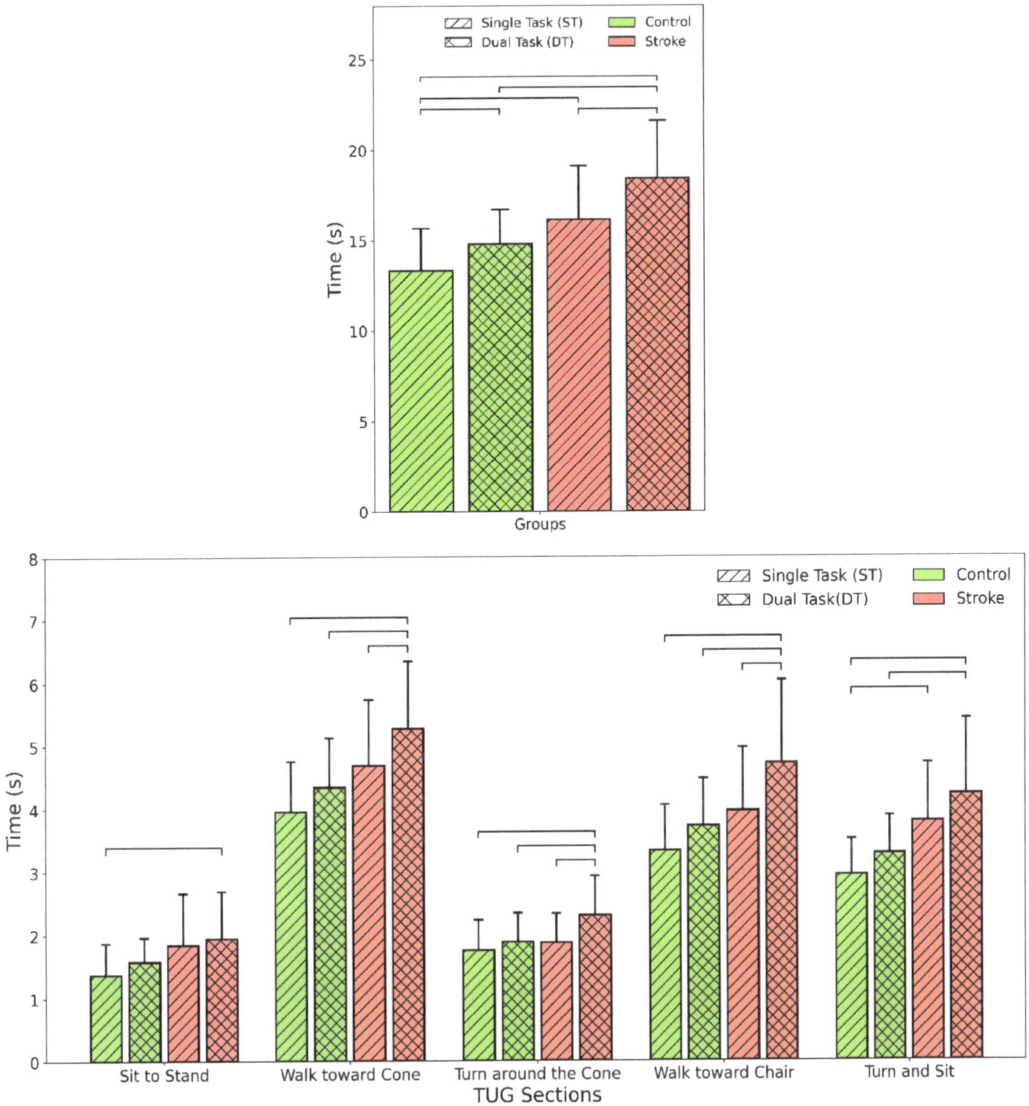

Figure 3. Comparison of the time taken to complete the tasks for different groups (stroke and control) and cognitive loads (single task and dual task). (Top) TUG total time and (bottom) times for the five subsections of TUG. The bars connected to the lines are significantly different (p-value < 0.05). Note: Each set of four bars (from left to right) in each section shows the results for the control group under ST and DT conditions and the stroke group under ST and DT conditions, respectively.

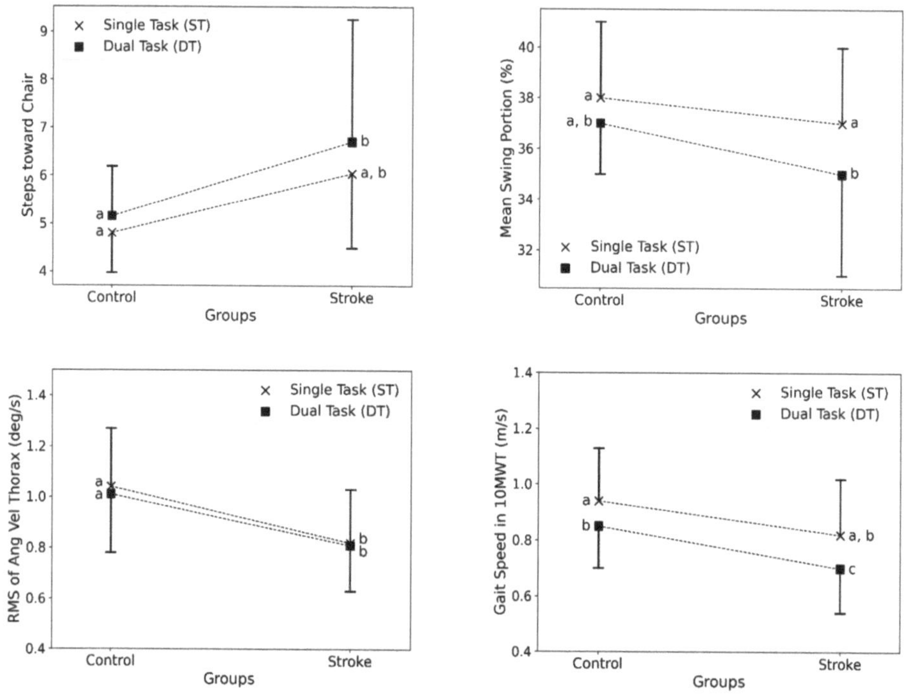

Figure 4. Comparison of key outcome measures between stroke survivors and healthy individuals in single-task (ST) and dual-task (DT) conditions. (Top left) Steps toward the chair in the TUG test, (top right) mean swing portion of strides in the 10MWT (%), (bottom left) RMS of thorax angular velocity in STS (deg/s), and (bottom right) gait speed in the 10MWT (m/s). Error bars indicate standard deviation. The points with no common letter labels (i.e., a, b, and c) were significantly different (p-value < 0.05) in the post-hoc analysis (lack of common letters between the control and stroke groups denotes statistical significance). The dotted lines connect the outcomes of the relevant cognitive loads in control and stroke groups.

4. Discussion

In this study, we examined the movement deficits in post-stroke patients by instrumenting the common clinical TUG, STS, and 10MWT assessments with the help of wearable sensors. The novelty of our approach was: (i) to enable detailed assessment of body movement as participants performed the tasks within these tests, and (ii) to assess the impact of cognitive loading while performing the tests by incorporating single and motor-cognitive DT conditions. This is the first study to conduct in-depth motion analysis to assess the impacts of DTs in common clinical tests by implementing multiple wearable motion sensors. The outcomes of the study revealed significant impairments in SS compared with HI as they performed tests that included a combination of ADLs, particularly sit/stand, walking, and turning. Notably, our findings (Table 3) also showed that the addition of cognitive loading in the DT conditions substantially impacted body movements in SS, highlighting the potential impact of divided attention on physical performance. Prior studies demonstrated that inducing additional cognitive loading can lead to decreased performance in completing tasks [15,18,20,34]. The findings of this study are consistent with earlier findings, and the introduction of a secondary cognitive task during the tests further worsened the motor impairments in SS.

Substantially greater variability was seen in the SS compared with the HI for several of the extracted kinematic features across the clinical tests (see Table 3). SS took 23% longer to complete the TUG test, reflecting significant challenges in the performance of sequential motor tasks. Segmentation of the TUG test into walking, turning, and sit/stand was beneficial for identifying differences between SS and HI, especially for assessing the impacts of cognitive loading [29]. As depicted in Figure 3, detailed segmentation revealed specific sections that were more demanding, such as the 29% increase in the time taken for the final turn and sit subsection of the TUG test. The TUG test best detected mobility limitations in stroke survivors, with total time increased by 23% compared with controls. Further analysis of TUG subsections provided insights into which activities were specifically challenging (Figure 3). Turning ability in SS was objectively evaluated by Hollands et al., 2014, wherein the mean time to turn during TUG was higher in the SS than in the HI (2.12 vs. 1.99 s). This study yielded similar outcomes, with the turn around the cone section being 15% longer in SS compared with HI, indicating turning deficiencies. The final turn and sit subsection had an even greater 29% increase, suggesting transitional movements like sitting are especially demanding. SS also took 20% longer for the straight walk sections. Detailed TUG analysis using wearable sensors, as conducted in this study, demonstrates the value of segmenting clinical tests to reveal specific problematic activities compared with using total time alone.

Consistent with the findings of our earlier study [49], cognitive loading through DT in this study impacted TUG performance in both groups, but more so in SS. DTs increased total TUG time by 20% in SS versus 11% in HI. In a previous study, cognitive loading in TUG increased the duration of the test by ~3.7–5.9 s (total time ~17–21 secs with cognitive loading), similar to the findings (Figure 3) reported in our study [20]. Examining the subsections showed that turn time was disproportionately affected, increasing by 31% under DT conditions in stroke survivors. Turning requires the integration of multiple sensorimotor processes, including postural transitions, asymmetric limb coordination, and spatial navigation [35,38]. DTs may overwhelm these processes in SS in the regions of the brain that govern cognition, attention, and motor control. Targeted training focusing on turning and transitions under cognitive load may improve TUG performance. Overall, instrumented analysis of TUG with DTs allowed robust assessment of the motor-cognitive capabilities affected by stroke.

During the STS test, the total test time under DT conditions had a standard deviation of 6.83 s in SS vs. 4.25 s in HI. This highlights the considerable heterogeneity in functional disabilities and compensatory movement strategies adopted by the stroke population. This result could be due to several factors, such as the location or extent of the injury (brain lesion), severity of the stroke, and duration since stroke occurrence [16]. The obtained kinematic features that showed pronounced deficits and variability in SS vs. HI show potential as sensitive metrics for tracking longitudinal recovery and improvements during rehabilitation. For example, the STS trunk angular velocity detected mobility limitations related to transitional movements [20,34]. Clinically, these objective instrumented measures could inform individualized prognoses and therapies tailored to each patient's specific functional challenges revealed during standardized tests. Furthermore, the pronounced deficits in trunk angular velocity during sit-to-stand in stroke survivors point to weakened core muscles and postural instability after stroke. STS transitions require dynamic balance capabilities, which may be further compromised by divided attention demands [37,50]. It is also possible that the greater attentional cost for maintaining stability could manifest in the form of affected kinematic parameters, such as reduced trunk velocity.

The absence of significant main effects of group or cognitive load on metrics like total time and phase durations during the STS test may be attributed to several factors. High inter-individual variability in transitional movements between participants likely contributed to considerable variability in STS performance that may have masked group differences. The straightforward sit-to-stand motion may have posed less of a motor challenge compared with more dynamic tasks like turning and walking for this cohort of stroke survivors. Furthermore, multiple repetitions of the STS task could have allowed participants to optimize their performance through practice effects, and the counting DT may not have sufficiently challenged cognitive-motor integration during the primarily motor sit-stand activity. Importantly, the limited sample size of stroke survivors may have reduced the statistical power to detect group or cognitive load effects. These factors potentially explain the lack of significant differences and indicate that STS test parameters may not differentiate stroke survivors from controls as well as other mobility metrics in certain populations and testing conditions.

The 10MWT revealed considerable variability in spatiotemporal gait metrics, including cadence, stride length, stance/swing ratios, and gait speed in stroke survivors. Similar outcomes were reported in [15], wherein cognitive loading affected gait speed, stride time, average stride length, and cadence during the 10MWT test. This could mean that stroke can disturb locomotor rhythmicity and symmetry owing to affected brain functions that map to leg motor functions, as seen in the form of alterations in gait parameters [27,51,52]. SS exhibited impaired gait speed, reduced cadence, shortened swing phase, and longer double support time compared with HI, as depicted in Table 3. DT likely shifts priority toward maintaining safe walking patterns, resulting in conservative alterations in cadence, swing time, and support phases in SS [52,53]. The changes in gait metrics may signify the recruitment of cognitive resources for conscious control and monitoring of movements. DT costs were higher across spatiotemporal gait parameters in stroke survivors, reiterating their limited cognitive reserve.

Among the studied clinical tests, 10MWT provided unique insights into spatiotemporal gait impairments in SS, such as reduced gait speed, cadence, stride length, and swing ratio. These gait metrics capture characteristics that may not be evident from total time alone during short walks in the TUG test. However, the 10MWT lacks the multi-tasking components involved in activities of daily living. Therefore, integrating the 10MWT straight path walking into an instrumented TUG test could offer advantages for both measures. According to the results of this study, we propose a modified TUG starting with the patient already standing, then walking 10 m, turning 180 degrees around a cone, walking 10 m back, and ending by sitting down. This would allow detailed quantification of gait parameters over an extended distance along with an assessment of turning under cognitive load. The sit-to-stand transition may provide less added value based on our findings. Eliminating this portion would shorten test time and reduce the complexity of the analysis while still capturing the key markers of mobility deficits. Thus, a modified instrumented TUG incorporating gait analysis over 10 m and a DT turn component may optimally assess motor-cognitive capabilities affected by stroke. Further research is needed to develop and validate such an integrated clinical assessment.

The instrumented metrics obtained in this study hold potential for translation to clinical practice to inform individualized rehabilitation. For instance, the disproportionate DT deficits in turn time could indicate the need for targeted training focused on transitional movements and navigation under divided attention in stroke survivors [54]. Likewise, reduced trunk velocity during repeated sit–stand transitions points to weakened core muscles requiring specific strengthening. Quantifying such functional limitations through sensor-based assessment during standardized tests allows customized therapies tailored to patient weaknesses. The metrics showing pronounced stroke-related impairments like TUG turn time and 10MWT gait speed could also serve as sensitive outcomes to track subtle longitudinal improvements resulting from interventions. Collaborations spanning engineers, clinicians, and data scientists would allow the development of predictive models leveraging these metrics to forecast fall risk or functional prognosis [55,56]. Overall, the methodology presented contributes to future technologically enabled precision rehabilitation paradigms and could optimize the quality of care. Integration of quantified instrumented clinical tests into immersive biofeedback platforms might actively engage patients in self-rehabilitation. Further research should explore such clinical translation while addressing practical challenges involving patient compliance, clinician adoption, and ethical considerations around data privacy.

This study, while demonstrating a detailed evaluation of motor-cognitive capabilities impacted by stroke through the integration of standardized clinical tests, motion analysis, and DT paradigms, is not without its constraints. The relatively small sample size, though a common limitation in studies involving clinical populations, may affect the generalizability of the findings. In addition, there was a marginally significant difference between the ages of the two groups that should be addressed in future studies to allow more reliable results. Furthermore, the potential benefits of incorporating additional sensor modalities such as electromyography to gain insights into neuromuscular control warrant consideration. To further enhance our understanding, future research endeavors should explore longitudinal studies that follow patients through rehabilitation, offering a dynamic perspective on how the observed metrics evolve over time and their therapeutic implications. These endeavors have the potential to refine rehabilitation strategies and, in turn, contribute to an improved quality of life for SS.

5. Conclusions

In this study, we used wearable sensors to perform a detailed study of movement alterations in stroke survivors during common clinical tests (TUG, STS, and the 10MWT). Using motion analysis, we effectively segmented the tests into specific activities and identified critical deficits in SS, such as reduced trunk velocity in STS, altered gait patterns in TUG, and asymmetric strides in the 10MWT. The outcomes of our study show that SS took significantly longer to complete the clinical tests, especially the TUG total time and turning sections, compared with HI. When DT conditions were imposed, impairments in SS were exacerbated across all tests, indicating the impact of divided attention on physical performance. This showed that the combination of instrumented clinical tests and DT paradigms allows robust assessment of motor-cognitive capabilities affected by stroke and provides insights into functional disabilities post-stroke. Overall, detailed quantification of specific task segments and DT conditions can inform targeted rehabilitation strategies. Future work should explore predictive modeling of fall risks and prognoses based on these instrumented metrics. The proposed methods can be translated to clinical practice for the objective evaluation of patient status and recovery. Overall, this research lays the groundwork for technologically enabled, precision rehabilitation to improve outcomes in stroke survivors.

Author Contributions: Conceptualization, M.A. and E.R.; methodology, M.A. and S.J.; software, M.A.; validation, M.A. and B.K.; formal analysis, M.A.; investigation, M.A.; resources, E.R. and M.D.; data curation, M.A., P.M.K. and S.J.; writing—original draft preparation, M.A. and P.M.K.; writing—review and editing, M.A., S.J. and P.M.K.; visualization, M.A. and P.M.K.; supervision, E.R., N.A.-F. and M.D.; project administration, M.A. All authors have read and agreed to the published version of the manuscript.

Funding: This research received no external funding.

Institutional Review Board Statement: The study was conducted in accordance with the Declaration of Helsinki and approved by the Institutional Review Board of Rochester Regional Health.

Informed Consent Statement: Informed consent was obtained from all subjects in the study.

Data Availability Statement: The raw data supporting the conclusions of this article will be made available by the corresponding authors upon reasonable request.

Conflicts of Interest: The authors declare no conflicts of interest.

References

1. Tsao, C.W.; Aday, A.W.; Almarzooq, Z.I.; Anderson, C.A.; Arora, P.; Avery, C.L.; Baker-Smith, C.M.; Beaton, A.Z.; Boehme, A.K.; Buxton, A.E.; et al. Heart disease and stroke statistics—2023 update: A report from the American Heart Association. *Circulation* **2023**, *147*, e93–e621. [PubMed]
2. Yang, Q.; Tong, X.; Schieb, L.; Vaughan, A.; Gillespie, C.; Wiltz, J.L.; King, S.C.; Odom, E.; Merritt, R.; Hong, Y.; et al. Vital signs: Recent trends in stroke death rates—United States, 2000–2015. *Morb. Mortal. Wkly. Rep.* **2017**, *66*, 933. [CrossRef] [PubMed]
3. Walsh, M.E.; Horgan, N.F.; Walsh, C.D.; Galvin, R. Systematic review of risk prediction models for falls after stroke. *J. Epidemiol. Community Health* **2016**, *70*, 513–519. [CrossRef]

4. Nowak, D.A. The impact of stroke on the performance of grasping: Usefulness of kinetic and kinematic motion analysis. *Neurosci. Biobehav. Rev.* **2008**, *32*, 1439–1450. [CrossRef]
5. Abdollahi, M.; Whitton, N.; Zand, R.; Dombovy, M.; Parnianpour, M.; Khalaf, K.; Rashedi, E. A Systematic Review of Fall Risk Factors in Stroke Survivors: Towards Improved Assessment Platforms and Protocols. *Front. Bioeng. Biotechnol.* **2022**, *10*, 910698. [CrossRef] [PubMed]
6. Abdollahi, M.; Kuber, P.M.; Hoang, C.; Shiraishi, M.; Soangra, R.; Rashedi, E. Kinematic Assessment of Turning and Walking Tasks Among Stroke Survivors by Employing Wearable Sensors and Pressure Platform. In Proceedings of the 2021 43rd Annual International Conference of the IEEE Engineering in Medicine & Biology Society (EMBC), Mexico, NA, USA, 1–5 November 2021; pp. 6635–6638.
7. Bonnyaud, C.; Pradon, D.; Vaugier, I.; Vuillerme, N.; Bensmail, D.; Roche, N. Timed Up and Go test: Comparison of kinematics between patients with chronic stroke and healthy subjects. *Gait Posture* **2016**, *49*, 258–263. [CrossRef] [PubMed]
8. Dennis, A.; Dawes, H.; Elsworth, C.; Collett, J.; Howells, K.; Wade, D.T.; Izadi, H.; Cockburn, J. Fast walking under cognitive-motor interference conditions in chronic stroke. *Brain Res.* **2009**, *1287*, 104–110. [CrossRef] [PubMed]
9. Plummer, P.; Zukowski, L.A.; Feld, J.A.; Najafi, B. Cognitive-motor dual-task gait training within 3 years after stroke: A randomized controlled trial. *Physiother. Theory Pract.* **2022**, *38*, 1329–1344. [CrossRef] [PubMed]
10. Cramer, S.C.; Wolf, S.L.; Adams, H.P.; Chen, D.; Dromerick, A.W.; Dunning, K.; Ellerbe, C.; Grande, A.; Janis, S.; Lansberg, M.G.; et al. Stroke Recovery and Rehabilitation Research: Issues, Opportunities, and the National Institutes of Health StrokeNet. *Stroke* **2017**, *48*, 813–819. [CrossRef] [PubMed]
11. Chan, P.P.; Tou, J.I.S.; Tse, M.M.; Ng, S.S. Reliability and Validity of the Timed Up and Go Test With a Motor Task in People With Chronic Stroke. *Arch. Phys. Med. Rehabil.* **2017**, *98*, 2213–2220. [CrossRef] [PubMed]
12. Kim, G.Y.; Han, M.R.; Lee, H.G. Effect of dual-task rehabilitative training on cognitive and motor function of stroke patients. *J. Phys. Ther. Sci.* **2014**, *26*, 1–6. [CrossRef] [PubMed]
13. Persson, C.U.; Danielsson, A.; Sunnerhagen, K.S.; Grimby-Ekman, A.; Hansson, P.-O. Timed Up & Go as a measure for longitudinal change in mobility after stroke—Postural Stroke Study in Gothenburg (POSTGOT). *J. NeuroEng. Rehabil.* **2014**, *11*, 83.
14. Sibley, K.M.; Straus, S.E.; Inness, E.L.; Salbach, N.M.; Jaglal, S.B. Balance assessment practices and use of standardized balance measures among Ontario physical therapists. *Phys. Ther.* **2011**, *91*, 1583–1591. [CrossRef] [PubMed]
15. Plummer-D'amato, P.; Altmann, L.J.; Saracino, D.; Fox, E.; Behrman, A.L.; Marsiske, M. Interactions between cognitive tasks and gait after stroke: A dual task study. *Gait Posture* **2008**, *27*, 683–688. [CrossRef] [PubMed]
16. Muir-Hunter, S.; Wittwer, J. Dual-task testing to predict falls in community-dwelling older adults: A systematic review. *Physiotherapy* **2016**, *102*, 29–40. [CrossRef] [PubMed]
17. Mori, T.; Takeuchi, N.; Izumi, S.-I. Prefrontal cortex activation during a dual task in patients with stroke. *Gait Posture* **2018**, *59*, 193–198. [CrossRef]
18. Subramaniam, S.; Hui-Chan, C.W.-Y.; Bhatt, T. Effect of dual tasking on intentional vs. reactive balance control in people with hemiparetic stroke. *J. Neurophysiol.* **2014**, *112*, 1152–1158. [CrossRef]
19. Hwang, W.J.; Cho, M.K.; Chung, Y. Relationship between anticipatory postural adjustment of the trunk, dual tasks and physical performance with chronic stroke survivors: A pilot test. *Phys. Ther. Rehabil. Sci.* **2015**, *4*, 44–48. [CrossRef]
20. Sasaki, A.C.; Pinto, E.B.; de Matos Sousa, M.C.; Monteiro, M.; Oliveira-Filho, J.; Júnior, A.D. Dual Task Performance Between Adults and Elderly Post Stroke Patients. *Braz. J. Med. Hum. Health* **2010**, *24*, 783–790. [CrossRef]
21. Yang, L.; Lam, F.M.H.; Liao, L.R.; Huang, M.Z.; He, C.Q.; Pang, M.Y.C. Psychometric properties of dual-task balance and walking assessments for individuals with neurological conditions: A systematic review. *Gait Posture* **2017**, *52*, 110–123. [CrossRef]
22. Zhou, Q.; Yang, H.; Zhou, Q.; Pan, H. Effects of cognitive motor dual-task training on stroke patients: A RCT-based meta-analysis. *J. Clin. Neurosci.* **2021**, *92*, 175–182. [CrossRef] [PubMed]
23. Patterson, K.K.; Gage, W.H.; Brooks, D.; Black, S.E.; McIlroy, W.E. Changes in gait symmetry and velocity after stroke: A cross-sectional study from weeks to years after stroke. *Neurorehabilit. Neural Repair* **2010**, *24*, 783–790. [CrossRef] [PubMed]
24. Bae, J.; Kong, K.; Byl, N.; Tomizuka, M. A mobile gait monitoring system for abnormal gait diagnosis and rehabilitation: A pilot study for Parkinson disease patients. *J. Biomech. Eng.* **2011**, *133*, 041005. [CrossRef] [PubMed]
25. Choi, W.; Lee, G.; Lee, S. Effect of the cognitive-motor dual-task using auditory cue on balance of surviviors with chronic stroke: A pilot study. *Clin. Rehabil.* **2015**, *29*, 763–770. [CrossRef] [PubMed]
26. Deblock-Bellamy, A.; Lamontagne, A.; McFadyen, B.J.; Ouellet, M.C.; Blanchette, A.K. Dual-Task Abilities During Activities Representative of Daily Life in Community-Dwelling Stroke Survivors: A Pilot Study. *Front. Neurol.* **2022**, *13*, 855226. [CrossRef] [PubMed]
27. Baek, C.Y.; Yoon, H.S.; Kim, H.D.; Kang, K.Y. The effect of the degree of dual-task interference on gait, dual-task cost, cognitive ability, balance, and fall efficacy in people with stroke: A cross-sectional study. *Medicine* **2021**, *100*, e26275. [CrossRef] [PubMed]
28. Denneman, R.P.M.; Kal, E.C.; Houdijk, H.; van der Kamp, J. Over-focused? The relation between patients' inclination for conscious control and single- and dual-task motor performance after stroke. *Gait Posture* **2018**, *62*, 206–213. [CrossRef] [PubMed]
29. Pumpho, A.; Chaikeeree, N.; Saengsirisuwan, V.; Boonsinsukh, R. Selection of the Better Dual-Timed Up and Go Cognitive Task to Be Used in Patients With Stroke Characterized by Subtraction Operation Difficulties. *Front. Neurol.* **2020**, *11*, 499761. [CrossRef] [PubMed]

30. Rice, J.; Corp, D.T.; Swarowsky, A.; Cahalin, L.P.; Cabral, D.F.; Nunez, C.; Koch, S.; Rundek, T.M.; Gomes-Osman, J. Greater Cognitive-Motor Interference in Individuals Post-Stroke during More Complex Motor Tasks. *J. Neurol. Phys. Ther.* **2022**, *46*, 26–33. [CrossRef] [PubMed]
31. Tang, P.F.; Yang, H.J.; Peng, Y.C.; Chen, H.Y. Motor dual-task Timed Up & Go test better identifies prefrailty individuals than single-task Timed Up & Go test. *Geriatr. Gerontol. Int.* **2015**, *15*, 204–210. [CrossRef] [PubMed]
32. Tasseel-Ponche, S.; Roussel, M.; Toba, M.N.; Sader, T.; Barbier, V.; Delafontaine, A.; Meynier, J.; Picard, C.; Constans, J.-M.; Schnitzler, A.; et al. Dual-task versus single-task gait rehabilitation after stroke: The protocol of the cognitive-motor synergy multicenter, randomized, controlled superiority trial (SYNCOMOT). *Trials* **2023**, *24*, 172. [CrossRef] [PubMed]
33. Yang, L.; He, C.; Pang, M.Y.C. Reliability and validity of dual-task mobility assessments in people with chronic stroke. *PLoS ONE* **2016**, *11*, e0147833. [CrossRef] [PubMed]
34. Bowen, A.; Wenman, R.; Mickelborough, J.; Foster, J.; Hill, E.; Tallis, R. Dual-task effects of talking while walking on velocity and balance following a stroke. *Age Ageing* **2001**, *30*, 319–323. [CrossRef] [PubMed]
35. Hollands, K.L.; Agnihotri, D.; Tyson, S.F. Effects of dual task on turning ability in stroke survivors and older adults. *Gait Posture* **2014**, *40*, 564–569. [CrossRef] [PubMed]
36. Plummer-D'Amato, P.; Altmann, L.J.; Behrman, A.L.; Marsiske, M. Interference between cognition, double-limb support, and swing during gait in community-dwelling individuals poststroke. *Neurorehabilit. Neural Repair* **2010**, *24*, 542–549. [CrossRef] [PubMed]
37. Bower, K.; Thilarajah, S.; Pua, Y.-H.; Williams, G.; Tan, D.; Mentiplay, B.; Denehy, L.; Clark, R. Dynamic balance and instrumented gait variables are independent predictors of falls following stroke. *J. NeuroEng. Rehabil.* **2019**, *16*, 3. [CrossRef] [PubMed]
38. Abdollahi, M.; Kuber, P.M.; Shiraishi, M.; Soangra, R.; Rashedi, E. Kinematic Analysis of 360○ Turning in Stroke Survivors Using Wearable Motion Sensors. *Sensors* **2022**, *22*, 385. [CrossRef] [PubMed]
39. Mao, Y.R.; Wu, X.Q.; Zhao, J.L.; Lo, W.L.A.; Chen, L.; Ding, M.H.; Xu, Z.Q.; Bian, R.H.; Huang, D.F.; Li, L. The crucial changes of sit-to-stand phases in subacute stroke survivors identified by movement decomposition analysis. *Front. Neurol.* **2018**, *9*, 185. [CrossRef] [PubMed]
40. Kirkland, M.C.; Wallack, E.M.; Rancourt, S.N.; Ploughman, M. Comparing three dual-task methods and the relationship to physical and cognitive impairment in people with multiple sclerosis and controls. *Mult. Scler. Int.* **2015**, *2015*, 650645. [CrossRef] [PubMed]
41. Petrillo, K.; Javed, B.; Toosizadeh, N. Association between dual-task function and neuropsychological testing in older adults with cognitive impairment. *Exp. Gerontol.* **2023**, *178*, 112223. [CrossRef] [PubMed]
42. Muir, S.W.; Speechley, M.; Wells, J.; Borrie, M.; Gopaul, K.; Montero-Odasso, M. Gait assessment in mild cognitive impairment and Alzheimer's disease: The effect of dual-task challenges across the cognitive spectrum. *Gait Posture* **2012**, *35*, 96–100. [CrossRef]
43. Fok, P.; Farrell, M.; McMeeken, J. Prioritizing gait in dual-task conditions in people with Parkinson's. *Hum. Mov. Sci.* **2010**, *29*, 831–842. [CrossRef] [PubMed]
44. Karvekar, S.; Abdollahi, M.; Rashedi, E. Smartphone-based human fatigue level detection using machine learning approaches. *Ergonomics* **2021**, *64*, 600–612. [CrossRef] [PubMed]
45. Karvekar, S.; Abdollahi, M.; Rashedi, E. A Data-Driven Model to Identify Fatigue Level Based on the Motion Data from a Smartphone. *bioRxiv* **2019**, 796854. [CrossRef]
46. Lee, J.B.; Mellifont, R.B.; Burkett, B.J. The use of a single inertial sensor to identify stride, step, and stance durations of running gait. *J. Sci. Med. Sport* **2010**, *13*, 270–273. [CrossRef] [PubMed]
47. Benson, L.C.; Clermont, C.A.; Watari, R.; Exley, T.; Ferber, R. Automated accelerometer-based gait event detection during multiple running conditions. *Sensors* **2019**, *19*, 1483. [CrossRef] [PubMed]
48. Cohen, J. *Statistical Power Analysis for the Behavioral Sciences*; Academic Press: Cambridge, MA, USA, 2013.
49. Abdollahi, M.; Kuber, P.M.; Pierce, M.; Cristales, K.; Dombovy, M.; LaLonde, J.; Rashedi, E. Motor-Cognitive Dual-Task Paradigm Affects Timed Up & Go (TUG) Test Outcomes in Stroke Survivors. In Proceedings of the 2023 11th International IEEE/EMBS Conference on Neural Engineering (NER), Baltimore, MD, USA, 24–27 April 2023; pp. 1–4.
50. Cho, K.; Lee, G. Impaired dynamic balance is associated with falling in post-stroke patients. *Tohoku J. Exp. Med.* **2013**, *230*, 233–239. [CrossRef] [PubMed]
51. An, H.-J.; Kim, J.-I.; Kim, Y.-R.; Lee, K.-B.; Kim, D.-J.; Yoo, K.-T.; Choi, J.-H. The effect of various dual task training methods with gait on the balance and gait of patients with chronic stroke. *J. Phys. Ther. Sci.* **2014**, *26*, 1287–1291. [CrossRef] [PubMed]
52. Haggard, P.; Cockburn, J.; Cock, J.; Fordham, C.; Wade, D. Interference between gait and cognitive tasks in a rehabilitating neurological population. *J. Neurol. Neurosurg. Psychiatry* **2000**, *69*, 479–486. [CrossRef] [PubMed]
53. Hyndman, D.; Ashburn, A.; Yardley, L.; Stack, E. Interference between balance, gait and cognitive task performance among people with stroke living in the community. *Disabil. Rehabil.* **2006**, *28*, 849–856. [CrossRef]
54. Heinzel, S.; Maechtel, M.; Hasmann, S.E.; Hobert, M.A.; Heger, T.; Berg, D.; Maetzler, W. Motor dual-tasking deficits predict falls in Parkinson's disease: A prospective study. *Park. Relat. Disord.* **2016**, *26*, 73–77. [CrossRef] [PubMed]

55. Abdollahi, M.; Rashedi, E.; Jahangiri, S.; Kuber, P.M.; Azadeh-Fard, N.; Dombovy, M. Fall Risk Assessment in Stroke Survivors: A Machine Learning Model Using Detailed Motion Data from Common Clinical Tests and Motor-Cognitive Dual-Tasking. *Sensors* **2024**, *24*, 812. [CrossRef] [PubMed]
56. Jahangiri, S.; Abdollahi, M.; Patil, R.; Rashedi, E.; Azadeh-Fard, N. An inpatient fall risk assessment tool: Application of machine learning models on intrinsic and extrinsic risk factors. *Mach. Learn. Appl.* **2024**, *15*, 100519. [CrossRef]

Disclaimer/Publisher's Note: The statements, opinions and data contained in all publications are solely those of the individual author(s) and contributor(s) and not of MDPI and/or the editor(s). MDPI and/or the editor(s) disclaim responsibility for any injury to people or property resulting from any ideas, methods, instructions or products referred to in the content.

bioengineering

MDPI

Article

Comparison of sEMG Onset Detection Methods for Occupational Exoskeletons on Extensive Close-to-Application Data

Stefan Kreipe [1,2], Thomas Helbig [2], Hartmut Witte [2], Nikolaus-Peter Schumann [1] and Christoph Anders [1,*]

1 FB Motorik und Pathophysiologie, Klinik für Unfall-, Hand- und Wiederherstellungschirurgie, Universitätsklinikum Jena, 07740 Jena, Germany
2 Fachgebiet Biomechatronik, Institut für Mechatronische Systemintegration, Fakultät für Maschinenbau, Technische Universität Ilmenau, 98693 Ilmenau, Germany
* Correspondence: christoph.anders@med.uni-jena.de or biomechatronik@tu-ilmenau.de; Tel.: +49-(0)-3641-9-328960

Abstract: The design of human-machine interfaces of occupational exoskeletons is essential for their successful application, but at the same time demanding. In terms of information gain, biosensoric methods such as surface electromyography (sEMG) can help to achieve intuitive control of the device, for example by reduction of the inherent time latencies of a conventional, non-biosensoric, control scheme. To assess the reliability of sEMG onset detection under close to real-life circumstances, shoulder sEMG of 55 healthy test subjects was recorded during seated free arm lifting movements based on assembly tasks. Known algorithms for sEMG onset detection are reviewed and evaluated regarding application demands. A constant false alarm rate (CFAR) double-threshold detection algorithm was implemented and tested with different features. Feature selection was done by evaluation of signal-to-noise-ratio (SNR), onset sensitivity and precision, as well as timing error and deviation. Results of visual signal inspection by sEMG experts and kinematic signals were used as references. Overall, a CFAR algorithm with Teager-Kaiser-Energy-Operator (TKEO) as feature showed the best results with feature SNR = 14.48 dB, 91% sensitivity, 93% precision. In average, sEMG analysis hinted towards impending movements 215 ms before measurable kinematic changes.

Keywords: surface EMG; occupational exoskeletons; onset detection; myoelectric control; constant-false-alarm rate; Teager-Kaiser-Energy-Operator; electromechanical delay

Citation: Kreipe, S.; Helbig, T.; Witte, H.; Schumann, N.-P.; Anders, C. Comparison of sEMG Onset Detection Methods for Occupational Exoskeletons on Extensive Close-to-Application Data. *Bioengineering* **2024**, *11*, 119. https://doi.org/10.3390/bioengineering11020119

Academic Editor: Carlo Albino Frigo

Received: 24 November 2023
Revised: 19 January 2024
Accepted: 23 January 2024
Published: 25 January 2024

Copyright: © 2024 by the authors. Licensee MDPI, Basel, Switzerland. This article is an open access article distributed under the terms and conditions of the Creative Commons Attribution (CC BY) license (https://creativecommons.org/licenses/by/4.0/).

1. Introduction

Powered or active exoskeletons are wearable robots designed to assist human movements by enhancing the physical capacity of a person [1–4]. They are currently developed for applications in medical therapy [4–6], support of patients with movement disorders [1,4], as well as in occupational contexts [2,3]. In the latter case, it is hypothesized that occupational exoskeletons can effectively prevent musculoskeletal diseases by assisting the workers in handling excessive loads or by increasing their endurance, thereby limiting muscular fatigue and associated diseases [1,3,7].

For all these applications, a main challenge is the design of a proper human-machine-interface between the powered exoskeleton and its user. Synchronizing the forces generated by the exoskeleton with the muscular efforts of the user is of utmost importance to react properly and timely adequate to the movement intention, and thus provide the required assistance [1,3,4]. This is necessary to ensure biocompatible joint kinematics and dynamics and allow the user to adapt to the system, and gain confidence in its use [3,8–11].

Surface electromyography (sEMG) is a biosensoric technique, which allows to effectively measure and analyse muscular activity. sEMG signal processing has been investigated for the control of powered exoskeletons [1,4,6,10,12–15] and in the related field of prosthesis control [11,16–18]. Complete decoding of motion intention is not in reach due to the high complexity of the human motor system, its time-varying nature and the high inter-subject variability [4,17]. Still, a broad range of techniques and algorithms has been developed to analyse sEMG with significant success within the corresponding scientific setup. These methods vary strongly in complexity, with the more complex approaches giving better results by accounting for the variability and stochastic nature of sEMG signals [18], but also having higher demands towards computational power and the robustness of measurements [4]. These demands proved as a hindrance to application of more complex sEMG analysis in real life situations: Almost all commercially available exoskeletons for occupational use are passive, and although recent hand prosthesis controllers allow for multiple class pattern recognition, most amputees are only able to reliably use as little as four classes simultaneously.

Considering an active, occupational exoskeleton whose basic control loop is based on inertia and force sensors, follow-up control and limiting of contact forces is possible even without biosensoric information. The disadvantage of these sensor modalities is their inherent time latency, being only able to detect the user motion after its onset. This method, although currently the most common control technique for active exoskeletons, leads to the user necessarily needing to work against the exoskeleton for a short period of time at the beginning of each movement, which is contrary to the goal to reduce muscular effort. Additionally, such an active exoskeleton would not be able to deliver support already at the beginning of the movement, which is important to alleviate stationary weight and friction forces as well as acceleration. sEMG onset detection allows to determine the onset of muscular activity which itself is the origin of all human motion. Due to the electromechanical delay, there is a time span of 20–80 ms between the rise of sEMG and measurable generation of muscular force [19]. This characteristic can be used to eliminate the time latencies of the exoskeleton control loop. In contrast to other sEMG analysis methods, onset detection has low demands towards computational power and robustness of measurements, due to its low complexity [4,20,21].

Many studies on sEMG analysis techniques used simulated data or laboratory investigations far from real-life application cases (i.e., [22,23]). Often, isometric, isokinetic or very simple tasks were used, diminishing the variability of sEMG signals and thereby overestimating the algorithm performance. Also, small groups of test subjects are common. Considering that the variability of the sEMG signal is known to be a main hindrance towards its application, re-evaluation of established methods on a close-to-application dataset of sufficient size is necessary. To this aim, shoulder sEMG of a heterogeneous group of 55 healthy subjects was recorded during seated free arm lifting movements based on assembly tasks. Several state-of-the-art sEMG onset detection techniques are reviewed and compared according to reported performance, computational demands and signal-to-noise-ratio (SNR). A short-list of promising algorithms is implemented and tested on the dataset. The results are compared with kinematic onset detection and onset timings determined through visual inspection of sEMG signals by experts.

The transfer of scientifically developed sEMG signal analysis techniques into real-life applications was rarely achieved up to date. To improve man-machine interaction towards a more synchronous, i.e., physiologically driven behaviour of active exoskeletons, it is of great importance to investigate the established techniques under realistic conditions. In this paper we investigate the performance and reliability of different methods for sEMG onset detection for the use case of an active, occupational shoulder exoskeleton under close-to-application conditions.

2. Materials and Methods

2.1. Experimental Protocol

The experimental protocol of this study was designed to resemble circumstances close to a real-life application of occupational exoskeletons. Exemplifying seated assembly work, the subjects were asked to perform arm lifts with a low weight (1 kg) while seated in front of a desk. The movements were performed alternating with both arms. The working space was reproduced by using different movement directions (Figure 1) and heights (Placing weight on the table, holding it just above the table, at shoulder height and above head height). All movements have been conducted freely and without any physical restrictions or direct feedback. The subjects were asked to lift the weight with their hand, move it to the target position, stop for at least one second, and return to the starting position, resting there for one second. On each height level, five repetitions of all direction have been performed with each side. Between each height level, a short break was included to avoid muscular fatigue. In total, 55 subjects (29 male, 26 female, 40 "young" (27.2 ± 4.7 years), 15 "old" (52.2 ± 4.2 years), 38 right-handed, 17 left-handed) were included in the study, deliberately chosen to achieve a diverse subject population in terms of sex, age, and handedness. The experimental protocol has been approved by the ethics committee of the Jena University Hospital, Germany (No. 2019-1350_1 BO).

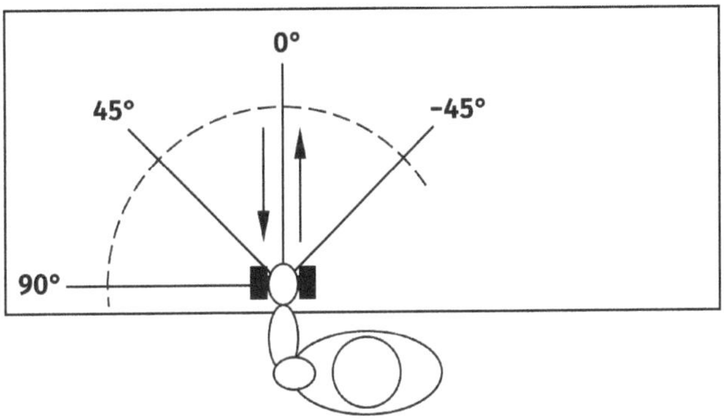

Figure 1. Movement directions, projected onto table plane. Movements for left arm are shown, but movement were done with both arms.

Bipolar sEMG was measured for multiple shoulder muscles using disposable adhesive electrodes (contact diameter 1.6 cm, H93SG, Covidien, Germany). Regarding the electrode placement SENIAM [24] recommendations were followed. Before application of the electrodes the skin was shaved and prepared with abrasive paste. The signal was preconditioned with an analogue 10–700 Hz band pass filter and digitized at 2048 Samples/s and a resolution of 6 nV/bit, using the ToM (Tower of Measurement, Demetec, Germany). Later, the sEMG was conditioned by a digital 20–500 Hz band pass filter and obvious movement were removed semiautomatically. Simultaneously, movement kinematics were recorded using a motion tracking system (Qualisys®, Göteborg, Sweden). Both measurements were synchronized using optical trigger signals.

For the sEMG analysis, the data of the anterior deltoids *[M. deltoideus pars anterior]* on both sides was used. Although in a real application of sEMG-based exoskeleton control, probably sEMG of multiple muscles will be included in the decision making, we here focused only on the prime mover muscle for the conducted tasks. To investigate the performance of sEMG onset detection methods, data that reliably contains sEMG activity at movement onsets is needed. Only the anterior deltoid can be expected to provide such muscular activity during the investigated movement. We generally assume that given a proper parameter calibration, any findings achieved for this particular muscle can be transferred to other skeletal muscles. For the kinematic analysis, we utilized the position of the subjects' hand, indicated by a motion tracking marker.

2.2. sEMG Onset Detection Techniques

When muscles are activated, their myoelectric signals were found to show a systematic structure during the phase of initial contraction [11]. Still, according to Drapala et al. [25], as well as others [4,22], precise determination of EMG onset was found to be a challenging task due to the stochastic nature of EMG, smooth gradual transitions from rest to movement, background noise and inter-subject variability. Over the years, several different approaches have been developed to achieve sEMG onset detection.

The task of sEMG onset detection can be split into the steps of signal pre-conditioning, feature extraction, detection and post-processing (Figure 2). Early works addressed the task with the straightforward approach of applying a single or double threshold algorithm, together with a simple time-domain feature, such as mean average value, squared, or rectified and low-pass filtered EMG [26–29]. Further studies investigated more sophisticated features, such as Teager-Kaiser-Energy-Operator (TKEO) [20,30–33], sets of optimized time-domain (TD) features [4], sample entropy (SampEn) [23] and wavelet transform (WT)-based time-frequency-domain features [34,35]. In terms of more advanced detectors likelihood-based methods [22,36], Bayesian changepoint analysis [37], Gaussian-mixture-models (GMM) [4,25] and constant false alarm rate (CFAR) adaptive, double thresholding [18,21] have been applied. These methods are summarized in Table 1 .

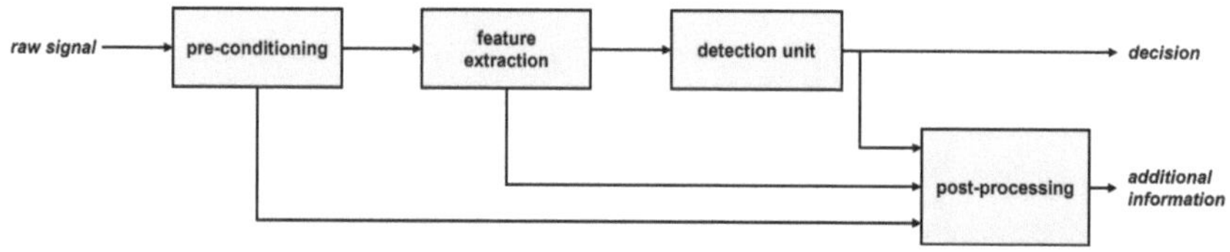

Figure 2. sEMG onset detection processing structure.

Table 1. sEMG onset detection methods.

Type	Method	Reference(s)
Features	TKEO	[20,30–33]
	optimized TD feature set	[4]
	SampEn	[23]
	WT features	[34,35]
Detectors	likelihood-based methods	[22,36]
	Bayesian changepoint analysis	[37]
	GMM	[4,25]
	CFAR adaptive tresholding	[18,21]

Although processing capacities have increased significantly, computational complexity remains a constraint for online sEMG analysis in real-life applications, especially when considering simultaneous analysis of multiple signal channels such as in the application of an occupational exoskeleton. Therefore, we excluded WT-based features from the comparisons due to their high computational effort [35], even though their performance is among the best reported. Further, Bayesian changepoint analyses, 2-step search algorithms and GMM cannot be used in online applications, because they incorporate the signal after the EMG onset into their decision-making. Likelihood methods are in principle real-time applicable but require training, which is not in the scope of this work.

To assess the performance of sEMG onset detection techniques for real-life applications of occupational exoskeletons, in the following we use a CFAR adaptive, double thresholding algorithm as detector. We compared its performance in combination with different features, namely TKEO, SampEn, the time domain feature set proposed by Trigili et al. [4] (Integrated Absolute Value (IAV), Simple Square Integral (SSI), Waveform Length (WL), Logarithm (LOG)) and variance (VAR) according to Tabie's and Kirchner's [20] results. When Trigili et al. [4] evaluated the time domain feature set, they used a GMM with multiple inputs, one for each feature. Since the CFAR algorithm has only one input, we fuse the feature set by taking the average of all four features, each weighted by their mean value to compensate for different numerical dimensions. Root-mean-square amplitude was not investigated separately due to its similarity with VAR. The methods applied are described in detail in Appendix A. The features have been calculated on 50 ms windows with 90% Overlap.

2.3. Onset Reference and Evaluation Criteria

When evaluating onset detection algorithm performance, the need for an appropriate reference arises. The gold standard is visual inspection of the sEMG signals by trained experts [21,22,25,35,37]. This method is known to provide the most accurate results, although with some subjective influence of the sEMG examiners. On the other hand, as a manual task, visual inspection requires a huge effort, being unfeasible for larger datasets. The alternative is to reference the detected sEMG onset towards the movement onset which is derived by kinematic analysis [4,20]. From the motion tracking data, we were able to automatically determine the moment, when within the course of one movement the distance covered excesses a certain percentage of the total distance between movement start and end point. We used 5% of within-movement distance of the subject's hand marker as kinematic onset and as reference point for ground truth for the sEMG analysis. In order to assess the differences to the gold standard method, we visually determined onsets in the sEMG of 4 subjects.

In sEMG onset detection studies, methods are evaluated according to sensitivity, specificity, influence of SNR, and detection latency or timing error [4,18,20–23,25,34,35]. As in the studies cited above, the algorithm sensitivity is defined as the rate of the movement onsets that are detected by sEMG analysis. We consider a detection as correct, if it is within 500 ms before the kinematic onset reference. In our opinion, specificity is not a feasible criterion here, because in the present experimental context the definition of negative detection is not intuitive. Confronted with the same problem, Tabie and Kirchner [20] used the number of time periods between onsets without detection events as true negatives. Another approach can arise from the fact that it is among the main design goals of an occupational exoskeleton to provide trust to the user. This would be contradicted by every single false positive detection, which in the worst case could initiate a movement not intended by the user. Therefore, we propose the usage of algorithm precision instead of specificity, calculated as 100% minus false positive rate, with the latter being the rate of detected onsets in the sEMG which are not within 500 ms before the kinematic onset reference. In terms of SNR, no additional noise was added artificially (as i.e., in [22]). The SNR was calculated as the feature amplitude ratio in dB between signal areas with and without muscular activation in general and more in detail before movement onset. Therefore, the visually inspected and marked data was used. Finally, as already stated by Trigili et al. [4], in control applications the time delay between detected sEMG onset and movement onset is not a bug, but a feature. Where in clinical applications, the aim of automated sEMG onset detection is to replace manual, visual inspection and therefore be as close to the manual detection as possible, this is not relevant for the application in the control of powered exoskeletons. There, the algorithm is only required to detect muscular activity early enough before the start of a movement to implement exoskeleton pre-control. No fixed minimum time difference can be given a priori, but rather has the exoskeleton control to be fast enough to react in the remaining time. We therefore rate higher onset delays as better, as long as they are within a reasonable range of up to 500 ms before movement onset. However, as long as the onset detection exhibits a sufficient time lag towards the kinematic reference the exact timing of an algorithm is secondary to its sensitivity and precision.

3. Results

For the example of one movement onset, the sEMG as well as the feature curves are shown in Figure 3. There, the area visually found to contain muscular activation is highlighted in grey. The dotted line with the '5%' mark shows the kinematic onset reference timestamp.

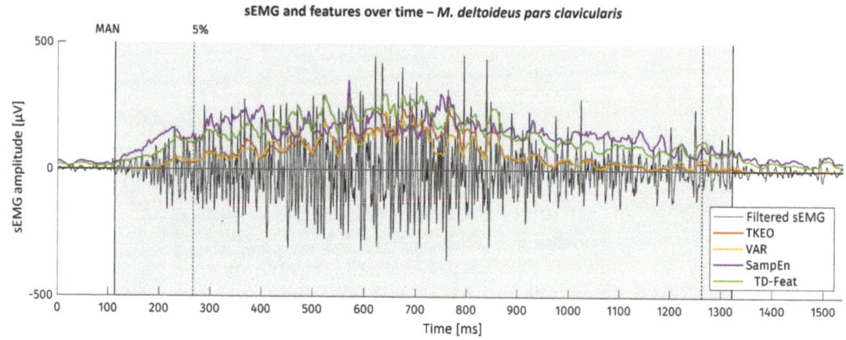

Figure 3. sEMG at movement onset and feature curves, exemplarily shown for one signal section. Area marked in grey: Visual detection of muscular activity. Dotted line with '5%' mark: Timestamp of kinematic onset reference.

The SNR of all features was calculated as their amplitude ratio in dB between signal areas manually labelled as with muscular activity and without. Additionally, contrast values in windows 300 ms before and after the onset of muscular activity were taken into consideration. The results are displayed in Table 2. It is obvious that TKEO shows the highest SNR values in all categories. The variance feature shows slightly increased values compared to the mean of the time-domain features proposed by Trigili [4]. In contrast, SampEn shows very low SNR values, but more consistently, reaching its maximum value almost directly after onset.

Table 2. sEMG onset detection feature SNR in decibel.

	TKEO	VAR	SampEn	TD-Features
Total SNR	14.84 dB	13.64 dB	4.44 dB	13.42 dB
Onset SNR	10.59 dB	10.29 dB	4.15 dB	10.10 dB

The combinations of the CFAR adaptive, double threshold and each feature were implemented and tested separately. To ensure optimal detector performance for each feature, the thresholding algorithms' parameters were optimized for each feature to reach a maximum mean value of Sensitivity and Precision (MSP) on a data subset.

All movement onsets of all subjects were included into the evaluation (total: approx. 11,000 onsets). The accuracy of each feature-detector combination is displayed in Figure 4. As comparison, the accuracy of the visual inspection is shown for both experts (MAN1, MAN2). The TKEO + CFAR method shows the highest accuracy values of all tested algorithms. Its sensitivity and precision is about 4% lower than both visual detections.

The onset detection latencies for each method are shown in histogram form in Figure 5. It needs to be remembered that different data sizes amounts are displayed for the time differences of each method. The visual inspection covered only approx. 900 onsets due to its limitation on 4 subjects. In contrast, the algorithms have been tested on approx. 11,000 onsets, but since time differences could only be calculated for successful detections, the amount of onsets taken into account vary between the algorithms by their sensitivity.

Most visually detected onsets showed a time difference of more than 150 ms to the kinematic reference point. The automatically detected onsets exhibited a larger variability and less time difference with most detections before 100 ms towards the kinematic reference. The onset time differences of all methods were statistically tested using a mixed linear model. Although the onset time differences of the methods show significant differences to each other, but with only small to medium effect sizes.

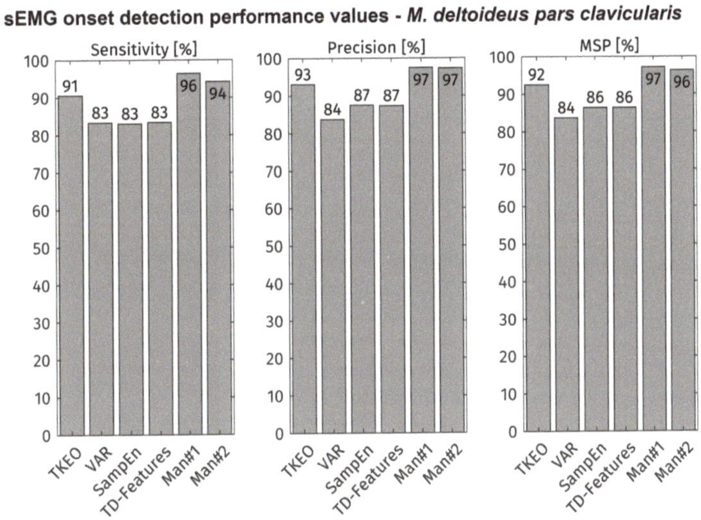

Figure 4. sEMG onset detection performance values. Sensitivity is true positive rate, precision is 100% minus false positive rate and MSP is the mean of sensitivity and precision. 'Man#1' and 'Man#2' depict the results of the manual, visual inspection by experts. Manual inspection: ca. 900 movement onsets under investigation; Automatic detection: ca. 11,000 movement onsets under investigation.

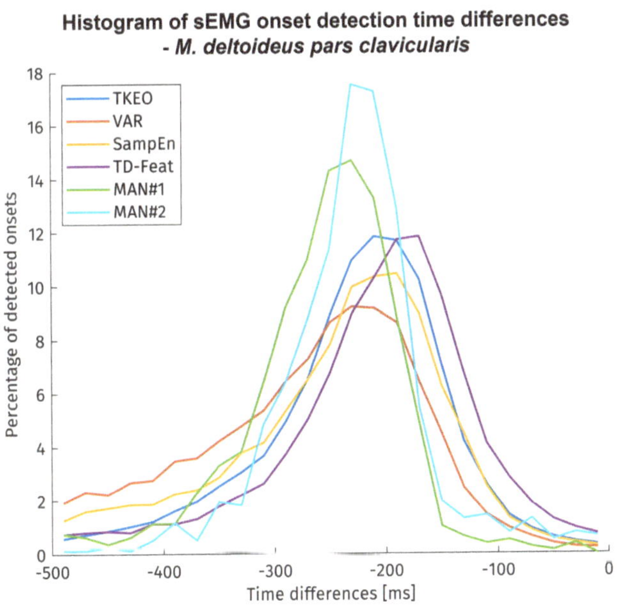

Figure 5. Histogram of the sEMG onset detection time differences to the '5%' kinematic reference. 'MAN#1' and 'MAN#2' depict the results of the manual, visual inspection by experts. Manual inspection: ca. 900 movement onsets under investigation; Automatic detection: ca. 11000 movement onsets under investigation.

4. Discussion

Due to the electromechanical processes during muscle activation, there are time delays between EMG onset, muscle force generation, and movement onset. The delay between EMG and force generation was named electromechanical delay (EMD) by Cavanagh and

Komi [38] and is in the order of 20–80 ms [19]. The movement onset (kinematic reference) might be delayed even further from the EMG onset because of inertia and forces opposing the movement, up to several hundred milliseconds [4,20,35]. This is represented by the results in Figure 5, showing an average time delay of about 200 ms to 5% of the distance covered during the movement. Since the manual inspection can be assumed to be as accurate as possible, its time difference towards the kinematic reference is close to the real time difference between the onset of muscular activation and the movement onset. Therefore, regarded inversely, the histograms of the manual, visual inspection represent the spread in time delay introduced by the kinematic reference method. Factors contributing to this variability may be other muscles initiating the movement or muscular co-contractions overlaying the onset. Compared to similar studies [4,20,21,37], the time differences here are found to be even slightly higher, possibly due to the use of an additional weight in the experimental protocol. Concerning the algorithms for sEMG onset detection, the onset time differences and their deviation are in the expected range. Their statistical indifference means that all methods under investigation exhibit no relevant performance differences regarding the onset detection timing error.

The visual inspection method reaches only 95% sensitivity and 97% precision although commonly perceived to detect all muscular activity present in the sEMG signal. This finding is probable to be influenced by the same causes as the variability of the onset time differences and shows the difficulty of non-laboratory sEMG onset detection, influenced by temporally low muscle activation, combinations of multiple muscles contributing to a certain movement, and the presence of noise in the signal. This becomes clear when reviewing the high sensitivity values reported by other studies with simulated data or signals measured under laboratory conditions [18,22,23]. Therefore, 100% onset detection performance is not to be expected when regarding a single muscle, even if it is the prime mover such as the *M. deltoideus pars clavicularis* for the movements investigated here.

Regarding the automated onset detection methods under investigation, the TKEO + CFAR combination performed best at sEMG onset detection under close-to-application. Its sensitivity and precision were 91% and 93% respectively, only 4% less than the average of visual inspection results by sEMG experts. The significance of algorithm sensitivity and precision becomes especially clear, when regarding the error rates: 91% sensitivity means 9% of movement onsets not being detected, while 83% sensitivity means 17% of movement onsets not being detected, which is almost the double. In the same perspective, 93% precision means 7% of the detected onsets not being in relation with an upcoming movement, resulting in possible unwanted actions of the exoskeleton. This highlights the value of the higher performance of the TKEO + CFAR technique versus the other methods. However, with continuous detection regular false alarms are still to be expected. In applications of occupational exoskeletons, this issue needs to be dealt with by post-processing, considering multiple muscles or a robust control scheme in order to avoid unintended movements and to ensure trust by the user in the assistive device.

5. Conclusions

The aim of this study was to investigate the performance of sEMG onset detection methods under close-to-application conditions of active exoskeletons. Therefore, free and unconstrained movements representing the workspace during seated manual assembly tasks have been investigated on a large and heterogeneous group of subjects. This unique dataset with approx. 11,000 movement onsets allowed testing of known and promising sEMG onset detection algorithms regarding their performance in a real application case unlike the common, limited laboratory setting.

For a reduced dataset, the gold standard for onset detection, manual, visual inspection allowed to evaluate the possible onset detection sensitivity and precision when only regarding the prime mover muscle, showing that before 95% of movement onsets sEMG activity can be found and that 97% of sEMG onsets happen shortly before a movement onset. It thereby also permitted evaluation of kinematic reference as an automated method to generate a ground truth for onset detection in larger datasets and possibly even during online testing.

The computationally simple approach with TKEO + CFAR showed good results for sEMG onset detection under close-to-application condition (91% sensitivity, 93% precision), performing best within the set of applied methods. Still, reliable sEMG onset detection in applications remains a challenging task, with accuracies not reaching 100% even if manual visual inspection, still serving as the gold standard is applied. Anyhow, time-frequency-features based algorithms and data of multiple muscles bear the potential to improve the presented results, but at the cost of necessarily higher computing cost. No matter the actual technique used in a real application, the results presented here show that the sEMG contains sufficient information for timely pre-detection of movement onsets and therefore exoskeleton pre-control even under close-to-application conditions.

Author Contributions: Conceptualization, S.K., H.W. and N.-P.S.; Data curation, S.K.; Formal analysis, S.K.; Funding acquisition, T.H., H.W. and N.-P.S.; Investigation, S.K.; Methodology, all authors; Project administration, all authors; Resources, S.K. and T.H.; Software, S.K.; Supervision, H.W., N.-P.S. and C.A.; Validation, all authors; Visualization, S.K.; Writing—original draft, S.K.; Writing—review and editing, all authors. All authors have read and agreed to the published version of the manuscript.

Funding: This research was funded by German Ministry for Education and Research grant numbers 16SV8003 and 16SV8004. (project: *"Muskelgesteuertes Exoskelett zur Kraftunterstützung—LEVIAKTOR"*—especially the partial projects *"Myoelektrische Aktivierungsmuster der Schulter- und Armmuskulatur"* and *"Biomechatronisches Systemkonzept und Gestaltung"*).

Institutional Review Board Statement: The study was conducted in accordance with the Declaration of Helsinki, and approved by the Ethics Committee of Universitätsklinikum Jena (protocol code 2019-1350_1 BO, 9.4.2019).

Informed Consent Statement: Informed consent was obtained from all subjects involved in the study.

Data Availability Statement: The data presented in this study are available on request from the corresponding author.

Acknowledgments: We would like to thank Elke Mey for her support with the experiments and data structuring, Roland Graßme for the synchronization between EMG and kinematic measurements, and Stefan Lutherdt for coordinating joint research efforts.

Conflicts of Interest: The authors declare no conflict of interest. The funders had no role in the design of the study; in the collection, analyses, or interpretation of data; in the writing of the manuscript; or in the decision to publish the results.

Abbreviations

The following abbreviations are used in this manuscript:

CFAR	Constant False Alarm Rate Thresholding
EMD	electromechanical delay
EMG	electromyography
GMM	Gaussian-mixture-model
IAV	Integrated Absolute Value
LOG	Logarithm
MSP	Mean value of Sensitivity and Precision
SampEn	Sample Entropy
sEMG	surface electromyography
SNR	signal-to-noise-ratio
SSI	Simple Square Integral
TKEO	Teager-Kaiser-Energy-Operator
ToM	Tower of Measurement
VAR	variance
WT	wavelet transform
WL	waveform length

Appendix A. Feature and Detector Description

Appendix A.1. Teager-Kaiser-Energy-Operator (TKEO)

Teager-Kaiser-Energy-Operator was first proposed by Teager and later detailed, explained and published by Kaiser [30]. It approximates the energy of a time-discrete signal in the time instant k as:

$$E_k = x_k^2 - x_{k-1} \cdot x_{k+1}$$

Appendix A.2. Variance (VAR)

Variance is a well known parameter in signal processing. For application on time-discrete EMG signals the following formula was proposed [20,39]:

$$V_k = \frac{1}{N-1} \sum_{i=-m}^{m} x_{k+i}^2 - \left(\frac{1}{N-1} \sum_{i=-m}^{m} x_{k+i} \right)^2$$

$N = 2m + 1$ is the width of the sliding window.

Appendix A.3. Sample Entropy (SampEn)

In information theory, entropy can be described as a measure of the signals variability and complexity in the sense of its probability to show an unpredictable change. It is therefore strongly connected to changes in the signals information content. Richman and Moorman [40] introduced the Sample Entropy as feature for processing of cardiovascular data and Zhang and Zu [23] transferred it to EMG analysis. It is based on the probability $P^n(r)$, that within a window of length N, two sequences $(x_{k+p})_{p=0}^{p=n}$ with length $n < N$ match. This probability can be estimated by the average number of sequences with a total difference lower than the threshold r. Sample Entropy is then calculated as:

$$SampEn_k(n,r) = -ln\left(\frac{P^{n+1}(r)}{P^n(r)} \right)$$

Appendix A.4. Time Domain Features (TD-Features)

The time domain feature set is calculated according to the work by Trigili et al. [4]. They used each of the features as a seperate input to the GMM algorithm. Since the CFAR algorithm accepts only one input signal without further changes, hence in this work the mean of the four features SSI, IAV, LOG and WL is taken as a single feature. The features are calculated according to the formulations by [4,41].

Appendix A.4.1. Simple Square Integral (SSI)

Simple Square Integral calculates the sum of the signal energy within a window of width N as:

$$SSI_k = \sum_{i=-m}^{m} x_{k+i}^2$$

Appendix A.4.2. Integrated Absolute Value (IAV)

Integrated Absolute Value is the sum of the absolute values of the signal within a window of length N:

$$IAV_k = \sum_{i=-m}^{m} |x_{k+i}|$$

Appendix A.4.3. Logarithm (LOG)

The logarithm feature is calculated as the mean of the decadic logarithm of the signals absolute value within a window of length N:

$$LOG_k = \frac{1}{N} \sum_{i=-m}^{m} log_{10}(|x_{k+i}|)$$

Appendix A.4.4. Waveform Length (WL)

The waveform length is the sum of the absolute values of the discrete derivative of the signal within a window of length N:

$$WL_k = \sum_{i=-m}^{m-1} |x_{k+i+1} - x_{k+i}|$$

It can also be understood as the cumulative length of the waveform.

Appendix A.5. Constant False Alarm Rate Thresholding (CFAR)

The constant false alarm rate algorithm is essentially an advanced adaptive double thresholding method. Whilst a single threshold algorithm signifies a change once the threshold is crossed, a double threshold method does only so if the value exceeds the threshold in m out of n samples. Being adaptive means that the threshold is not a fixed value, but is set based on the signals value in a reference window in the past. This threshold can be further adapted to move between a minimum and maximum value to take special cases into account. All those adaptions aim towards reducing the algorithms false alarm rate. If the signals statistical properties are known *a priori*, the adaptive thresholds gain can be set in order to achieve a certain constant false alarm rate [18,21]. We here did not set the adaptive threshold gain towards a defined false alarm rate, since we did not assume *a priori* knowledge on the signal properties. The parameters were indeed optimized towards maximum combined sensitivity and precision. Still, the original structure of the CFAR algorithm was kept. The algorithm can be described as follows.

The signal is separated backwards into a detection window of length D whose values are compared to the threshold, a "guard" window of length G whose values are not taken into account and a reference window of length R from whose values the threshold base is calculated. The adaptive threshold is here defined using the reference windows median

$$Ta_k = K_a \cdot median(x_i),\ i = k - D - G - R + 1, \ldots, k - D - G$$

with K_a being the adaptive thresholds gain. The threshold is limited as described above. If the signal value x_k exceeds the threshold Ta_k in C out of D times, an onset is detected:

$$Onset = \left(\sum_{i=k-D}^{k} (x_i > Ta_k) \right) > C$$

References

1. Lenzi, T.; de Rossi, S.M.M.; Vitiello, N.; Carrozza, M.C. Intention-based EMG control for powered exoskeletons. *IEEE Trans. Biomed. Eng.* **2012**, *59*, 2180–2190. [CrossRef] [PubMed]
2. de Looze, M.P.; Bosch, T.; Krause, F.; Stadler, K.S.; O'Sullivan, L.W. Exoskeletons for industrial application and their potential effects on physical work load. *Ergonomics* **2015**, *59*, 671–681. [CrossRef] [PubMed]
3. Theurel, J.; Desbrosses, K. Occupational Exoskeletons: Overview of Their Benefits and Limitations in Preventing Work-Related Musculoskeletal Disorders. *IISE Trans. Occup. Ergon. Hum. Factors* **2019**, *7*, 264–280. [CrossRef]
4. Trigili, E.; Grazi, L.; Crea, S.; Accogli, A.; Carpaneto, J.; Micera, S.; Vitiello, N.; Panarese, A. Detection of movement onset using EMG signals for upper-limb exoskeletons in reaching tasks. *J. Neuroeng. Rehabil.* **2019**, *16*, 45. [CrossRef] [PubMed]
5. Sale, P.; Franceschini, M.; Waldner, A.; Hesse, S. Use of the robot assisted gait therapy in rehabilitation of patients with stroke and spinal cord injury. *Eur. J. Phys. Rehabil. Med.* **2012**, *48*, 111–121. [PubMed]

6. Vitiello, N.; Lenzi, T.; Roccella, S.; de Rossi, S.M.M.; Cattin, E.; Giovacchini, F.; Vecchi, F.; Carrozza, M.C. NEUROExos: A Powered Elbow Exoskeleton for Physical Rehabilitation. *IEEE Trans. Robot.* **2013**, *29*, 220–235. [CrossRef]
7. Huysamen, K.; Bosch, T.; de Looze, M.P.; Stadler, K.S.; Graf, E.; O'Sullivan, L.W. Evaluation of a passive exoskeleton for static upper limb activities. *Appl. Ergon.* **2018**, *70*, 148–155. [CrossRef] [PubMed]
8. Beck, O.N.; Sheperd, M.K.; Rastogi, R.; Martino, G.; Ting, L.H.; Sawicki, G.S. Exoskeletons need to react faster than physiological responses to improve standing balance. *Sci. Robot.* **2023**, *8*, eadf1080. [CrossRef]
9. Hébert, L.J.; Moffet, H.; McFadyen, B.J.; Dionne, C.E. Scapular behavior in shoulder impingement syndrome. *Arch. Phys. Med. Rehabil.* **2002**, *83*, 60–69. [CrossRef]
10. Fleischer, C.; Wege, A.; Kondak, K.; Hommel, G. Application of EMG signals for controllingexoskeleton robots. *Biomed. Eng.* **2006**, *51*, 314–319. [CrossRef]
11. Hudgins, B.; Parker, P.; Scott, R.N. A new strategy for multifunction myoelectric control. *IEEE Trans. Biomed. Eng.* **1993**, *40*, 82–94. [CrossRef]
12. Mefoued, S.; Belkhiat, D.E.C. A Robust Control Scheme Based on Sliding Mode Observer to Drive a Knee–Exoskeleton. *Asian J. Control* **2019**, *21*, 439–455. [CrossRef]
13. Kinnaird, C.R.; Ferris, D.P. Medial gastrocnemius myoelectric control of a robotic ankle exoskeleton. *IEEE Trans. Neural Syst. Rehabil. Eng. A Publ. IEEE Eng. Med. Biol. Soc.* **2009**, *17*, 31–37. [CrossRef]
14. Anam, K.; Al-Jumaily, A.A. Active Exoskeleton Control Systems: State of the Art. *Procedia Eng.* **2012**, *41*, 988–994. [CrossRef]
15. Moreno, J.C.; Brunetti, F.; Navarro, E.; Forner-Cordero, A.; Pons, J.L. Analysis of the human interaction with a wearable lower-limb exoskeleton. *Appl. Bionics Biomech.* **2009**, *6*, 245–256. [CrossRef]
16. Asghari Oskoei, M.; Hu, H. Myoelectric control systems—A survey. *Biomed. Signal Process. Control* **2007**, *2*, 275–294. [CrossRef]
17. Zecca, M.; Micera, S.; Carrozza, M.C.; Dario, P. Control of Multifunctional Prosthetic Hands by Processing the Electromyographic Signal. *Crit. Rev. Biomed. Eng.* **2002**, *30*, 459–485. [CrossRef] [PubMed]
18. Kontunen, A.; Rantanen, V.; Vehkaoja, A.; Ilves, M.; Lylykangas, J.; Mäkelä, E.; Rautiainen, M.; Surakka, V.; Lekkala, J. Low-latency EMG Onset and Termination Detection for Facial Pacing. *IFMBE Proc.* **2018**, *65*, 1016–1019. [CrossRef]
19. Norman, R.W.; Komi, P.V. Electromechanical delay in skeletal muscle under normal movement conditions. *Acta Physiol. Scand.* **1979**, *106*, 241–248. [CrossRef] [PubMed]
20. Tabie, M.; Kirchner, E.A. EMG Onset Detection: Comparison of Different Methods for a Movement Prediction Task based on EMG. In *Proceedings ofthe International Conference on Bio-Inspired Systems and Signal Processing (BIOSIGNALS-2013)*; SciTePress: Setúbal, Portugal, 2013; pp. 242–247. [CrossRef]
21. Kang, K.; Rhee, K.; Shin, H.C. Event Detection of Muscle Activation Using an Electromyogram. *Appl. Sci.* **2020**, *10*, 5593. [CrossRef]
22. Staude, G.; Flachenecker, C.; Daumer, M.; Wolf, W. Onset Detection in Surface Electromyographic Signals: A Systematic Comparison of Methods. *Eurasip J. Adv. Signal Process.* **2001**, *2001*, 867853. [CrossRef]
23. Zhang, X.; Zhou, P. Sample entropy analysis of surface EMG for improved muscle activity onset detection against spurious background spikes. *J. Electromyogr. Kinesiol. Off. J. Int. Soc. Electrophysiol. Kinesiol.* **2012**, *22*, 901–907. [CrossRef]
24. Hermens, H.; Hägg, G.; Freriks, B. (Eds.) *SENIAM: European Applications on Surface ElectroMyoGraphy: Proceedings of the Second General SENIAM Workshop*; Roessingh Research and Development: Enschede, The Netherlands, 1997.
25. Drapala, J.; Brzosowski, K.; Szpala, A.; Rutowska-Kucharska, A. Two stage EMG onset detection method. *Arch. Control Sci.* **2012**, *LVIII*, 427–440. [CrossRef]
26. Hodges, P.W.; Bui, B.H. A comparison of computer-based methods for the determination of onset of muscle contraction using electromyography. *Electroencephalogr. Clin. Neurophysiol.* **1996**, *101*, 511–519. [CrossRef] [PubMed]
27. Bonato, P.; D'Alessio, T.; Knaflitz, M. A statistical method for the measurement of muscle activation intervals from surface myoelectric signal during gait. *IEEE Trans. Biomed. Eng.* **1998**, *45*, 287–299. [CrossRef] [PubMed]
28. Lidierth, M. A computer based method for automated measurement of the periods of muscular activity from an EMG and its application to locomotor EMGs. *Electroencephalogr. Clin. Neurophvstol.* **1986**, *64*, 378–380. [CrossRef] [PubMed]
29. Abbink; der Bilt, V.; der Glas, V. Detection of onset and termination of muscle activity in surface electromyograms. *J. Oral Rehabil.* **1998**, *25*, 365–369. [CrossRef] [PubMed]
30. Kaiser, J.F. On a simple algorithm to calculate the 'energy' of a signal. In Proceedings of the International Conference on Acoustics, Speech, and Signal Processing, Albuquerque, NM, USA, 3–6 April 1990; pp. 381–384. [CrossRef]
31. Li, X.; Aruin, A.S. Muscle Activity Onset Time Detection Using Teager-Kaiser Energy Operator. In Proceedings of the 2005 IEEE Engineering in Medicine and Biology 27th Annual Conference, Shanghai, China, 1–4 September 2005.
32. Lauer, R.T.; Prosser, L.A. Use of the Teager-Kaiser Energy operator for muscle activity detection in children. *Ann. Biomed. Eng.* **2009**, *37*, 1584–1593. [CrossRef] [PubMed]
33. Solnik, S.; Rider, P.; Steinweg, K.; DeVita, P.; Hortobágyi, T. Teager-Kaiser energy operator signal conditioning improves EMG onset detection. *Eur. J. Appl. Physiol.* **2010**, *110*, 489–498. [CrossRef] [PubMed]
34. Merlo, A.; Farina, D.; Merletti, R. A fast and reliable technique for muscle activity detection from surface EMG signals. *IEEE Trans. Biomed. Eng.* **2003**, *50*, 316–323. [CrossRef] [PubMed]
35. Vannozzi, G.; Conforto, S.; D'Alessio, T. Automatic detection of surface EMG activation timing using a wavelet transform based method. *J. Electromyogr. Kinesiol. Off. J. Int. Soc. Electrophysiol. Kinesiol.* **2010**, *20*, 767–772. [CrossRef] [PubMed]

36. Micera, S.; Sabatini, A.M.; Dario, P. An algorithm for detecting the onset of muscle contraction by EMG signal processing. *Med. Eng. Phys.* **1998**, *20*, 211–215. [CrossRef] [PubMed]
37. Tenan, M.S.; Tweedell, A.J.; Haynes, C.A. Analysis of statistical and standard algorithms for detecting muscle onset with surface electromyography. *PLoS ONE* **2017**, *12*, e0177312. [CrossRef] [PubMed]
38. Cavanagh, P.R.; Komi, P.V. Electromechanical Delay in Human Skeletal Muscle Under Concentric and Eccentric Contractions. *Eur. J. Appl. Physiol. Occup. Physiol.* **1979**, *42*, 159–163. [CrossRef] [PubMed]
39. Nikolic, M.; Krarup, C. EMGTools, an adaptive and versatile tool for detailed EMG analysis. *IEEE Trans. Biomed. Eng.* **2011**, *58*, 2707–2718. [CrossRef]
40. Richman, J.S.; Moorman, J.R. Physiological time-series analysis using approximate entropy and sample entropy. *Am. J. Physiol. Heart Circ. Physiol.* **2000**, *278*, H2039–H2049. [CrossRef]
41. Phinyomark, A.; Limsakul, C.; Phukpattaranont, P. A Novel Feature Extraction for Robust EMG Pattern Recognition. *arXiv* **2009**, arXiv:0912.3973. [CrossRef]

Disclaimer/Publisher's Note: The statements, opinions and data contained in all publications are solely those of the individual author(s) and contributor(s) and not of MDPI and/or the editor(s). MDPI and/or the editor(s) disclaim responsibility for any injury to people or property resulting from any ideas, methods, instructions or products referred to in the content.

bioengineering

MDPI

Article

Optimizing RNNs for EMG Signal Classification: A Novel Strategy Using Grey Wolf Optimization

Marcos Aviles [1,*], José Manuel Alvarez-Alvarado [1], Jose-Billerman Robles-Ocampo [2,3], Perla Yazmín Sevilla-Camacho [2,4] and Juvenal Rodríguez-Reséndiz [1,*]

1 Facultad de Ingeniería, Universidad Autónoma de Querétaro, Santiago de Querétaro 76010, Mexico; jmalvarez@uaq.mx
2 Programa de Postgrado en Energías Renovables, Universidad Politécnica de Chiapas, Suchiapa 29150, Mexico; jrobles@upchiapas.edu.mx (J.-B.R.-O.); psevilla@upchiapas.edu.mx (P.Y.S.-C.)
3 Departamento de Ingeniería Energética, Universidad Politécnica de Chiapas, Suchiapa 29150, Mexico
4 Departamento de Ingeniería Mecatrónica, Universidad Politécnica de Chiapas, Suchiapa 29150, Mexico
* Correspondence: marcosaviles@ieee.org (M.A.); juvenal@uaq.edu.mx (J.R.-R.)

Abstract: Accurate classification of electromyographic (EMG) signals is vital in biomedical applications. This study evaluates different architectures of recurrent neural networks for the classification of EMG signals associated with five movements of the right upper extremity. A Butterworth filter was implemented for signal preprocessing, followed by segmentation into 250 ms windows, with an overlap of 190 ms. The resulting dataset was divided into training, validation, and testing subsets. The Grey Wolf Optimization algorithm was applied to the gated recurrent unit (GRU), long short-term memory (LSTM) architectures, and bidirectional recurrent neural networks. In parallel, a performance comparison with support vector machines (SVMs) was performed. The results obtained in the first experimental phase revealed that all the RNN networks evaluated reached a 100% accuracy, standing above the 93% achieved by the SVM. Regarding classification speed, LSTM ranked as the fastest architecture, recording a time of 0.12 ms, followed by GRU with 0.134 ms. Bidirectional recurrent neural networks showed a response time of 0.2 ms, while SVM had the longest time at 2.7 ms. In the second experimental phase, a slight decrease in the accuracy of the RNN models was observed, standing at 98.46% for LSTM, 96.38% for GRU, and 97.63% for the bidirectional network. The findings of this study highlight the effectiveness and speed of recurrent neural networks in the EMG signal classification task.

Keywords: RNN; GRU; LSTM; bidirectional recurrent neural networks; GWO; metaheuristic algorithms; EMG

Citation: Aviles, M.; Alvarez-Alvarado, J.M.; Robles-Ocampo, J.-B.; Sevilla-Camacho, P.Y.; Rodríguez-Reséndiz, J.. Optimizing RNNs for EMG Signal Classification: A Novel Strategy Using Grey Wolf Optimization. *Bioengineering* **2024**, *11*, 77. https://doi.org/10.3390/bioengineering11010077

Academic Editors: Chiara Giulia Fontanella and Carlo Albino Frigo

Received: 6 December 2023
Revised: 8 January 2024
Accepted: 11 January 2024
Published: 13 January 2024

Copyright: © 2024 by the authors. Licensee MDPI, Basel, Switzerland. This article is an open access article distributed under the terms and conditions of the Creative Commons Attribution (CC BY) license (https://creativecommons.org/licenses/by/4.0/).

1. Introduction

The classification and analysis of electromyographic (EMG) signals have emerged as essential research fields in biomechanics and neuroscience. By reflecting the electrical activity produced by the muscles, these signals offer a detailed overview of muscle functionality and potential associated pathologies [1,2]. However, extracting meaningful information from these signals for practical applications requires advanced and efficient processing techniques. Traditionally, EMG analysis has relied on feature extraction to interpret the information, that could help improve the quality of life of people, in different applications [2,3]. However, recurrent neural networks (RNNs) have opened doors to new analysis methods [4].

Using the segmented signal directly instead of extracting features to feed into the network offers a series of advantages that are crucial to the efficiency and quality of the analysis [4,5]. Firstly, working with the totality of the information in the signal allows the preservation of details that, although they may be subtle, are essential. Feature extraction often involves dimensional reduction that could omit important aspects of the signal. Furthermore, this direct approach minimizes complexity in the preprocessing stage by avoiding a meticulous process based on specific domain knowledge [6]. While a set of selected features might be adequate for one application, it may be insufficient for another. In that sense, the segmented signal-based model becomes more flexible. Likewise, neural networks, particularly RNNs, have a remarkable ability to detect complex patterns in data. Feeding the network with the raw signal allows it to identify and learn patterns that may not be evident in a manual extraction process [7].

Despite the promising benefits they present, RNNs are not without significant challenges. One of the main obstacles is the proper selection of hyperparameters, which significantly influence network performance [8]. The adequate selection of hyperparameters in deep learning models is a critical task, but at the same time, it is highly challenging. Hyperparameters, unlike parameters, are not learned during training but are set beforehand. An incorrect choice leads to problems such as overfitting, where the model performs exceptionally well on the training data but fails when faced with unseen data, or underfitting, where the model fails to capture the underlying complexity of the data [9]. Manually searching for these values is notoriously laborious and dependent on the knowledge of the researcher. Although there are automatic techniques, such as grid search or random search, these are computationally expensive [10]. In this context, the Grey Wolf Optimization (GWO) algorithm emerges as a promising solution for hyperparameter selection. This is a metaheuristic algorithm inspired by the social and hunting behavior of gray wolves. The advantages of using GWO lie in its ability to explore and exploit the hyperparameter search space simultaneously [11] and its fast convergence compared with alternative optimization algorithms [12].

Based on the above, this work proposes that through using a combination of an RNN and GWO to analyze EMG signals, an accurate classification is achieved, and traditional limitations in the analysis of these signals are overcome. Long short-term memories (LSTMs), gated recurrent units (GRUs), and bidirectional recurrent neural networks, each with their unique characteristics, are used to capture the complex and sequential nature of EMG signals. LSTMs are notable for their ability to learn long-term dependencies, which is crucial given that EMG signals can contain classification-relevant information over long periods. GRUs, on the other hand, offer an efficient and less computationally intensive option, ideal for real-time applications where resources may be limited. Additionally, bidirectional recurrent networks provide a more complete view of the data by processing information in both directions, ensuring that the context of the entire sequence is taken into account for more accurate classification [13–15]. By integrating these advanced RNN methods with the GWO technique, the ability of the model to identify patterns in the data is further improved, resulting in superior performance and greater accuracy in classifying movements based on EMG signals. The contributions of this work are the following:

- Implement a novel methodology using the GWO algorithm to extract features from EMG signals using recurrent neural networks, thereby improving accuracy.
- This approach compares three RNN structures by establishing a solid baseline. This provides a rigorous foundation for evaluating the improvements that each structure can bring to the system performance by integrating a GWO algorithm.

The document is structured as follows. Section 2 reviews previous research addressing issues similar to those discussed here. Section 3 reviews the theoretical foundation, providing a vision of the techniques used. Later, in Section 4, the implemented methodology is detailed. The experiments and their respective results are detailed in Section 5, while Section 6 focuses on an in-depth analysis of the findings. Finally, the study is concluded in Section 7.

2. Related Works

Xie et al. [16] developed an advanced neural network model, Bi-ConvGRU, to recognize hand gestures from EMG signals, allowing detailed measurement of muscle activity. This model was evaluated by considering 18 hand gestures from the Ninapro dataset performed by both amputee and non-amputee individuals. The results highlight the potential of this approach for a bio-intuitive and non-invasive control of upper limb prostheses with a physiologically acceptable latency. In [17], the researchers developed a gesture classifier using an RNN model with LSTM layers specifically for hand control in prosthetics. A notable contribution of the authors was enhancing the model's adaptability for embedded systems by reducing the number of EMG channels.

Metaheuristic algorithms are already used in the field of machine learning. Ref. [18] introduced variations in the Artificial Bee Colony algorithm, which they applied in a KNN classification system to discern hand movements. Likewise, they highlight that this proposal can have applications in physical activities and physiotherapy therapies thanks to its notable performance. In [19], an LSTM model was used to classify the gestures from the forearm muscles. The authors compared the proposed neural network against GRU and demonstrated great performance during online classification. The work of Xiong et al. [20] compared techniques based on machine learning and four RNN configurations, such as GRU, LSTM, and the bidirectional alternative of these two. The models were run to classify eight different gestures from a dataset. The results showed that the bidirectional LSTM configuration obtained the best performance compared to the other RNN configurations and the machine learning models. Aviles et al. [10] developed an SVM classifier incorporating genetic algorithms for feature extraction. Two sets of data were used: the first referring to the right upper extremity and the second composed of movements of the right lower extremity. Likewise, Particle Swarm Optimization (PSO) was implemented to compare both algorithms. The SVM-GA approach significantly improves classification, efficiency, and provides a reduction in the number of parameters compared to the PSO-based approach.

To classify flexion, extension, and ramp walking movements, the authors of [21] employed an LSTM due to its strong suitability for processing nonlinear time-series data. Additionally, to enhance accuracy of the model, they integrated a PSO algorithm for fine-tuning the parameters of LSTM. The PSO-LSTM model significantly improved performance compared to the randomly initialized traditional LSTM. Li et al. [6] employed a methodology based on CNN for classification tasks and RNN for handling timing issues. This approach excelled in real-time recognition, accurately classifying 20 distinct hand movement patterns. A hybrid approach for classifying EMG signals was implemented by [22], utilizing a CNN-LSTM model integrated with a kernel-based PCA technique. The findings demonstrate that the PCA-CNN-LSTM method effectively recognizes lower limb activities from the signals. The overview of the related works is presented in Table 1. However, there is a need for continued research and development to create even more effective algorithms that improve the classical models in this field.

Table 1. Overview of EMG classification methods in related studies.

Reference	Classification Method	Tuning Algorithm	Dataset	Accuracy (%)	No. of Channels
[16]	Bidirectional Convolutional-GRU	-	Ninapro DB5 Field	88.70	16
[17]	RNN-LSTM	Manual hyperparameter tuning	500 samples from 5 different gestures	87	4
[18]	KNN	ABC improved	Ninapro	97.06	12
[19]	LSTM	-	UC2018 DualMyo and the Ninapro DB5	95 for DualMyo and 91 for Ninapro DB5	16
[20]	LSTM	-	CapMyo	98.57	8 × 16 electrode array
[10]	SVM	PSO	Healthy muscular limbs collection	91.00	4
[21]	LTSM	PSO	Lower limb muscle	98.58	4
[6]	1D CNN-RNN	-	gForce EMG Armband	91	8
[22]	CNN-LSTM	PCA	Codamotion system acquisition	98.50	16

3. Materials and Methods

This section discusses and analyzes the main concepts of the theoretical foundation and the materials used to develop this work. The development of the RNN and GWO was carried out in Python using TensorFlow. On the other hand, the filtering and segmentation of the EMG signals was carried out in MATLAB 2018b. The equipment used was a laptop with a 12th-generation i7 processor with an RTX 3060 GPU.

3.1. Database

The database presented in [10] was used for this study, which is focused on the muscles specified in Table 2. In the analog filtering phase, a combination of a low-pass filter and a Butterworth high-pass filter was used, both implemented with a second-order Sallen–Key topology and with cut-off frequencies of 600 Hz and 10 Hz, respectively. Additionally, a second-order Bainter notch filter was applied to eliminate 60 Hz interference caused by the power supply. Digitization of the signals was performed using a USB-6002 data acquisition device (DAQ).

Table 2. Action of the muscles whose action potentials are used for movement classification.

Muscle	Action
Biceps brachii (long head)	Flexes the forearm at the elbow joint, supinates the forearm at the radioulnar joints, and flexes the arm at the shoulder joint.
Triceps brachii (long head)	Extends the forearm at the elbow joint and extends the arm at the shoulder joint.
Superficial flexor of the fingers	Flexes the middle phalanx of each finger at the proximal interphalangeal joint, the proximal phalanx of each finger at the metacarpophalangeal joint, and the hand at the wrist joint.
Finger extensor	Extends the distal and middle phalanges of each finger at the interphalangeal joints, the proximal phalanx of each finger at the metacarpophalangeal joint, and the hand at the wrist joint.

The study population consisted of 9 participants, aged between 23 and 27 years: five men and four women. All participants were free of pathologies related to the locomotor system and nervous system and did not have amputation conditions or obesity problems. Five different types of arm and hand movements were recorded, including flexion and extension of the arm at the elbow joint, flexion and extension of the fingers, and a resting state. For this purpose, four bipolar channels placed directly over the muscles of interest were used, which are shown in Table 2. Additionally, a reference electrode was placed on the wrist. Each movement was performed for 6 s, followed by a 2 s relaxation period, and was repeated 20 times using a sampling rate of 1.5 kHz.

The Surface Electromyography for the Non-Invasive Assessment of Muscles (SENIAM) recommendations were followed. The SENIAM project is a European initiative focused on superficial electromyography. It seeks to standardize aspects such as electrode placement and signal processing for EMG. SENIAM recommends locating sensors in 30 individual muscles to obtain quality and stable EMG signals. The recommendations include details on the location, orientation, and distance between electrodes, as well as advice for fixation on the skin and the location of the reference electrode [23].

For the placement of the electrodes, a separation of 20 mm between them was ensured, and attention was paid to the specific characteristics of each muscle. Initially, the reference electrode was placed on the dorsal area of the wrist of the right hand. Subsequently, the reading electrodes were placed on the selected muscles, connecting them to the conditioning and acquisition equipment to begin data capture. During acquisition, participants were asked to perform the indicated movements, ensuring rest periods between each repetition and after each type of movement to avoid muscle fatigue.

3.2. Recurrent Neural Network

RNNs are a category of neural networks explicitly designed to work with data sequences, especially useful in natural language processing tasks and time-series analysis [24]. This study explored three types of RNNs: LSTM, GRU, and bidirectional recurrent neural networks.

LSTM is a variant of an RNN designed to address the problem of gradient disappearance, a challenge that occurs in traditional RNNs when processing long data sequences. This is achieved through a cell structure containing entry, exit, and forget gates, allowing the network to have long- and short-term memory. LSTMs learn and remember over long sequences and are, therefore, less sensitive to gaps in data sequences [13]. A GRU is another variant of an RNN that, like LSTM, seeks to solve the problem of gradient disappearance. However, unlike LSTM, GRU simplifies the cell structure by merging the input and forget gates into a single update gate. This reduces the computational complexity and, in specific contexts, offers performance comparable to or even superior to LSTM with a shorter training time [14]. Bidirectional neural networks take advantage of sequence information in both directions (past and future) to improve accuracy in classification and prediction tasks. This is achieved by running two traditional RNNs: one that moves forward through the sequence and one that moves backward. Both outputs combine to provide a more informed perspective on the sequence, which can result in better accuracy on specific tasks [15]. Algorithm 1 shows the programming logic to implement RNNs with GWO.

Algorithm 1 Optimization of bidirectional neural network, LSTM, and GRU with GWO.

Inputs: Training data, validation data
Initial hyperparameters:
Learning rate: $lr \in [0.00001, 0.1]$
Neurons in layer 1: $n_1 \in [10, 150]$
Neurons in layer 2: $n_2 \in [10, 150]$
Batch size: $batch\_size \in [128, 512]$
Training epochs: $epochs \in [10, 100]$
procedure CREATEMODEL(lr, n_1, n_2)
    Initialize RNN model with defined structure
    Add RRN layer units of n_1 neurons
    Add RRN layer units of n_2 neurons
    Add dense layer for classification
    Compile model with learning rate lr
    return model
end procedure
procedure FITNESSFUNCTION($hyperparameters$)
    $lr, n_1, n_2, batch\_size, epochs \leftarrow hyperparameters$
    $model \leftarrow$ CREATEMODEL(lr, n_1, n_2)
    Train $model$ with $epochs$ and $batch\_size$
    Evaluate $model$ on validation data
    return validation error
end procedure
Optimize with GWO:
Define search space with defined ranges
$optimal\_hyperparameters \leftarrow$ GWO(FITNESSFUNCTION)
$optimal\_lr, optimal\_n_1, optimal\_n_2, optimal\_batch\_size, optimal\_epochs \leftarrow optimal\_hyperparameters$
Train and validate with selected hyperparameters:
$optimal\_model \leftarrow$ CREATEMODEL($optimal\_lr, optimal\_n_1, optimal\_n_2$)
Train $optimal\_model$ with $optimal\_epochs$ and $optimal\_batch\_size$

3.3. *Hyperparameters*

A hyperparameter is a parameter not intrinsically derived from the data but set before training the model. Hyperparameters guide how the neural network learns and how the model optimizes. Ensuring the appropriate selection of these hyperparameters is essential to achieving exceptional model performance [25].

When working with neural networks such as GRU, LSTM, and bidirectional neural networks, various hyperparameters are essential and drastically influence the behavior of the model. These hyperparameters cover aspects such as the number of units or neurons in the layers, the activation function used, the learning rate, and the total number of training epochs. For example, the number of units in the layers largely determines the ability of the network to model complex interactions in the data. Increasing this number can allow the network to understand more sophisticated patterns but also runs the risk of overfitting [26].

The activation function introduces nonlinearity into the network, thus its ability to model nonlinear relationships. Although the sigmoid function is recognized, in networks such as LSTM or GRU, functions such as the scaled exponential linear unit (SELU) are frequently used [27]. Regarding the learning rate, this regulates the magnitude of adjustment of the weights in each training cycle. Too high a rate can cause oscillations in the network, preventing convergence, while an excessively low rate can cause slow convergence, trapping the model in local minima. On the other hand, the number of epochs establishes how often the entire dataset is used during training, which is crucial to avoid overfitting or underlearning [8].

Table 3 shows the hyperparameters adjusted using GWO to determine the most suitable values in the GRU, LSTM, and bidirectional recurrent neural networks.

Table 3. Hyperparameters considered for adjustment using GWO.

Hyperparameter
Number of neurons per layer
Batch size
Training epoch
Learning rate

3.4. *Grey Wolf Optimizer*

GWO is a metaheuristic optimization algorithm proposed by [11], inspired by the social and hunting behavior of gray wolves. Its design emulates the hierarchical structure and hunting tactics these creatures deploy in nature. Hierarchy in gray wolves: Gray wolves have a very marked hierarchical social structure in the wild. Within this hierarchy, four main types of wolves stand out:

- Alpha (α): They are the leaders of the pack, usually a couple (male and female). They make all the critical decisions, from the time of hunting to the time of migrating or resting.
- Beta (β): They are second in command. If both alphas die, the beta would assume leadership. They help the alphas make decisions and act as an "advisor".
- Delta (δ): They act as guardians of the pack. They protect the wolves from external threats and maintain order within the group.

The GWO uses this hierarchy to update the wolves' (solutions) positions in the search space. The rankings are updated based on the position of the top three wolves (α, β, δ). The position of the rest of the wolves is updated based on these three best positions, emulating hunting and tracking behavior. The hunting process is modeled mathematically using equations that represent the pursuit, encirclement, and attack of prey. These equations are based on the distance between the wolf and its prey and are adjusted according to the hierarchy. One of the main advantages of GWO is its balance between exploration (looking for new areas in the solution space) and exploitation (refining a solution in a specific area). This is due to hierarchical and cooperative behavior of the wolves when hunting, which allows the algorithm to evade local optima and converge towards a suitable global solution.

Hunting behavior is imitated using hunting coefficients. For each wolf (except alpha), the following coefficients are used [11]:

- A_1, A_2, A_3: These coefficients define the magnitude of the attraction towards the leading wolves. They control the ability of the wolves to explore and exploit.
- C_1, C_2, C_3: These coefficients are random vectors obtained for each iteration and wolf. They help in adjusting the position of each wolf concerning the leading wolves.

The hunting coefficients are typically calculated as follows:

$$A_{ij} = 2a \times r_1 - a \tag{1}$$

$$C_{ij} = 2 \times r_2 \tag{2}$$

where r_1 and r_2 are random numbers in [0,1], and a decreases linearly from 2 to 0 over the iterations. For each wolf in the group (except the leading wolves), the distances to the three leading wolves are calculated as follows:

$$D_\alpha = |C_{1j} \times X_\alpha - X_i| \tag{3}$$

$$D_\beta = |C_{2j} \times X_\beta - X_i| \tag{4}$$

$$D_\delta = |C_{3j} \times X_\delta - X_i| \tag{5}$$

where X_α, X_β, and X_δ are the positions of the alpha, beta, and delta wolves, respectively, and X_i is the position of the current wolf. These distances are then used to adjust the position of each wolf based on the positions of the leading wolves. The goal is to bring the wolves closer to the best solutions in the search space, guiding the pack toward possible optimal solutions.

Algorithm 2 shows the programming logic to implement GWO.

Algorithm 2 Grey Wolf Optimizer.

```
Inputs: Objective function f(x), Population size N, Maximum iterations T
Initialization:
for i = 1 to N do
    Initialize wolf position X_i randomly
    Calculate fitness f(X_i)
end for
for t = 1 to T do
    Update coefficients:
    a = 2 - t × 2/T
    Update alpha, beta, and delta wolves:
    Identify the top three wolves X_α, X_β, X_δ based on fitness
    for i = 1 to N do
        for j = 1 to dimension do
            Calculate hunting coefficients A_1, C_1, A_2, C_2, A_3, C_3
            Calculate distances D_α, D_β, D_δ
            Update position using X_α, X_β, X_δ
        end for
        Apply boundary conditions if necessary
        Update the fitness of X_i
    end for
end for
Output: Best solution X_α
```

3.5. Windowing

The windowing technique is widely used in signal processing and time series to divide a continuous dataset into more manageable segments called windows. This technique is advantageous when analyzing data that undergo temporal variations, such as electromyographic or electrocardiographic signals. A common and notable variant of windowing is the use of overlapping windows. Unlike segmentation into discrete, non-overlapping windows, overlapping windows allow an overlap between consecutive windows by a given number of points. Overlapping windows offer several advantages: an improvement in temporal resolution, a reduction in edge error, and an increase in data density. The improvement in temporal resolution is because the overlap between windows allows us to detect events or features in the data that could go unnoticed or not be clearly defined in a segmentation without overlap. Reducing edge error is an essential benefit since, in some applications, the start and end of a window can introduce artifacts or errors. These errors can be minimized by overlapping windows since data at the edges of a window are also analyzed in the context of the adjacent window. Finally, the increase in data density refers to the fact that segmentation with overlapping windows generates a more significant number of segments for the same dataset compared to segmentation without overlap, which can benefit machine learning techniques by providing more examples to train and validate models [28,29].

When implementing overlapping windows, it is crucial to consider the degree of overlap, generally defined as a percentage of the window size. It should be noted that a more significant overlap increases the correlation between consecutive windows, which can be beneficial for detecting subtle transitions in the data. However, it can also introduce redundancy [30].

4. Methodology

This section outlines the methodological steps undertaken to implement this work.

4.1. Signal Processing

The first step in processing consisted of filtering the signals to attenuate noise. Since the original data sequence of the EMG signal was used, the classification of the signals may have been susceptible to interference and artifacts. Therefore, it was essential to perform filtering before proceeding with window segmentation. For this purpose, a second filtering stage was used in addition to the analog filtering of the database. In this case, it corresponded to a second-order digital Butterworth bandpass filter with cut-off frequencies between 10 and 500 Hz, which were the frequencies of interest, using the "Butter" and "filtfilt" functions of the MATLAB 2018b software [31].

After filtering, the signals were segmented into 250 ms windows, overlapping by 190 ms [28,29]. It is important to note that the EMG signal contains 2 s of rest before the start of the movement. Therefore, these were discarded to focus exclusively on the information generated by the movement of the arm. After removing these 2 s, the remaining signal was divided into 63 windows. The choice of using overlapping windows is due to their ability to continuously collect information during the operation of the classification algorithm, which is essential for its real-time application. Additionally, using this approach increases the cadence of classification decisions since each analysis window requires less data to complete, in this case, 250 ms.

After extracting the windows, the information was organized to be introduced into the neural networks in a three-dimensional matrix of dimensions i, j, and k. Here, i represents the total number of windows for each acquisition in the database, calculated as $9 \text{ people} \times 20 \text{ trials} \times 5 \text{ movements} \times 63 \text{ windows}$, resulting in $i = 56{,}700$. The dimension j is related to the number of sensors used in each acquisition, which is four. Meanwhile, k represents the total number of data points found in each window, with a total of 375 points per window. This value reflects the data collected in a time interval of 250 ms, with a sampling rate of 1.5 kHz.

The experimentation was developed in two stages. In the first, the applicability of the method to EMG signals was validated, involving most volunteers in the training and validation phases. In contrast, the second stage was designed to evaluate the robustness of the methodology, using a higher percentage of individuals in the testing phase.

In the first stage, the generated matrix was organized so that the data of the first eight people were allocated to the training and validation phases, reserving the information of the ninth individual exclusively for testing. Of the set of 8 people, 80% of their data were used for training and the remaining 20% for validation. It is vital to highlight that these subsets were mixed randomly to prevent any possibility of overfitting in the network.

A second experimental round was carried out to check the efficiency and viability of the proposed method. In this second round, a ratio of 5 to 4 of the people was used. This means that five individuals were used for the training and validation round, while the remaining four were used to testing the models. On this occasion, the 80–20 division for training and validation was also respected.

It is relevant to note that the validation accuracy is used for hyperparameter tuning. This is evident in Algorithm 1, specifically in lines 20 and 21. On the other hand, the testing accuracy is used to confirm the model results. The testing set is made up exclusively of subjects not included in the training and validation sets. It is also important to mention that the models were trained from scratch in both stages. The accuracy calculation is presented in Equation (6). This equation defines accuracy as the ratio of correct predictions to the total number of predictions [10].

$$Accuracy = \frac{TP + TN}{TP + TN + FP + FN} \tag{6}$$

where TP represents true positives, the cases in which the model correctly predicts the positive class. TN refers to true negatives, cases where the model correctly identifies the negative class. FP indicates false positives, which occur when the model incorrectly predicts a positive outcome for a case that is negative. Finally, FN are false negatives, in which the model fails to recognize the positive class, erroneously classifying it as harmful.

Sensitivity, also known as true positive rate, measures the proportion of correctly predicted positive instances to all actual positive instances. It focuses on the ability of the model to capture all positive instances and avoid false negatives. Equation (7) shows the equation that defines the sensitivity.

$$Sensitivity = \frac{TP}{TP + FN} \tag{7}$$

Specificity, or true negative rate, measures the proportion of correctly predicted negative instances to all true negative instances. It indicates the ability of the model to identify negative examples correctly and is crucial for its discriminative power. Equation (8) shows the equation that defines the specificity.

$$Specificity = \frac{TN}{TN + FP} \tag{8}$$

4.2. Recurrent Neural Networks

Within the framework of this work, three variants of recurrent neural network architectures were designed and implemented, namely, LSTM, GRU, and a bidirectional neural network. These architectures were implemented using Python, relying on the TensorFlow library. Two recurrent layers are included in each of these architectures, and the SELU activation function is used. The Adam optimizer was selected to adjust the weights, while, to evaluate the performance of the model, the accuracy metric was adopted. The cross-entropy loss function was used during the training phase to calculate the error.

As the output of the model, a dense layer composed of five neurons was added, one for each movement, using the softmax activation function. Furthermore, to stabilize the activations and facilitate training, a normalization layer, specifically LayerNormalization, was included before the recurrent layers.

The early_stopping callback was integrated to optimize training time, which stops training if no improvement in accuracy is perceived in the validation dataset for five consecutive iterations. The number of iterations executed during training served as feedback for the GWO algorithm, allowing the number of epochs needed to achieve the best results to be adjusted.

Bidirectional networks can be built using GRU or LSTM structures as a base. In our particular case, LSTM layers were used for the bidirectional configuration. Additionally, the computational efficiency of the different architectures was assessed, identifying advantages and disadvantages in terms of training time, memory use, and precision.

4.3. Grey Wolf Optimizer

GWO algorithm was implemented in Python, using the numpy and pandas libraries. A population of 20 wolves was established for this optimization, and the algorithm iterated over ten cycles. Table 4 shows the ranges of the hyperparameters to optimize.

Table 4. Hyperparametervalues optimized using GWO.

Parameter	Considered Range
Number of neurons per layer	10 to 150
Batch size	128 to 512
Training epochs	10 to 100
Learning rate	0.00001 to 0.1

At the end of the ten iterations of the GWO, various data of interest were recorded for the best solutions found. These included the best position (representing the suggested hyperparameters), the associated cost (indicating the validation classification error in the neural network), the structure of the obtained neural model, and the corresponding training and prediction times. This process allowed us to fine-tune the configuration of the neural networks, searching for the best combinations of hyperparameters that would minimize the classification error while optimizing the performance and efficiency of the model.

5. Results

This section details the results obtained for the two experimental stages described in Section 4.

5.1. First Stage

Table 5 shows the hyperparameters achieved for each of the recurrent networks optimized using GWO.

Table 5. Hyperparameter values selected using GWO for LSTM, GRU, and bidirectional neural network.

Hyperparameter	LSTM	GRU	Bidirectional
Number of neurons in first layer	28	25	14
Number of neurons in second layer	74	18	13
Batch size	188	329	199
Training epochs	10	21	16
Learning rate	0.00346	0.00554	0.0117

A notable robustness is observed in the LSTM model, with 102 neurons distributed between two layers. It has a total of 76,861 trainable parameters. On the other hand, bidirectional networks have only 27 neurons in total but have 48,183 trainable parameters. On the other hand, the GRU is presented with only 43 neurons and 33,425 trainable parameters. The lighter nature of GRU may be the reason why it takes more epochs to reach convergence. Despite this difference in the density of neurons between GRU and bidirectional networks, there is no considerable disparity in complexity. This observation shows that a more significant number of neurons does not necessarily result in an intrinsically more complex network. Concerning learning rates, a high rate such as the one adopted by the bidirectional model (0.0117) suggests a faster adaptation of the weights, although with possible oscillations that may be experienced during the process. Meanwhile, more contained rates, such as those adopted by the LSTM (0.00346) and GRU (0.00554), suggest a more cautious approach toward convergence.

The batch size, which is another crucial hyperparameter, shows variations between architectures. In GRU, a considerable batch of 329 is used, probably to speed up training through simultaneous data processing. However, this benefit may be risky, as larger batch sizes may compromise convergence accuracy. Despite these risks, on all architectures, including LSTM with a batch size of 188 and bidirectional with 199, a flawless accuracy of 100% was achieved during testing. Figure 1 shows the final block diagram for each of the three trained models.

Table 6 shows the temporal analysis of the different architectures of the recurrent neural networks studied. A difference in time is observed between the different stages evaluated.

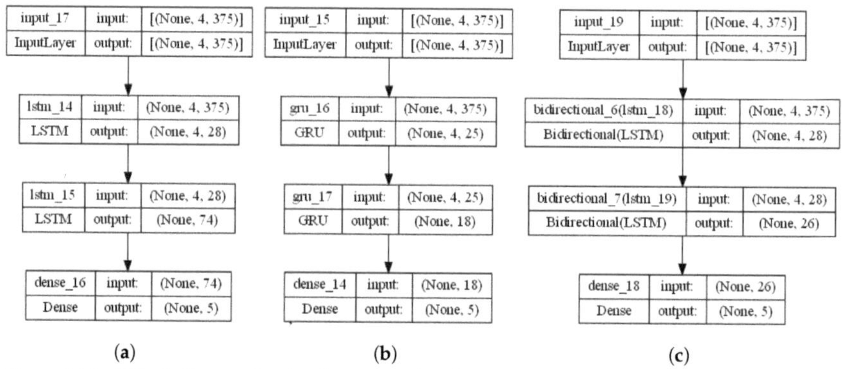

Figure 1. Final block diagram for the three trained and adjusted models, (**a**) LSTM network, (**b**) GRU, and (**c**) bidirectional, for first stage.

Table 6. Training, validation, and prediction times for LSTM, GRU, and bidirectional neural network.

Model	Training Time (s)	Validation Time (s)	Prediction Time (ms)
LSTM	31.47	0.81	0.12
GRU	51.28	0.85	0.134
Bidirectional	81.60	1.23	0.2

The LSTM network proved to be the most efficient in terms of training time, requiring only 31.47 s. This result is particularly interesting given its high neuronal density and relatively large number of trainable parameters. The moderate learning rate (0.00346) and batch size (188) could contribute to this rapid convergence and efficient training. Regarding validation time, the LSTM was also slightly faster than the GRU, needing only 0.81 s. LSTM was remarkably effective for prediction, with a time of only 0.12 ms.

On the other hand, the GRU, despite being less dense and having fewer parameters than the LSTM, required a longer training time of 51.28 s. Given its lighter architecture, this longer duration is related to the need for more epochs to converge. The validation time of the GRU was slightly longer than that of the LSTM, registering 0.85 s. Despite this marginal difference, it is relevant to mention that the GRU's prediction time, while still relatively fast, was slower than the LSTM, taking 0.134 ms.

Finally, the bidirectional architecture, which uses an underlying LSTM structure to process sequences in both directions, showed the longest training time of the three, at 81.60 s. This increase in time is associated with the bidirectional nature of the model, which processes forward and backward information, intrinsically increasing the computational load. Despite its compact configuration of neurons, its validation time was the longest, requiring 1.23 s. In terms of prediction, it also showed the longest time, at 0.2 ms.

Figure 2 presents the error evolution in different recurrent networks through the GWO optimization method. The GRU network, shown in Figure 2b, starts with the highest error, approximately 17.5%. However, its rapid convergence is notable, reaching an error of 0% in the third iteration. On the contrary, the bidirectional network, which can be seen in Figure 2c, starts with the lowest error, 1.6%, in its first iteration, thanks to an appropriate combination of hyperparameters obtained by the algorithm. Despite this, it requires six iterations to minimize the error to 0%, showing a more gradual reduction than the other architectures, a direct consequence of its low starting error. In the case of the LSTM, presented in Figure 2a, it starts with an error of 6% and shows a rapid decrease until the third iteration, after which its decrease becomes more gradual, reaching 0% in the eighth iteration.

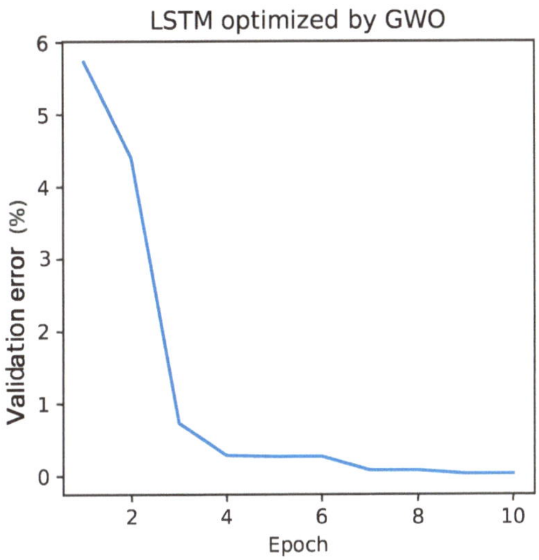

(a)

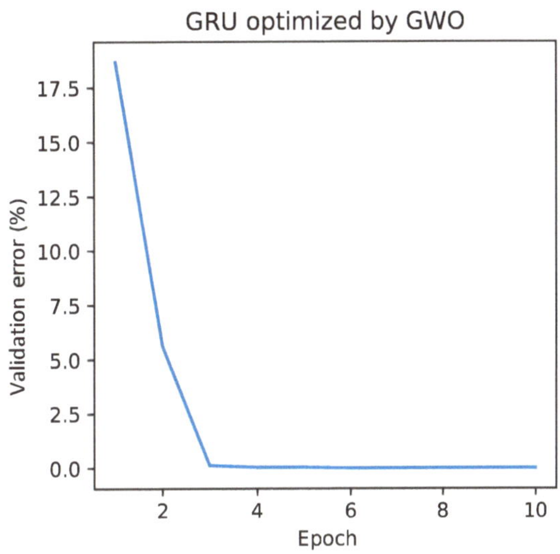

(b)

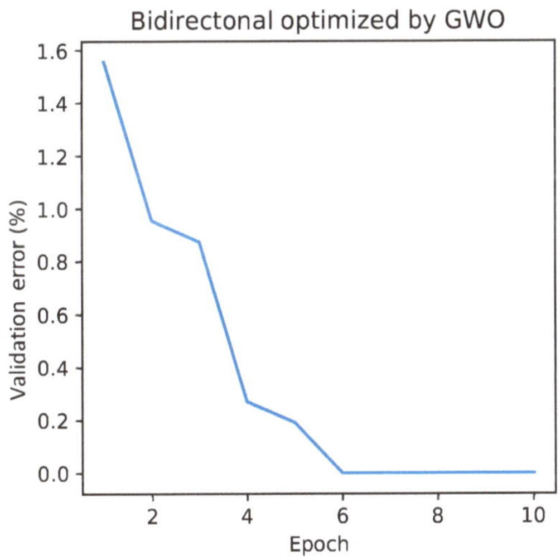

(c)

Figure 2. Reduction in the error classification due to the selection of hyperparameters by GWO. Where (**a**) represents the error in the LSTM, (**b**) in the GRU, and (**c**) in the bidirectional network.

Figure 3 illustrates the training and validation accuracy behavior of the three recurrent neural network models: LSTM, GRU, and bidirectional, each optimized with the GWO optimization algorithm. Consistently across all three models, an increase in classification across iterations is observed, indicative of an absence of overfitting. The LSTM model shows a rapid increase in accuracy that soon stabilizes, maintaining a slight advantage in training accuracy over validation, suggesting effective generalization without falling into memorization. On the other hand, although the GRU model follows a similar trend in increasing precision, it presents a distinctive peak in the validation curve that could be attributed to temporal overfitting or variations in the test data. However, this model also stabilizes its precision, demonstrating its ability to adapt and generalize with the advancement of time. The bidirectional network maintains the general behavior observed in LSTM and GRU, with the training and validation accuracy curves advancing in close formation throughout the process.

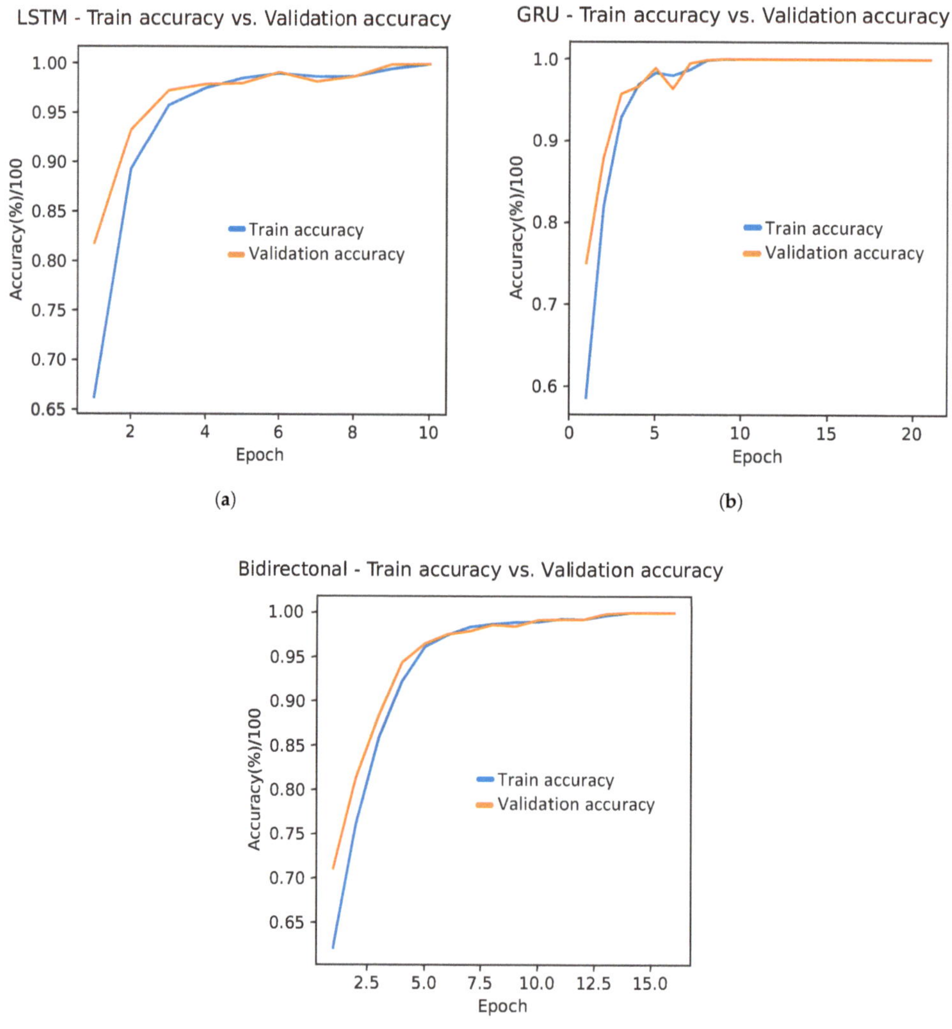

(**a**) (**b**) (**c**)

Figure 3. Evolution of training and validation accuracy with hyperparameters defined by GWO in the first stage. Where (**a**) represents the accuracy evolution in the LSTM, (**b**) in the GRU, and (**c**) in the bidirectional network.

Figure 4 presents the evolution of the average error in the wolf population throughout the iterations, illustrating how the global solutions improve as they advance. A distinctive feature of metaheuristic algorithms is their ability to offer multiple solutions at the end of the iterative process. Each solution, corresponding to an individual in the population, can be adapted to the desired objective but with different properties. At the end of iteration

10, several RNN configurations reported an error of less than 1%, each with different sets of hyperparameters. For this study, those networks with faster response times in the evaluation stage of each topology were chosen. However, it is possible to select networks according to other criteria, such as the minimum number of neurons or the shortest training time, depending on the specifics of the problem addressed.

Figure 4a, corresponding to LSTM, reveals a start with the highest average error, approximately 65%. Furthermore, it shows a convergence to the lowest error in iteration 9, characterized by a gradual decrease. This behavior suggests a constant and balanced optimization of the prediction for the LSTM population. In contrast, Figure 4b, corresponding to GRU, exhibits a more irregular evolution, with an initial error close to 60%, reaching the minimum average error at iteration 8. This slightly oscillating behavior in GRU suggests that the GWO algorithm faces challenges in finding solutions that significantly reduce the error. Finally, Figure 4c shows that bidirectional neural networks start with a lower average error, around 41%. These networks reach faster convergence, achieving the minimum error in iteration 5. Their smooth and rapid trajectory suggests that GWO has a better facility to identify favorable solutions in this topology.

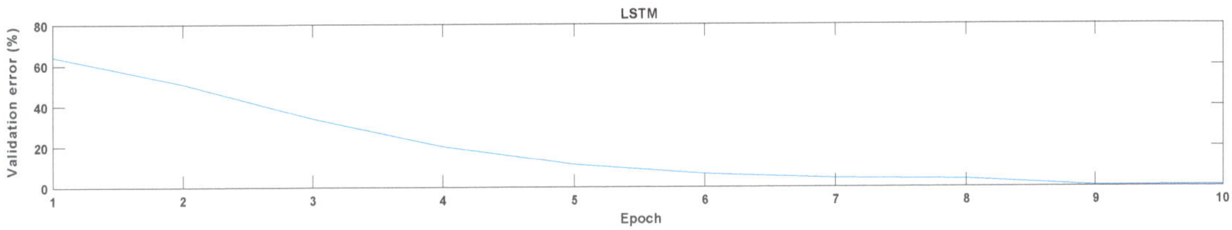

(a)

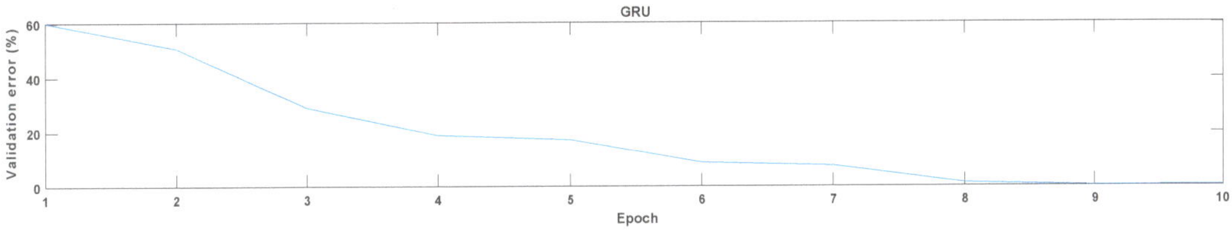

(b)

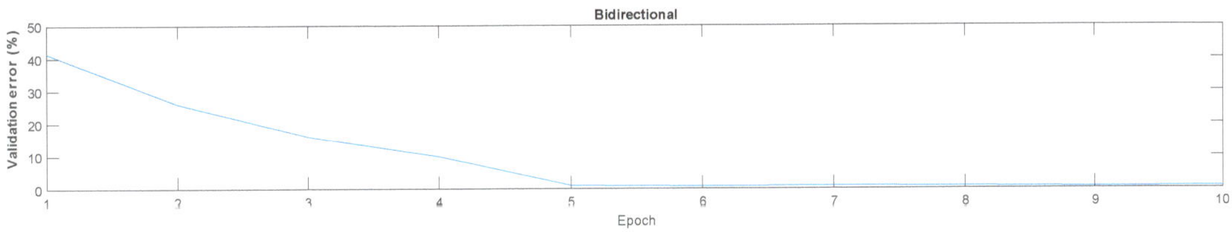

(c)

Figure 4. Evolution of mean validation error for LSTM, GRU, and bidirectional recurrent neural networks. Where (**a**) represents the mean validation error in the LSTM, (**b**) in the GRU, and (**c**) in the bidirectional network.

This study used SVM with a Gaussian kernel as a reference model. Since SVM does not allow direct processing of raw signals, performing a proper characterization of these signals was imperative. The characteristics proposed in ref. [10] were used for this. The features used are shown in Table 7.

Table 7. Features selected by the sensor for classification using SVM.

Sensor	Feature
1	Wavelength
2	Log detector
3 and 4	Shannon entropy
2 and 4	Myopulse percentage rate
3	Modified mean absolute value type 1
3	Zero crossings

It is relevant to highlight that the dataset and features used in this study are the same as those used in [10]. These features were carefully selected for this database in the previously mentioned work. By implementing the above-mentioned features, the SVM model achieved an efficiency of 93%. Table 8 presents the fundamental comparisons between the models based on RNN and SVM.

Table 8. Training, prediction times, and testing accuracy for LSTM, GRU, bidirectional neural network, and SVM.

Model	Training Time (s)	Prediction Time (ms)	Accuracy
LSTM	31.47	0.81	100%
GRU	51.28	0.85	100%
Bidirectional	81.60	1.23	100%
SVM	14	2.7	93%

Table 9 provides a detailed analysis of the performance of an SVM classifier in the testing stage for the five moves. Regarding sensitivity, class 1 shows the best performance with 85.2%, closely followed by class 2 with 81.9%. Class 3 also performs well, with 80.2%. However, the sensitivity decreases noticeably for classes 4 and 5, with 63.9% and 72.1%, respectively, indicating that the SVM classifier has difficulty correctly identifying these classes compared to the first three. Regarding specificity, which evaluates the classifier's ability to correctly identify negatives, a generally high performance is observed in all classes. Class 1 achieves a specificity of 95.6%, and classes 2 and 3 also exhibit high specificity, 93.2% and 95.9%, respectively. Although classes 4 and 5 present lower specificity, 83.2% and 82.9%, these values are still relatively high. It is important to contrast these results with the performance achieved by the LSTM, GRU, and bidirectional models which, by achieving 100% accuracy, also achieve 100% sensitivity and specificity. The lower performance of the SVM, particularly in sensitivity for classes 4 and 5, could indicate limitations in its ability to handle certain characteristics of these data or require more specific tuning of the model.

Table 9. Classifier performance for the testing step for SVM.

	Class 1 (%)	Class 2 (%)	Class 3 (%)	Class 4 (%)	Class 5 (%)
Sensitivity	85.2	81.9	80.2	63.9	72.1
Specificity	95.6	93.2	95.9	83.2	82.9

Table 8 shows an interesting comparison concerning the training and response times of the models. Noteworthy is the fact that SVM has the shortest training time. However, this efficiency is offset by a longer response time in the classification phase. This behavior is attributed to extracting features from the data before entering them into the classifier. This additional step imposes a delay that affects its performance in terms of response time. In contrast, RNN networks have the advantage of working directly with the raw data, eliminating the need for a feature extraction step and offering faster response.

Another relevant aspect is classification efficiency. Even though all models are trained using the same database and identical preprocessing, SVM has a lower classification rate. This discrepancy is due to the added complexity of selecting appropriate features. While the effectiveness of RNNs focuses on the quality and complexity of the input data, SVM has the particularity of depending not only on the selected features but also on the interaction and synergy between them. This analysis highlights the fundamental differences between feature-based approaches and those based on raw data, highlighting the strengths and limitations inherent to each methodology in the context of EMG signal classification.

5.2. Second Stage

Table 10 shows the hyperparameters achieved for each of the recurrent networks optimized using GWO in the second stage.

Table 10. Hyperparameter values for LSTM, GRU, and bidirectional neural network for the second stage.

Hyperparameter	LSTM	GRU	Bidirectional
Number of neurons in first layer	31	22	15
Number of neurons in second layer	13	101	15
Batch size	206	339	206
Training epochs	14	19	15
Learning rate	0.00502	0.00731	0.0177

For the LSTM model, an increase in the number of neurons in the first layer is observed from 28 to 31, suggesting a need for greater capacity to adapt to the variability in the data in the second stage, where a more significant number of individuals was included in the testing set. However, there was a significant reduction in the number of neurons in the second layer, going from 74 to 13, which could indicate an attempt to simplify the model to prevent overfitting. The batch size increased from 188 to 206, while the training epochs increased from 10 to 14, indicating that the model required more iterations on the data to reach convergence. Additionally, there was a slight increase in the learning rate.

In this second stage, a significant change is observed in the configuration of the GRU model. The number of neurons in the first layer was slightly reduced to 22, while it was increased to 101 in the second layer. This redistribution in model capacity suggests a change in modeling strategy, possibly due to differences in variability. By having a more significant number of individuals in the testing set in the second stage, the model may have needed to strengthen the internal layers to better generalize over the unseen data, thus avoiding overfitting the peculiarities of the training set. The batch size experienced a slight increase to 339, and the training epochs decreased to 19. These changes in the training hyperparameters suggest a search for balance between the stability and the speed of convergence of the model. A larger batch size may contribute to a more stable gradient estimation during training. At the same time, the reduction in the number of epochs suggests that the model was able to achieve an excellent fit to the data more efficiently. Finally, the learning rate increased from 0.00554 to 0.00731, indicating a more aggressive adjustment to the model weights during training. This increase can be interpreted as an attempt to speed up the training process.

In the second stage, the number of neurons in both layers experienced a slight increase, reaching 15 for both for the bidirectional model. This change suggests an adjustment of the model in response to the increased variability in the data introduced by the change in the distribution of the training, validation, and testing sets. It is important to note that a relatively simple structure is maintained despite this increase in the capacity of the model. The batch size was kept constant at 199, indicating that the data processed in each training iteration was adequate from the first stage. However, the training epochs decreased slightly to 15, suggesting that the model was able to fit the data more efficiently in the second stage despite potential additional complexities. One of the most notable changes was the learning rate, which increased from 0.0117 to 0.0177. This increase in the speed at which the model adjusts its weights is an effort to speed up the training process and achieve faster convergence. Figure 5 shows the final block diagram for each of the three trained models.

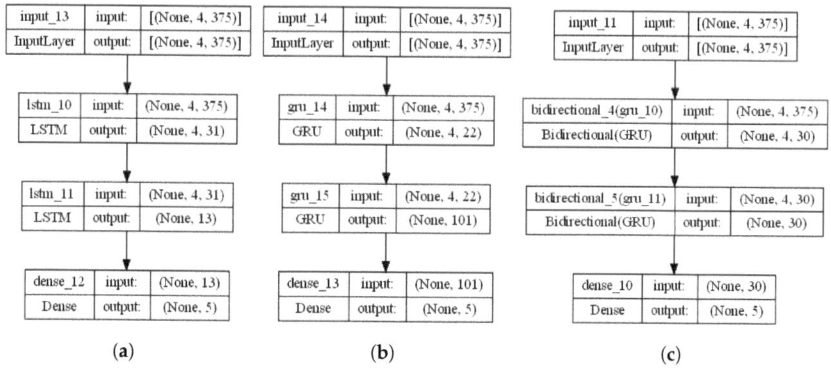

Figure 5. Final block diagram for the three trained and adjusted models, (**a**) LSTM network, (**b**) GRU, and (**c**) bidirectional, for second stage.

Table 11 shows the temporal analysis of the different architectures of recurrent neural networks studied in stage two.

Table 11. Training, validation, and prediction times for LSTM, GRU, and bidirectional neural network for the second stage.

Model	Training Time (s)	Validation Time (s)	Prediction Time (ms)
LSTM	52.76	1	0.21
GRU	57.90	1.2	0.24
Bidirectional	115.16	2.6	0.34

In evaluating the training, validation, and prediction times of the different recurrent neural network architectures, distinctive patterns and significant changes are observed between the two stages of the study. The LSTM model proved the most time efficient, with 31.47 s for training, 0.81 s for validation, and 0.12 ms for predictions. However, in the second stage, these times increased, recording 52.76 s, 1 s, and 0.21 ms, respectively. This increase can be attributed to increased training epochs, which implies a higher computational cost.

On the other hand, despite being generally slower than the LSTM, the GRU model maintained reasonable times and experienced a less pronounced increase between the two stages. In the first stage, the GRU recorded 51.28 s, 0.85 s, and 0.134 ms for training, validation, and prediction, respectively, and in the second stage, these times increased to 57.90 s, 1.2 s, and 0.24 ms. This behavior may be related to the adjustments to the number of neurons and the learning rate observed in the hyperparameters.

The bidirectional neural network, for its part, showed the highest times in both stages, underlining its computationally more intensive nature due to information processing in two directions. In the first stage, the times were 81.60 s for training, 1.23 s for validation, and 0.2 ms for predictions, while in the second stage, these increased dramatically to 115.16 s, 2.6 s, and 0.34 ms, respectively. This increase in times can be justified by the increase in the complexity of the model, reflected in the number of neurons and the learning rate.

Figure 6 presents the evolution of the best solution per iteration of the GWO optimization algorithm applied to the data from the second stage. In this instance, particular behaviors can be observed in each of the neural network architectures evaluated. In the case of the LSTM network, Figure 6a, it starts with an error close to 14%, which is higher than that recorded in the first stage. However, this network shows a remarkable ability to quickly adjust its parameters, resulting in an accelerated decrease in error. This phenomenon can be attributed to the reduction in the number of individuals used in the training and validation phases, decreasing the variability in these datasets and facilitating the network learning process. On the GRU network side, shown in Figure 6b, a similar initial behavior is observed in both stages, with a comparable starting error. However, during the second stage, the decrease in error manifests itself more gradually, reaching a minimum in the fourth iteration for both phases of the experiment. Finally, in Figure 6c, the bidirectional network presents a less abrupt error decay during the second stage, reaching a minimum error in iteration 9. This contrasts with the first phase, where the minimum error is achieved in iteration 6.

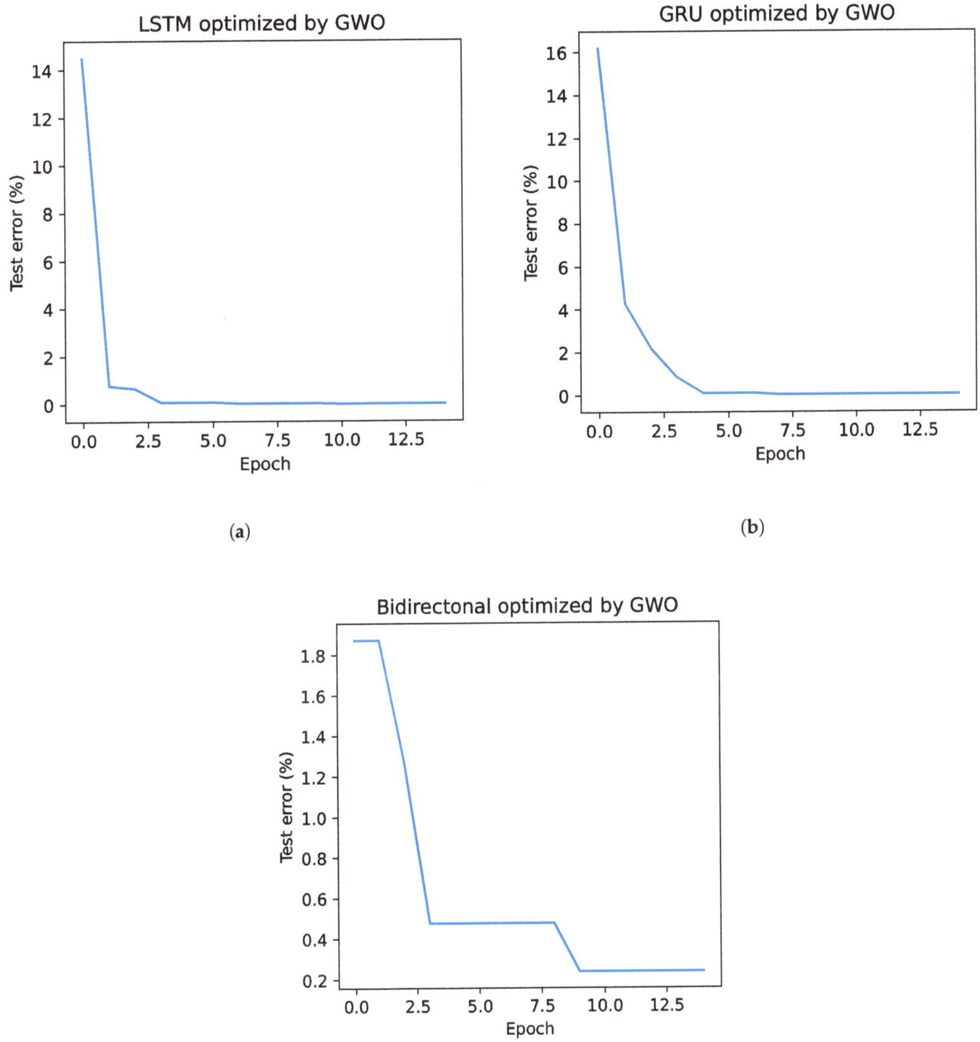

(a) (b) (c)

Figure 6. Reduction in the error due to the selection of hyperparameters by GWO in the second stage. Where (**a**) represents the error in the LSTM, (**b**) in the GRU, and (**c**) in the bidirectional network.

Figure 7 illustrates an encouraging behavior of the models during the training and validation phases, highlighting the absence of overfitting, since a concurrent increase in precision is observed for both phases. However, it is particularly interesting to note the peculiar behavior of the bidirectional network between iterations 8 and 11, where a brief decrease in percentage accuracy is experienced, as shown in Figure 7c. This small valley in accuracy could be attributed to a slightly high learning rate, which could have caused oscillations in model convergence. However, the crucial thing to highlight is the ability of the bidirectional network to recover, eventually achieving a classification close to 99%. This demonstrates the notable resilience and robustness of the model, highlighting its ability to overcome temporary setbacks in training and improve its accuracy.

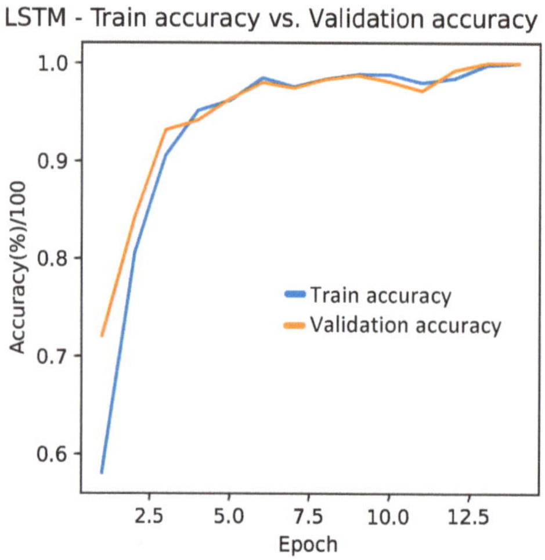

(a)

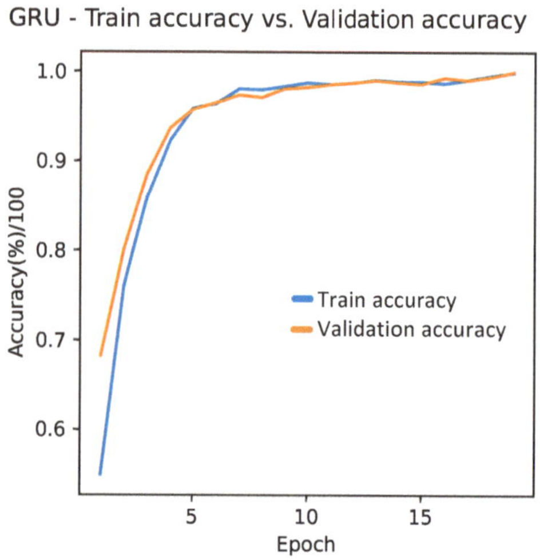

(b)

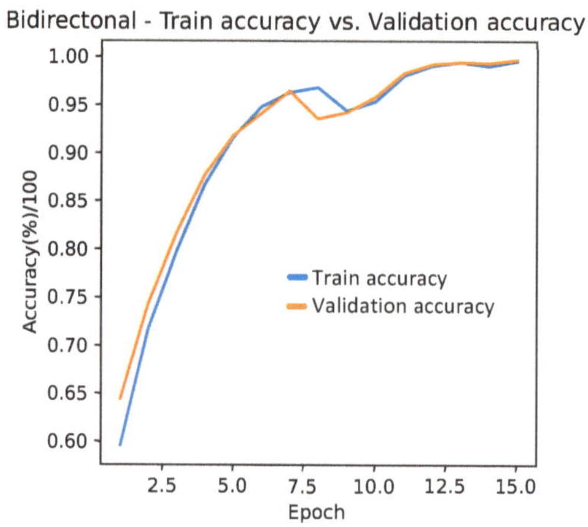

(c)

Figure 7. Evolution of training and validation accuracy with hyperparameters defined by GWO. Where (**a**) represents the evolution of the accuracy in the LSTM network, (**b**) in the GRU network, and (**c**) in the bidirectional network.

Table 12 summarizes the precision achieved by each model in the testing phase for each experimental stage. During the first stage, the LSTM, GRU, and bidirectional models achieved an impressive 100% accuracy, highlighting their ability to capture and learn from the complexity of arm movement patterns based on EMG signals. However, in the transition to the second stage, a slight decrease in the accuracy of all models was observed. The LSTM model, initially achieving classification perfection, experienced a slight drop, recording 98.46% accuracy. This decrease is attributed to the variability introduced by the new distribution of individuals in the training and testing phases. For its part, despite having maintained an accuracy of 100% in the first stage, the GRU model showed a more pronounced decrease in the second, reaching 96.38% accuracy. This reduction is due to its simpler structure compared to the LSTM, making it more susceptible to variations in the data. Despite its ability to process information in both directions and capture more complex contexts, the bidirectional network was not immune to variability between stages and experienced a decrease in accuracy, registering 97.63% in the second stage. Although this decrease is notable, the bidirectional network managed to maintain relatively high performance, demonstrating its robustness and ability to adapt.

Table 12. Accuracy of LSTM, GRU, and bidirectional neural network for testing. The accuracy for the first stage was 100% for all models.

Model	Second Stage Accuracy
LSTM	98.46%
GRU	96.38%
Bidirectional	97.63%

The results presented in Table 13 reveal the performance of the three implemented models regarding sensitivity and specificity across five different classes. In general, all models exhibit high sensitivity and specificity in all classes, with most values exceeding 95%. This demonstrates a strong ability of the models to identify instances of each class (sensitivity) correctly and to properly exclude instances that do not belong to that class (specificity). Furthermore, there is notable consistency in performance across different classes for each model, suggesting good generalization of the models across various classification conditions.

Analyzing each model individually, LSTM achieves the highest sensitivity and specificity rates in almost all classes for values greater than 97%. On the other hand, although achieving a sensitivity and specificity of 100% in classes 1 and 3, respectively, the GRU model shows slightly lower performance in other classes compared to the LSTM, being more notable in classes 4 and 5, where the sensitivity drops below 95%. The bidirectional model shows behaviors similar to LSTM and GRU. Regarding the analysis by class, classes 1 to 4 are those that the three models most accurately identify. However, class 5 is the most challenging regarding sensitivity, especially for the GRU model. This could suggest greater complexity or similarity to other classes that make their precise identification difficult.

Table 13. Sensitivity (sens) and specificity (spec) of LSTM, GRU, and bidirectional models for the different movements.

Model	Class 1		Class 2		Class 3		Class 4		Class 5	
	Sens (%)	Spec (%)	Sens (%)	Spec (%)	Sens (%)	Spec (%)	Sens (%)	Spec (%)	Sens (%)	Spec (%)
LSTM	99.6	99.9	98.5	99.7	99.9	99.9	98.0	99.5	97.5	99.2
GRU	100.0	99.8	98.0	98.3	99.8	100.0	94.6	99.4	91.4	98.3
Bidirectional	98.4	99.7	95.4	99.8	99.0	99.7	98.1	99.0	96.0	98.5

6. Discussion

This study conducted a meticulous comparative analysis between various recurrent neural network architectures, including LSTM, GRU, and bidirectional, evaluating crucial aspects such as accuracy, training times, and testing and prediction capabilities. Using the GWO optimization method to tune the hyperparameters, exceptional accuracy was achieved during the evaluation stage, reaching 100% on all RNN models during the first experimental phase. These results highlight the effectiveness of RNNs in processing EMG data with minimal preprocessing. However, when advancing to the second experimental phase, a decrease in precision was observed, obtaining 98.46% for LSTM, 96.38% for GRU, and 97.63% for bidirectional. Despite this level of reduction, the RNN models continued to demonstrate outstanding performance, underscoring their robustness and reliability in the classification task. Likewise, a nonlinear relationship is observed between the number of neurons and the computational complexity of the networks. Although intuitively this can be interpreted as, the greater the number of neurons, the greater the complexity, this study revealed that this is not always the case. Despite having the smallest number of neurons, bidirectional recurrent neural networks proved to be as complex as LSTM and more complex than GRU in terms of trainable parameters.

In this study, training, validation, and prediction times varied significantly between recurrent neural network architectures. The LSTM model stands out for its temporal efficiency, indicating more agile processing. On the other hand, the GRU and bidirectional models show longer times, which suggests a greater demand on processing resources, possibly due to more elaborate structures and adjustments in their hyperparameters. These differences reflect how each architecture handles tasks, providing insight into their operation and efficiency in different scenarios.

It is worth highlighting, however, some limitations of the present study. Although an extensive set of 56,700 data windows was available, these come from only nine individuals, raising questions about the generalization capacity of the models. This aspect highlights the need to increase the dataset by including information from a more diverse group of participants to strengthen the validity of the inferences made. Regarding the sensor configuration, only four sensors were used to differentiate five different movements. Based on previous work [10], this opens a field for future research, exploring how RNNs could behave under an even greater variety of movements with a reduced set of sensors. The discrimination capacity of RNNs in these circumstances constitutes a promising and highly relevant line of research.

Finally, expanding the database and considering the multiple solutions GWO can offer is essential. Although they converge towards a common goal, these solutions have unique characteristics, which could allow diversification in robustness, speed, and complexity, among others. Regarding SVM, a widely adopted technique in classification, it is essential to highlight its limitations. Despite its ease of use and adaptability, SVM requires a prior feature extraction stage, which can significantly lengthen the total classification time and affect its accuracy compared to RNNs. This work has not only shed light on the potential of recurrent neural networks in EMG data classification but has also pointed out important directions for future research, especially regarding the optimization and adaptability of the models according to the specific requirements of each application.

7. Conclusions

In this study, an in-depth analysis has been carried out on the effectiveness of recurrent neural networks, focusing on LSTM, GRU, and bidirectional architectures, for the EMG signal classification task. The applied methodology, complemented by optimizing hyperparameters using the GWO algorithm, has allowed us to achieve outstanding results, reaching 100% precision in the evaluation stage during the first experimental phase. RNNs, compared to traditional SVM models, show greater versatility in handling input data. This advantage is because recurrent neural networks can directly process data sequences after preprocessing, eliminating the need for specific feature extraction. This comparison highlights the advantage of using RNNs for EMG signal analysis, underscoring their ability to capture and learn from temporal sequences in the data, a limitation in models like SVM.

However, when moving to the second experimental phase, a slight decrease in the accuracy of the models was noticed: LSTM obtained 98.46%, GRU 96.38%, and bidirectional 97.63%. Although these results indicate a slight drop in performance, they are still remarkably high and demonstrate the robustness of recurrent neural networks in the task in question. This variation in results can be attributed to differences in training and validation settings between the two experimental phases, as well as the intrinsic nature of the data. It highlights the importance of the careful selection and tuning of hyperparameters specifically tailored to the characteristics of each dataset and stage of the experiment.

Despite the challenges and decreased accuracy observed in the second phase, RNN-based models have proven robust and practical tools for arm movement classification from EMG signals, maintaining outstanding performance throughout the experiment.

Author Contributions: Conceptualization, M.A.; methodology, M.A.; software, M.A. and J.M.A.-A.; validation, M.A., J.M.A.-A., J.-B.R.-O., J.R.-R. and P.Y.S.-C.; formal analysis, M.A.; investigation, M.A.; resources, J.R.-R.; data curation, M.A.; writing—original draft preparation, M.A. and J.M.A.-A.; writing—review and editing, J.M.A.-A., J.R.-R., J.-B.R.-O. and P.Y.S.-C.; visualization, M.A.; supervision, J.R.-R. All authors have read and agreed to the published version of the manuscript.

Funding: This research received no external funding.

Institutional Review Board Statement: Not applicable.

Informed Consent Statement: Not applicable.

Data Availability Statement: Data are contained within the article.

Acknowledgments: We sincerely thank the National Council of Science and Technology (CONACYT) for the invaluable support provided through the doctoral scholarship. This recognition and financial support have been fundamental for completing this research and the professional development of the beneficiary. We extend our gratitude for your commitment to promoting science and training new researchers in our country.

Conflicts of Interest: The authors declare no conflicts of interest.

References

1. Disselhorst-Klug, C.; Williams, S. Surface Electromyography Meets Biomechanics: Correct Interpretation of sEMG-Signals in Neuro-Rehabilitation Needs Biomechanical Input. *Front. Neurol.* **2020**, *11*, 603550. [CrossRef] [PubMed]
2. Song, T.; Yan, Z.; Guo, S.; Li, Y.; Li, X.; Xi, F. Review of sEMG for Robot Control: Techniques and Applications. *Appl. Sci.* **2023**, *13*, 9546. [CrossRef]
3. Suddaby, A.; Iqbal, J. EMG Controlled Modular Prosthetic Hand–Design and Prototyping. In Proceedings of the International Congress on Information and Communication Technology, London, UK, 20–23 February 2023; pp. 97–107.
4. Zhang, Z.; He, C.; Yang, K. A Novel Surface Electromyographic Signal-Based Hand Gesture Prediction Using a Recurrent Neural Network. *Sensors* **2020**, *20*, 3994. [CrossRef] [PubMed]
5. Xiu, K.; Xiafeng, Z.; Le, C.; Dan, Y.; Yixuan, F. EMG Pattern Recognition based on Particle Swarm Optimization and Recurrent Neural Network. *Int. J. Perform. Eng.* **2020**, *16*, 1404. [CrossRef]
6. Li, S.; Zhang, Y.; Tang, Y.; Li, W.; Sun, W.; Yu, H. Real-Time sEMG Pattern Recognition of Multiple-Mode Movements for Artificial Limbs Based on CNN-RNN Algorithm. *Electronics* **2023**, *12*, 2444. [CrossRef]
7. Song, Q.; Ma, X.; Liu, Y. Continuous online prediction of lower limb joints angles based on sEMG signals by deep learning approach. *Comput. Biol. Med.* **2023**, *163*, 107124. [CrossRef]
8. Aviles, M.; Rodríguez-Reséndiz, J.; Ibrahimi, D. Optimizing EMG Classification through Metaheuristic Algorithms. *Technologies* **2023**, *11*, 87. [CrossRef]
9. Gaspar, A.; Oliva, D.; Cuevas, E.; Zaldívar, D.; Pérez, M.; Pajares, G. Hyperparameter Optimization in a Convolutional Neural Network Using Metaheuristic Algorithms. In *Metaheuristics in Machine Learning: Theory and Applications*; Springer: Berlin/Heidelberg, Germany, 2021; pp. 37–59.
10. Aviles, M.; Sánchez-Reyes, L.M.; Fuentes-Aguilar, R.Q.; Toledo-Pérez, D.C.; Rodríguez-Reséndiz, J. A novel methodology for classifying EMG movements based on SVM and genetic algorithms. *Micromachines* **2022**, *13*, 2108. [CrossRef]
11. Mirjalili, S.; Mirjalili, S.M.; Lewis, A. Grey Wolf Optimizer. *Adv. Eng. Softw.* **2014**, *69*, 46–61. [CrossRef]
12. Rezoug, A.; Iqbal, J.; Tadjine, M. Extended grey wolf optimization–based adaptive fast nonsingular terminal sliding mode control of a robotic manipulator. *J. Syst. Control Eng.* **2022**, *236*, 1738–1754. [CrossRef]
13. Smagulova, K.; James, A.P. A survey on LSTM memristive neural network architectures and applications. *Eur. Phys. J. Spec. Top.* **2019**, *228*, 2313–2324. [CrossRef]
14. Jaouedi, N.; Boujnah, N.; Bouhlel, M.S. Deep Learning Approach for Human Action Recognition Using Gated Recurrent Unit Neural Networks and Motion Analysis. *J. Comput. Sci.* **2019**, *15*, 1040–1049. [CrossRef]
15. Chadha, G.S.; Panambilly, A.; Schwung, A.; Ding, S.X. Bidirectional deep recurrent neural networks for process fault classification. *ISA Trans.* **2020**, *106*, 330–342. [CrossRef] [PubMed]
16. Xie, B.; Meng, J.; Li, B.; Harland, A. Biosignal-based transferable attention Bi-ConvGRU deep network for hand-gesture recognition towards online upper-limb prosthesis control. *Comput. Methods Programs Biomed.* **2022**, *224*, 106999. [CrossRef] [PubMed]

17. Toro-Ossaba, A.; Jaramillo-Tigreros, J.; Tejada, J.C.; Peña, A.; López-González, A.; Castanho, R.A. LSTM Recurrent Neural Network for Hand Gesture Recognition Using EMG Signals. *Appl. Sci.* **2022**, *12*, 9700. [CrossRef]
18. Sahu, P.; Singh, B.K.; Nirala, N. An improved feature selection approach using global best guided Gaussian artificial bee colony for EMG classification. *Biomed. Signal Process. Control* **2023**, *80*, 104399. [CrossRef]
19. Simão, M.; Neto, P.; Gibaru, O. EMG-based online classification of gestures with recurrent neural networks. *Pattern Recognit. Lett.* **2019**, *128*, 45–51. [CrossRef]
20. Xiong, D.; Zhang, D.; Zhao, X.; Zhao, Y. Hand gesture recognition using instant high-density EMG graph via deep learning method. In Proceedings of the 2020 Chinese Automation Congress (CAC), Shanghai, China, 6–8 November 2020; pp. 5143–5148.
21. Sohane, A.; Agarwal, R. A single platform for classification and prediction using a hybrid bioinspired and deep neural network (PSO-LSTM). *Mapan* **2022**, *37*, 47–58. [CrossRef]
22. Zhu, M.; Guan, X.; Li, Z.; He, L.; Wang, Z.; Cai, K. sEMG-based lower limb motion prediction using CNN-LSTM with improved PCA optimization algorithm. *J. Bionic Eng.* **2023**, *20*, 612–627. [CrossRef]
23. Hermens, H.; Freriks, B.; Merletti, D.; Stegeman, J.; Blok, G.; Rau, C.; Disselhorst-Klug, C.; Hägg, G. *European Recommendations for Surface ElectroMyoGraphy, Results of the SENIAM Project*; Roessingh Research and Development: Enschede, The Netherlands, 1999; ISBN 90-75452-15-2.
24. Paramasivan, S.K. Deep Learning Based Recurrent Neural Networks to Enhance the Performance of Wind Energy Forecasting: A Review. *Rev. Intell. Artif.* **2021**, *35*, 1–10. [CrossRef]
25. Bakhashwain, N.; Sagheer, A. Online Tuning of Hyperparameters in Deep LSTM for Time Series Applications. *Int. J. Intell. Eng. Syst.* **2021**, *14*, 212–220. [CrossRef]
26. Fan, Y.; Zhang, Y.; Guo, B.; Luo, X.; Peng, Q.; Jin, Z. A Hybrid Sparrow Search Algorithm of the Hyperparameter Optimization in Deep Learning. *Mathematics* **2022**, *10*, 3019. [CrossRef]
27. Gong, X.; Zhi, Z.; Feng, K.; Du, W.; Wang, T. Improved DCNN Based on Multi-Source Signals for Motor Compound Fault Diagnosis. *Machines* **2022**, *10*, 277. [CrossRef]
28. Asogbon, M.G.; Samuel, O.W.; Jiang, Y.; Wang, L.; Geng, Y.; Sangaiah, A.K.; Chen, S.; Fang, P.; Li, G. Appropriate Feature Set and Window Parameters Selection for Efficient Motion Intent Characterization towards Intelligently Smart EMG-PR System. *Symmetry* **2020**, *12*, 1710. [CrossRef]
29. Ashraf, H.; Waris, A.; Gilani, S.O.; Kashif, A.S.; Jamil, M.; Jochumsen, M.; Niazi, I.K. Evaluation of windowing techniques for intramuscular EMG-based diagnostic, rehabilitative and assistive devices. *J. Neural Eng.* **2021**, *18*, 016017. [CrossRef]
30. Zhang, H.; Su, M. Hand Gesture Recognition of Double-Channel EMG Signals Based on Sample Entropy and PSO-SVM. *J. Phys. Conf. Ser.* **2020**, *1631*, 012001. [CrossRef]
31. Toledo-Pérez, D.C.; Rodríguez-Reséndiz, J.; Gómez-Loenzo, R.A.; Jauregui-Correa, J.C. Support Vector Machine-Based EMG Signal Classification Techniques: A Review. *Appl. Sci.* **2019**, *9*, 4402. [CrossRef]

Disclaimer/Publisher's Note: The statements, opinions and data contained in all publications are solely those of the individual author(s) and contributor(s) and not of MDPI and/or the editor(s). MDPI and/or the editor(s) disclaim responsibility for any injury to people or property resulting from any ideas, methods, instructions or products referred to in the content.

bioengineering

MDPI

Article

Walking with a Posterior Cruciate Ligament Injury: A Musculoskeletal Model Study

Lucia Donno [1,*], Alessandro Galluzzo [2,3], Valerio Pascale [2,4], Valerio Sansone [2,5] and Carlo Albino Frigo [1]

1 Movement Biomechanics and Motor Control Lab, Department of Electronics, Information and Bioengineering, Politecnico di Milano, I-20133 Milan, Italy; carlo.frigo@polimi.it
2 IRCCS Istituto Ortopedico Galeazzi, I-20161 Milan, Italy; alegalluzzo@gmail.com (A.G.); valerio.pascale@unimi.it (V.P.); valerio.sansone@unimi.it (V.S.)
3 Residency Program in Orthopaedics and Traumatology, University of Milan, I-20122 Milan, Italy
4 Department of Biomedical Sciences for Health, University of Milan, I-20122 Milan, Italy
5 Department of Biomedical, Surgical and Dental Sciences, University of Milan, I-20122 Milan, Italy
* Correspondence: lucia.donno@polimi.it

Abstract: The understanding of the changes induced in the knee's kinematics by a Posterior Cruciate Ligament (PCL) injury is still rather incomplete. This computational study aimed to analyze how the internal loads are redistributed among the remaining ligaments when the PCL is lesioned at different degrees and to understand if there is a possibility to compensate for a PCL lesion by changing the hamstring's contraction in the second half of the swing phase. A musculoskeletal model of the knee joint was used for simulating a progressive PCL injury by gradually reducing the ligament stiffness. Then, in the model with a PCL residual stiffness at 15%, further dynamic simulations of walking were performed by progressively reducing the hamstring's force. In each condition, the ligaments tension, contact force and knee kinematics were analyzed. In the simulated PCL-injured knee, the Medial Collateral Ligament (MCL) became the main passive stabilizer of the tibial posterior translation, with synergistic recruitment of the Lateral Collateral Ligament. This resulted in an enhancement of the tibial–femoral contact force with respect to the intact knee. The reduction in the hamstring's force limited the tibial posterior sliding and, consequently, the tension of the ligaments compensating for PCL injury decreased, as did the tibiofemoral contact force. This study does not pretend to represent any specific population, since our musculoskeletal model represents a single subject. However, the implemented model could allow the non-invasive estimation of load redistribution in cases of PCL injury. Understanding the changes in the knee joint biomechanics could help clinicians to restore patients' joint stability and prevent joint degeneration.

Keywords: musculoskeletal modeling; PCL injury; knee joint biomechanics; knee ligaments

Citation: Donno, L.; Galluzzo, A.; Pascale, V.; Sansone, V.; Frigo, C.A. Walking with a Posterior Cruciate Ligament Injury: A Musculoskeletal Model Study. *Bioengineering* **2023**, *10*, 1178. https://doi.org/10.3390/bioengineering10101178

Academic Editors: Massimiliano Pau and Philippe Gorce

Received: 2 September 2023
Revised: 5 October 2023
Accepted: 9 October 2023
Published: 11 October 2023

Copyright: © 2023 by the authors. Licensee MDPI, Basel, Switzerland. This article is an open access article distributed under the terms and conditions of the Creative Commons Attribution (CC BY) license (https://creativecommons.org/licenses/by/4.0/).

1. Introduction

The knee joint is a remarkably complex system from both the anatomical and functional points of view, with very specific biomechanical requirements. In the sagittal plane, it allows a wide range of flexion/extension in the presence of major loads resulting from body weight and inertia forces. In the coronal plane, it provides a high degree of stability throughout the full range of motion, and concerning the transverse plane it allows internal/external rotation when the knee is flexed. All these functional properties result from the specific morphology of the joint contact surfaces and the presence of ligaments.

In recent decades, plenty of literature was published regarding the role of cruciate ligaments and the effects of ligament deficiency in anterior–posterior stability, both in the native and the prosthetic knee. Most of the scientific attention was devoted to the Anterior Cruciate Ligament (ACL), probably due to the particular epidemiology of knee joint injuries, showing that the ACL is affected significantly more frequently than other fibrocartilaginous and ligamentous structures [1]. In contrast, the Posterior Cruciate Ligament (PCL) has been less investigated, despite being a pivotal structure within proper knee biomechanics. This ligament is an intra-articular extra-synovial cord of dense connective tissue, consisting of two distinct bundles, i.e., anterolateral and posteromedial bundles [2]. It plays a crucial role in resisting posterior tibial translation relative to the femur during knee motion, and recent literature enlightens the importance of this structure in maintaining the rotational stability of the knee, especially between 90° and 120° of flexion [3–6].

The role of the PCL is crucial also in the choice between cruciate-retaining and posterior-stabilized prosthetic implants [7]. If the cruciate-retaining implants could better preserve the physiological kinematics of the knee [8,9], their application becomes challenging in cases of PCL laxity due to the problematic balancing of the ligaments [10]. However, in the case of cruciate-retaining implants, recent findings showed that the mobile bearing design could ensure the stability of the knee joint even in case of PCL deficiency [11].

Although these findings are fairly well accepted in the scientific community, the understanding of the changes induced in the knee kinematics by a PCL injury is still rather incomplete. The large cross-sectional area of this ligament would lead us infer that it must bear significant loads, but in practice the functional impairments induced by an isolated injury of this ligament are essentially limited. In fact, many surgeons prefer to treat isolated injuries of the PCL conservatively. The PCL retention could influence not only the knee kinematics and stability, but also the shear forces on the tibial contact surface and the proprioception [12–14]. The surgical treatment of ligament reconstruction is therefore reserved for rare cases of significant and persistent functional limitations. The relative paucity of symptoms suggests that there are intrinsic forms of compensation that the joint can exploit when a ligament injury occurs. One of these could be the greater healing capacity of the PCL compared to the ACL as reported in the literature, probably due to a better synovial coverage and perfusion [15]. Patients with isolated PCL tears may have few functional problems in the short-medium term, but there is growing literature enlightening the detrimental effects of PCL deficiency in terms of acute or degenerative injury of the remaining joint structures and development of early arthrosis [16–18].

In general, the analysis of weightbearing activities (i.e., walking and climbing stairs) is limited in the literature. Specifically, for the task of walking, the effects of PCL lesion remain unclear and are worthy of further investigation [19]. Understanding the changes in the knee joint biomechanics could help clinicians to restore patients' joint stability and prevent joint degeneration. The inherent complexity of the knee joint and the close interdependence between the various elements makes the study of individual anatomical structures and their pathophysiology particularly difficult, and this applies specifically to PCL. In addition, this ligament injury is relatively infrequent, and the isolated rupture is rare. In an observational study of 85 PCL injuries, Fanelli G.C. et al. [20] reported that isolated PCL ruptures were no more than 3.5% (3 of 85), whereas 95.5% (82 of 85) of PCL injuries occurred in combination with other ligament lesions. This indicates the difficulty of analyzing the effect of this particular lesion on the knee biomechanics while also having regard to the dynamic interaction with other articular structures.

Computational musculoskeletal models effectively overcome this difficulty by enabling the researcher to observe the effect of a single-element modification on the overall joint motion and biomechanics. This is one of the main reasons why research on musculoskeletal models has grown exponentially (the number of scientific publications has increased by nearly 1500% in the last 20 years). Musculoskeletal models have proven to be fundamental tools for improving our understanding of the musculoskeletal system and related pathological conditions [21,22]. For example, in a previous study [23], the effects on knee joint kinematics and internal load distribution produced by removal of the ACL were simulated with reference to the walking cycle. In that study it was also possible to investigate the effects of different muscle contractions, the compensatory role of quadriceps and hamstrings in particular.

Kang et al. [24] simulated a PCL deficiency in a subject-specific knee model through a force-dependent kinematics method under gait and squat loading conditions. The cited authors found out that the PCL deficiency affects the contact forces of the tibial–femoral and patellar–femoral joints less during the gait cycle with respect to the squat loading condition. However, this study did not focus on the effects of PCL injury on the knee kinematics under dynamic conditions, but only on passive flexion and posterior drawer tests. Moreover, to our best knowledge, no other studies investigated on the impact of different degrees of PCL lesion on knee kinematics during walking.

In the present study, several dynamic simulations of the gait cycle were performed aimed at answering to two main questions: (i) how are the internal loads redistributed among the remaining joint ligaments when the PCL is lesioned at different degrees? (ii) is there a possibility to compensate for a PCL lesion by changing the hamstring's contraction in a particular phase of the stride cycle?

2. Materials and Methods

2.1. Modeling Approach

The dynamic simulations of the gait cycle were performed by means of a previously developed three-dimensional musculoskeletal model implemented on the SimWise-4D platform (Design Simulation Technologies, DST, Canton, MI, USA) [25]. This software performs both the forward and the inverse dynamics of complex articulated systems and solves the mechanical interaction between body surfaces. Figure 1 represents a schematic overview of the numerical procedure. The integration algorithm used to solve the dynamic equilibrium equations was based on the Kutta–Merson process. The following parameters were adopted: integration step 0.02 s, configuration tolerance 0.01 mm and 0.1°.

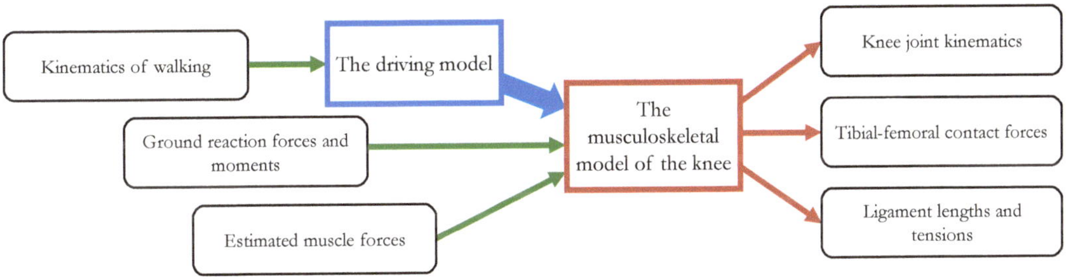

Figure 1. Schematic overview of the numerical research procedure.

The implemented model included a "driving model", composed of trunk, pelvis, thighs, shanks and feet, and a detailed "knee joint model" attached to it. The anatomical segments in the driving model, represented by rigid bodies, were linked together by rotational actuators (motors), by means of which it was possible to impose the desired kinematics, in our case the walking cycle (Figure 2). The pelvis was linked to the trunk by three rotational actuators controlling 3 degrees of freedom (d.o.f): anteversion/retroversion, frontal plane tilt and horizontal rotation. Three other motors were used to control the relative movement between the thigh and the pelvis (hip flexion/extension, adduction/abduction, internal/external rotation). Similar actuators controlled the 2 d.o.f between the shank and the thigh (knee flexion/extension and internal/external rotation) and the 2 d.o.f between the foot and the shank (ankle plantar/dorsiflexion and pronation/supination).

All these motors received as input the kinematic data from our repository, referring to the average of 14 strides recorded on 5 healthy male subjects aged between 24 and 36 years, walking barefoot at their self-selected speed. These data (trajectory of a reference point on the trunk and joint angles corresponding to the d.o.f of the model) were obtained by applying the SAFLo (Servizio di Analisi della Funzionalità Locomotoria) marker set protocol [26] and were normalized in time so as to obtain an ideal walking cycle lasting 1 s. To ensure that the results were obtained at steady state condition, two subsequent walking cycles were simulated, and only the second one was analyzed. Anthropometric tables [27] provided masses of body segments corresponding to a healthy subject with a height of 1.72 m and 70 kg of body mass.

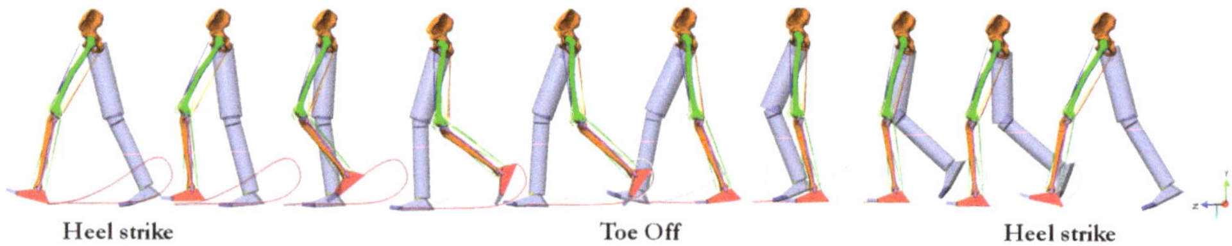

Figure 2. The walking model simulating the gait cycle (from right to left). Gait cycles events (heel strikes and toe off) are referred to the left lower limb, including the detailed knee joint model.

The knee joint model was composed of the digital models of the femur and tibia, obtained by digital segmentation of MRI images (Amira 5.3.3, Visage Imaging, Inc., San Diego, CA, USA) of a 42 year old Caucasian male who was 1.72 m tall and had 70 kg of body mass. The two femoral condyles were in contact with the tibial plateaus and could slide along the contact surfaces. In order to obtain a proper sliding between the contact surfaces, the portions of the distal femur and of the proximal tibia were separated from the rest of the bones and their contact surfaces were smoothed by using a commercial software (Meshmixer 3.5.47, Autodesk, San Rafael, CA, USA). Then, for the purpose of reducing the simulation time, the two portions were imported again in the software SimWise4D and connected to the respective bones by rigid constraints; thereby, the algorithms analyzing the surface interaction had to work only on these small portions of the bones. Specifically, a friction coefficient equal to 0.01 between the contact surfaces of the tibia and the femur was set [28] and their interaction was modeled as an inelastic collision. The two bones were connected by springs representing the knee joint ligaments. The knee joint model was introduced in the driving model by rigidly connecting the femur to the thigh. A specially designed device called a "Grood&Suntay mechanism (G&S)" (the name comes from the convention adopted for the definition of the knee joint d.o.f. [29]) was used to impose the flexion/extension movement of the knee while keeping the other 5 d.o.f. unconstrained. Thus, the movement of the tibia relative to the femur was determined by the geometrical interaction between the surfaces of femoral condyles and tibial plateau, the tension produced by the ligaments, and all the forces acting on this structure (muscle forces, ground reaction forces, weight and inertial forces of shank and foot). The relative movement between the tibia and femur was measured by the G&S mechanism in terms of adduction/abduction, internal/external rotation and proximal/distal, medial/lateral and anterior/posterior tibial displacements. As for the muscle forces, an estimate was obtained by implementing an optimization procedure for the minimization of the maximum force (Min/Max criterion [30]) in relation to the muscle's physiological cross-sectional area. The muscle lever arms were obtained by analyzing the moment produced by each muscle when it was activated with a predefined muscle force. Each muscle acting on the knee joint was modeled as a force actuator producing the respective estimated concentric force. The quadriceps group was represented by four actuators corresponding to the Vastus Lateralis, Vastus Medialis, Vastus Intermedius and Rectus Femoris. The hamstrings group was considered to be composed of the Semitendinosus, Semimembranosus, Biceps Femoris long head and Biceps Femoris short head. The Gastrocnemius Lateralis and Gastrocnemius Medialis were also implemented in the model, acting as knee flexors.

The ground reaction forces (anterior–posterior, medial–lateral and vertical components) measured during walking were applied to the knee model. Through an inverse dynamics analysis, the inertia forces and moments associated with the movement of the shank and foot were computed.

A detailed ligament structure was implemented in the knee model. Each ligamentous fascicle was represented by a straight spring with non-linear behavior (viscoelastic element).

Hence, for each ligament, several springs were included to represent the different fascicles that may be tensioned in different working conditions (Figure 3). The Anterior (ACL) and Posterior (PCL) Cruciate Ligaments were subdivided into two springs each. Three springs representing the anterior, intermediate and posterior fascicles were used both to model the Lateral Collateral Ligament (LCL) and the superficial Medial Collateral Ligament (MCL). The deep fascicles of the Medial Collateral Ligament (MCL deep) were modeled by two additional springs. The fibrous capsule was also included in the knee model and was represented by its antero-lateral (Cap-Ant-L), posterior-lateral (Cap-Post-L) and posterior-medial (Cap-Post-M) bundles. The springs representing the ligaments were attached to specific points identified with the support of many publications [31–35] and a joint physiology text [36]. Each spring was characterized by a quadratic rise of the force up to a predefined strain limit (assumed equal to 0.03) and a linear behavior for larger deformations, in accordance to [31,37,38]. As regards the stiffness parameter associated to the non-linear springs, the values reported by [33] were adopted. Specifically, for those ligaments appearing as single units in the cited study but split in more bundles in our model, the stiffness of each elastic element was obtained by dividing the reported value by the number of bundles constituting the ligament.

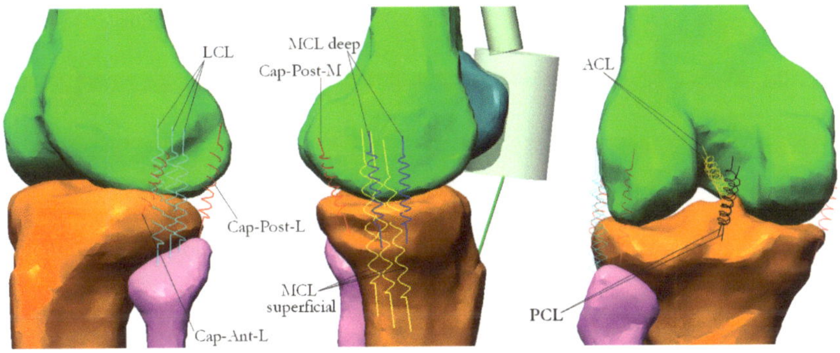

Figure 3. Springs with non-linear characteristics represented the ligaments connecting femur and tibia. On the left, Lateral Collateral Ligament and anterior and posterior lateral bundles of fibrous capsule. In the middle, the extensor mechanism, superficial and deep bundles of Medial Collateral Ligament and medial portion of fibrous capsule. On the right, Anterior and Posterior Cruciate Ligaments.

As for the extensor mechanism, to reduce the computational time the rotula was reproduced by a cylinder sliding on the femoral trochlea and having the lower extremity attached to the tibial tuberosity by a rope representing the patellar tendon (Figure 3). Its upper extremity was linked by a series of smaller cylinders reproducing the quadriceps tendon, which in turn were attached to the muscle actuators. The cylinders were connected to each other by a spherical joint. The cylinder representing the rotula was put in collision mode with the femur.

Interested readers can find an exhaustive description of the implemented model in our previous publication [25].

2.2. Model Validation

The performance of our intact knee model was assessed in detail in our previous publication [25] by looking at three major determinants of the knee function during the whole gait cycle. First of all, we considered the knee kinematics, since in our model only the flexion/extension movement was imposed and the remaining 5 d.o.f were completely free. A forward displacement of the tibia with respect to the femur was observed during each of the two knee flexion phases, the first during the stance phase and the second associated with the swing phase. This was consistent with the consolidated knowledge about the association of sliding and rotation in the tibial–femoral movement. Secondly, in the mentioned study, we observed the presence of the well-known "screw home mechanism": the tibia experienced an external rotation when approaching the full extension and, conversely, showed an internal rotation during the initial flexion of the knee.

Thirdly, we analyzed the tibial–femoral contact forces. In our model, the values obtained were larger than the force measured by instrumented prostheses [39–41], but the main features [42] were well reproduced: they exhibited two peaks during the stance phase (second peak higher than the first one) and a third peak at the end of the swing phase, ensuring the joint stability. To make a quantitative assessment, we made a comparison between our data and those provided as a reference for the Grand Challenge competition [41]: after downscaling our results, the RMS difference with respect to the reference data was 318 N. This result shows that the predicted time-course reproduces the force measured in vivo well. As for the amplitude, it must be considered that our input data for the model are probably different from those of the subjects analyzed in the reference study (operated by knee arthroplasty) in terms of anthropometric measures and walking speed. In our musculoskeletal model, the first peak reaches 3.5 BW and the second peak is equal to 4.8 BW. These values are similar to the results obtained by Hu et al. [43]: 4 BW and 4.7 BW for the first and the second peaks, respectively. Moreover, the value we predicted for the first peak lies within the range of many studies published in the literature: Winby et al. [44] reported 3.2–4.9 BW, Meireles et al. [45] obtained 3.3–4.8 BW, and in Morrison et al. [46] 2.1–4.0 BW was found.

Thus, in terms of amplitude, the tibial–femoral contact force estimated in our intact knee model was higher than the force measured in vivo using sensorized prostheses, but in agreement with several studies that estimate the amplitude through computational models.

Further comparisons were made with data in the literature considering the typical clinical functional tests aimed at assessing the ligament's condition. The knee model was subjected to the typical anterior–posterior drawer, varus–valgus and internal–external rotation tests performed under passive conditions and different flexion angles. The predicted stiffness and laxity of the knee joint were compared to the experimental results obtained by Markolf et al. [47] on cadaver specimens, and proved to be in relatively good agreement.

Finally, as for the muscle forces, our approach looked for the minimization of the maximum force referred to the muscular physiological cross-sectional area and our results were consistent with the well-known electromyographic patterns [48], as was the case for most of the published works [49–53].

2.3. *Simulation Conditions*

In this study, two sets of simulations were performed to respond to two main questions (Figure 4). The first objective was to quantify the effects of different degrees of PCL injury on the knee kinematics, ligaments load redistribution and tibial–femoral contact force. Starting from the 'IntactKnee' model, a progressive PCL injury was simulated by reducing gradually, by 10% steps, the stiffness of the two non-linear springs representing the PCL. Thus, the conditions of PCL residual stiffness from 90% to 20% were obtained. Then, in this study, the most critical condition was obtained assuming a residual stiffness of 15% of the original one. In each of the nine simulated conditions, a dynamic simulation of the gait cycle was run and the ligament's tension, contact force and knee kinematics in each PCL injury condition were analyzed. In particular, we quantified the changes occurring in the phase of maximum PCL recruitment, the second half of the swing phase, from 76% to 100% of the stride cycle.

The second objective of this study was to understand if the hamstrings muscle group could compensate for PCL injury. Hence, starting from the most critical condition (PCL residual stiffness being 15% of the original, named "PCL-15%"), the force produced by each hamstring's muscle (Semitendinosus, Semimembranosus, Biceps Femoris long head and Biceps Femoris short head) was progressively reduced by 10% steps. Four new dynamic simulations of walking were performed, corresponding to a hamstring residual force of 90%, 80%, 70% and 60% of the original, having a PCL residual stiffness at 15%. Knee kinematics, ligaments load and tibial–femoral contact force were analyzed in all these conditions as before. Specifically, the changes occurring in the time window from 76% to 100% of the gait cycle were quantified in relation to the reference condition of 15% PCL stiffness and hamstring force at 100%.

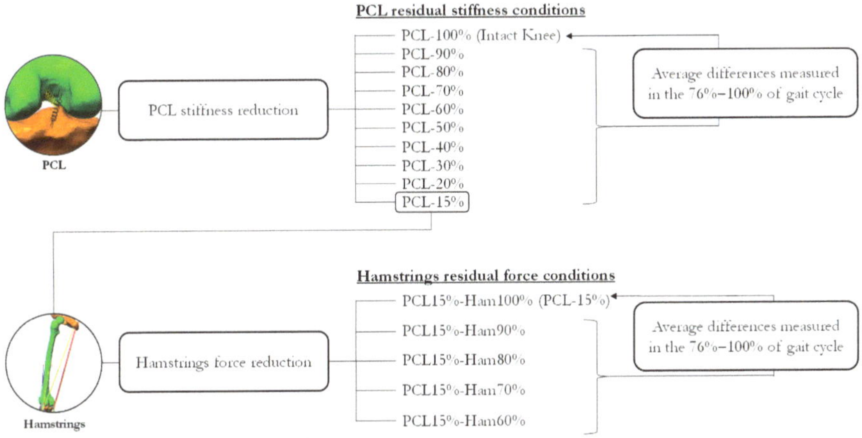

Figure 4. Workflow of the study.

3. Results

In a physiological knee joint (100% of PCL stiffness), the PCL expresses its maximum tension in the second half of the swing phase, between 76% and 100% of the stride cycle. To quantify the changes with respect to the intact knee, in each condition and for each variable, we calculated the average of the difference between the curve obtained in the altered condition and the curve computed for the intact knee in the time window 76% to 100% of the stride cycle.

Considering the posterior displacement of the tibia in relation to the femur, we realized that the most critical conditions were reached when the PCL residual stiffness was less than 50% of the physiological state. In fact, as reported in Figure 5, the posterior displacement did not increase significantly due to a decrease in stiffness up to that value.

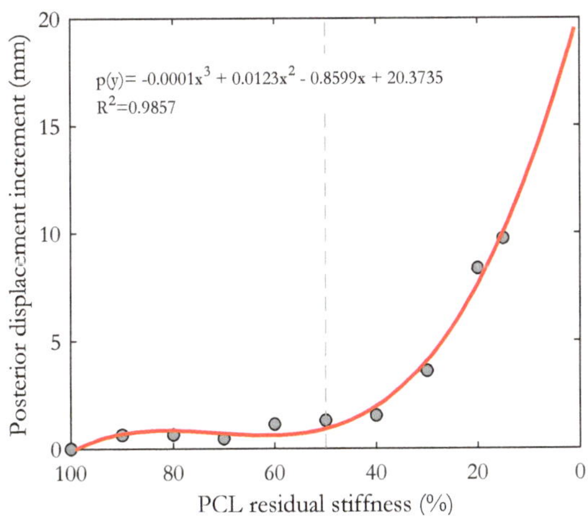

Figure 5. Mean increment of posterior tibial displacement calculated in the second half of the swing phase resulting from each simulated condition. In red, the polynomial regression function. Points to the right of the dashed line correspond to the most critical conditions.

In the following section, only the main findings obtained from the conditions of PCL residual stiffness lower than 50% will be outlined.

3.1. Residual PCL Stiffness from 50% to 15%

For the sake of clarity, in this section, the effects of PCL stiffness reduction on the knee kinematics, ligaments tension and tibial–femoral contact force will be outlined separately.

3.1.1. Effects on Anterior–Posterior Displacement of the Tibia

The progressive reduction in PCL stiffness consistently decreased the ligament tension all along the stride cycle, as shown in Figure 6a. The anterior–posterior displacement of the tibia during the gait cycle in the different PCL conditions is reported in Figure 6b. It appears that this variable was almost unaffected for the whole stance phase and the first half of the swing phase, while it exhibited a progressive increase in the backwards displacement in the second half of the swing phase, precisely when the PCL would express its maximum tension in a physiological knee (Figure 6a, black curve).

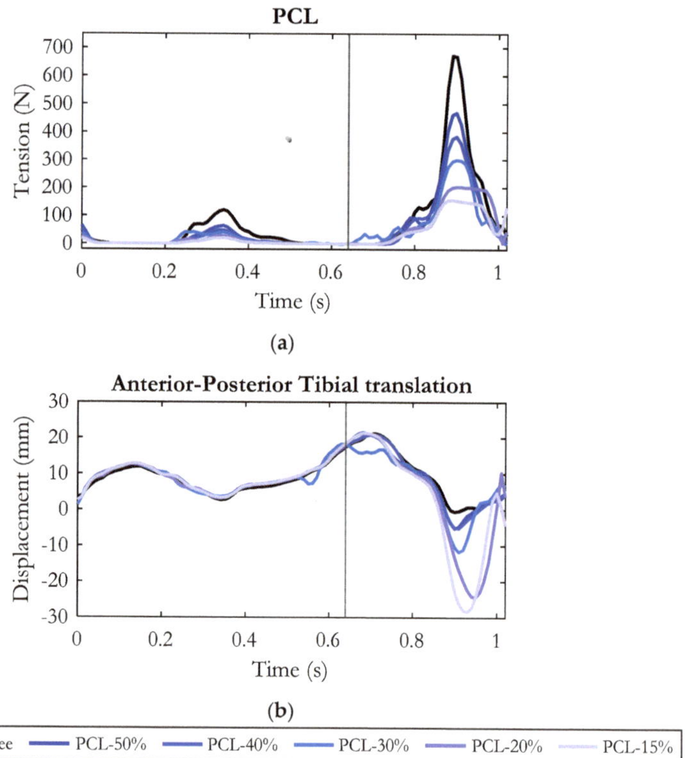

Figure 6. Tension of PCL in the simulated conditions (**a**). Anterior–posterior displacement of the tibia (**b**) during the gait cycle in the six simulated conditions. Values are positive for the anterior displacement. The gray vertical line refers to the toe–off event, marking the end of the stance phase and the beginning of the swing phase.

Specifically, in the time window considered (76% to 100% of the gait cycle), the average increase in the posterior displacement of the tibia was 1.3 mm and 1.5 mm with respect to the intact knee when the PCL residual stiffness was 50% and 40% of the healthy condition, respectively, but became 3.6 mm for a reduction in the PCL stiffness to 30% of the original one, 8 mm for a PCL stiffness of 20% and 10 mm for a PCL stiffness reduced to 15% of the original one (see also Figure 5). The peak of posterior displacement occurred at around 90% of the stride cycle, with a tendency towards lag for lower PCL stiffness.

3.1.2. Effects on the Tension of the Remaining Ligaments

Regarding the loads on the remaining ligaments, the collateral ligament structures were the most affected by the PCL injury (Figure 7). In particular, the gradual reduction in PCL stiffness resulted in a progressive increase in MCL (deep and superficial) and LCL tension specifically in the late swing phase. The average increase for a reduction in PCL stiffness to 50% and 40% compared to the intact knee condition was in both conditions about 30 N for LCL (Figure 7c). The peak of force was about 220 N and occurred at the same instant in which the injured PCL exhibited the peak of force and the tibia gained its maximum posterior displacement (90% point of the gait cycle, Figure 6).

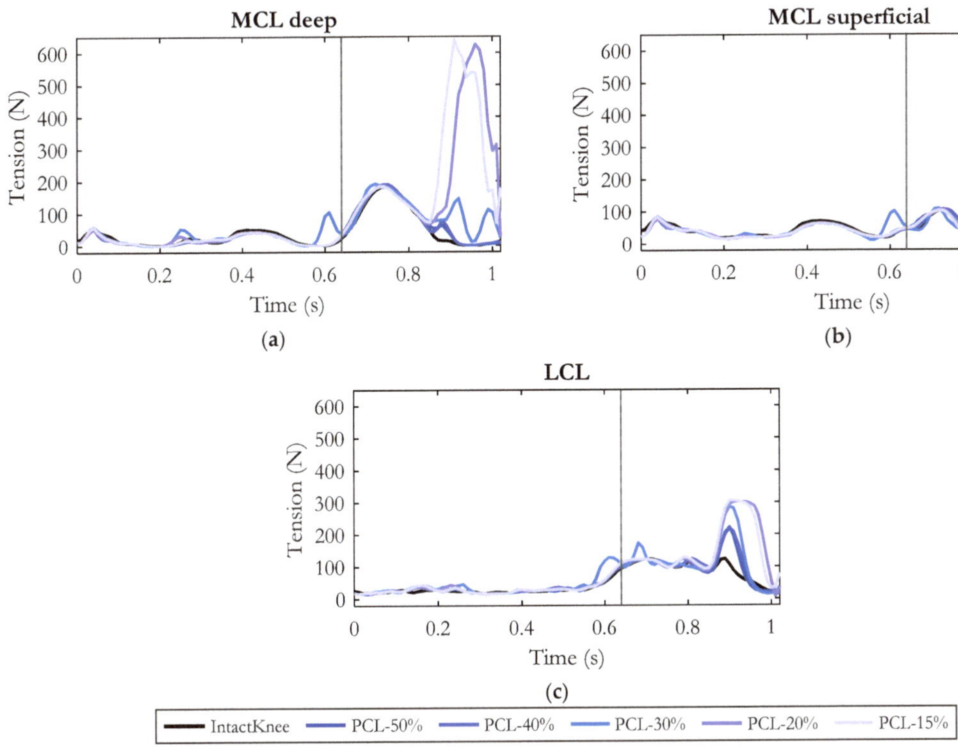

Figure 7. Tension of deep Medial Collateral Ligament (**a**), superficial Medial Collateral Ligament (**b**), Lateral Collateral Ligament (**c**) along the gait cycle in the six simulated conditions. The gray vertical line refers to the toe-off event, marking the end of the stance phase and the beginning of the swing phase.

Concerning the MCL (deep and superficial), no appreciable changes were observed until the PCL stiffness decreased to 30%. In this condition (see Figure 7a,c), a cooperation between the lateral and medial passive stabilizers was observed: the deep bundles of MCL (MCL-deep) and LCL increased their tension by approximately 50 N and 63 N on average compared to physiological conditions.

A further reduction in PCL stiffness (PCL-20%, PCL-15%) additionally demanded the recruitment of the deep and superficial layers of MCL, resulting in an average increase of 250 N and 50 N, respectively. In the same conditions, the LCL tension increased on average by 113 N.

Thanks to the combined recruitment of the passive stabilizers, at the end of the gait cycle the tibia recovered a physiological position (Figure 6b), except for the most critical condition (PCL-15%) in which the tibia remained less than 1 mm more posterior than the intact knee.

3.1.3. Effects on Tibial–Femoral Contact Force

Figure 8 provides an insight into the effects of the different degrees of PCL injury on the contact force between tibia and femur in the second half of the swing phase.

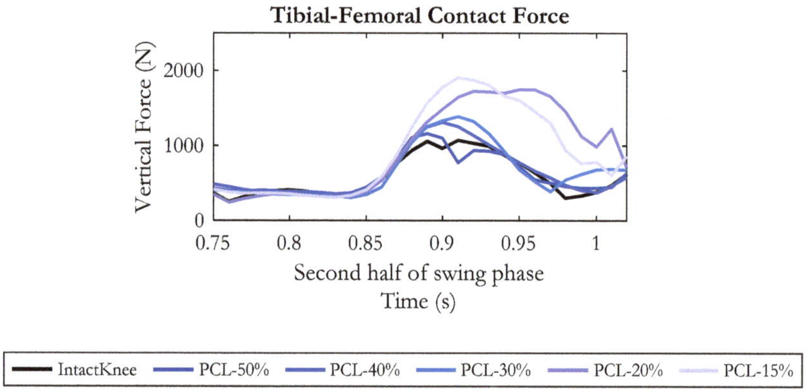

Figure 8. Tibial–femoral contact force in the second half of the swing phase, resulting from the six simulated conditions.

Except for some small oscillations, the contact force was unaffected by the tested conditions until the 87% point of the gait cycle. Then, in the late-swing phase, the contact force increased as far as the PCL residual stiffness was reduced. With respect to the intact knee, the average force increment was 70 N for PCL-50%, 90 N for PCL-40% and 190 N for PCL-30% conditions. The average increment became dramatic for PCL-20% and PCL-15%, achieving approximately 500 N.

3.2. *Effects of the Hamstrings on the PCL-Injured Knee (PCL-15%)*

Further dynamic simulations of the gait cycle were performed keeping the PCL residual stiffness at 15% of the healthy state and gradually reducing the hamstring's force from 100% down to 60% of the physiological condition. In this section, the main findings about the effects of hamstring force on a PCL-injured knee (PCL-15%) are reported.

3.2.1. Effects on Anterior–Posterior Displacement of the Tibia

The force applied by the hamstrings to the tibia was progressively reduced, as shown in Figure 9a. Figure 9b depicts the resulting anterior–posterior displacements of the tibia. Quite obviously, the main differences were noticed in the late-swing phase, when hamstrings are activated: the gradual reduction of hamstring force resulted in a progressive decrease in the posterior tibial displacement.

Specifically, as summarized in Figure 9c, with respect to the condition in which hamstrings expressed the physiological muscular activity ("PCL15%-Ham100%", reference condition), a force reduction to 90% lead to a decrease in the posterior displacement of the tibia by 2.5 mm on average. A further 10% drop in the hamstring's force ("PCL15%-Ham80%") produced a reduction of 3.5 mm on average with respect to the reference condition. When the hamstring force was reduced to 70%, the tibial posterior sliding decreased by an average of 7.3 mm; when it was reduced to 60%, the posterior displacement was reduced by 7.6 mm on average. As a consequence, the peak of backward displacement of the tibia decreased from 28 mm to 8 mm (see Figure 9b), very close to the displacement occurring in the intact knee.

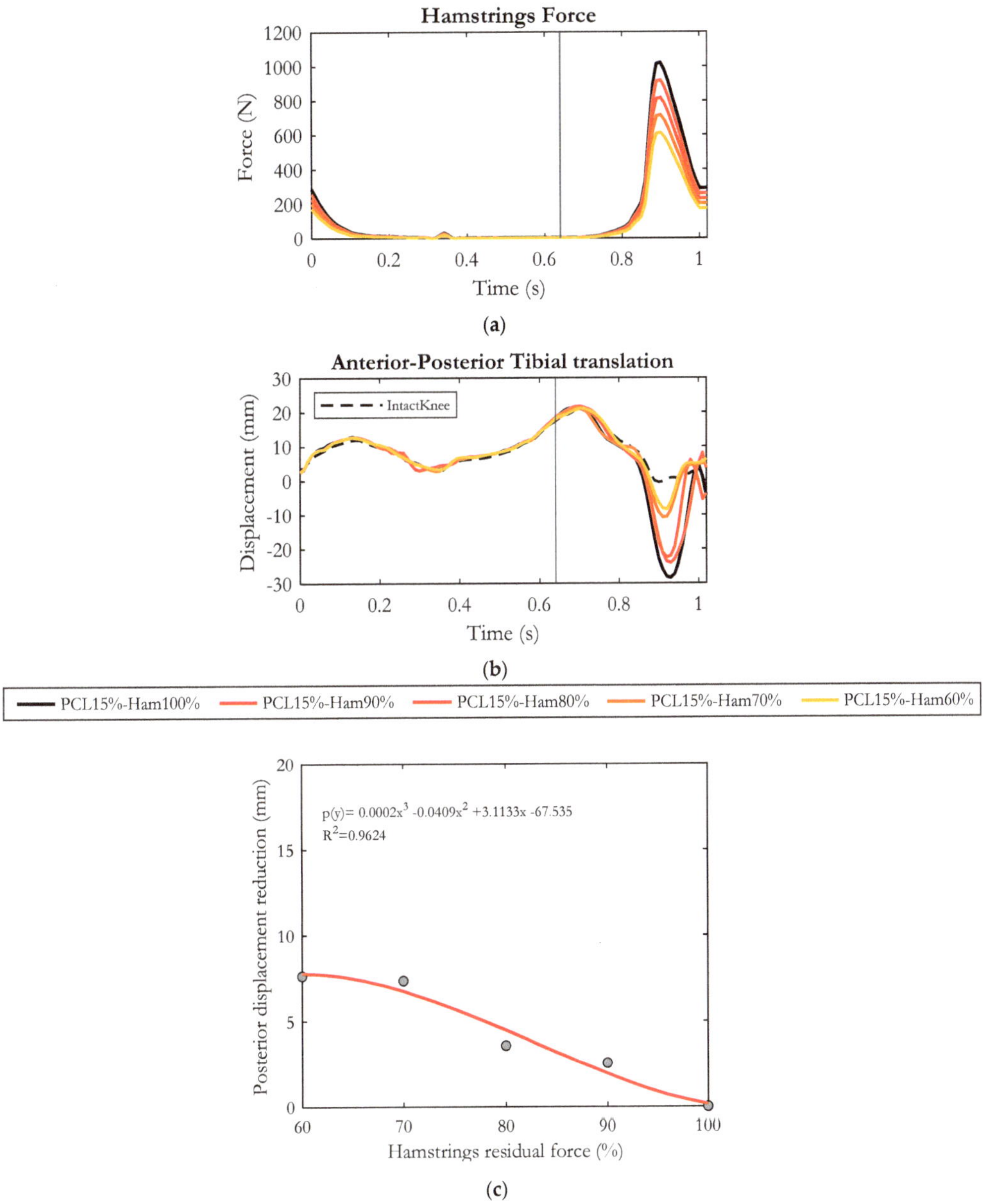

Figure 9. Progressive reduction of hamstring force (**a**). Anterior–posterior displacement of the tibia (**b**) during the gait cycle in the five simulated conditions. Values are positive for the anterior displacement. Dashed curve represents the intact knee condition. The gray vertical line refers to the toe–off event, marking the end of the stance phase and the beginning of the swing phase. Posterior tibial displacement average reduction with respect to "PCL15%-Ham100%" condition calculated in the second half of the swing phase (**c**), resulting from each simulated condition. In red, the polynomial regression function.

3.2.2. Effects on the Tension of the Remaining Ligaments

Figure 10 depicts the effects of hamstring force on the tension produced by the collateral ligaments in an injured knee (residual stiffness of PCL equal to 15%). The main differences in ligament tension were observed during the hamstring's activation phase (late-swing phase of the gait cycle). The progressive reduction of the muscular force resulted in a gradual relieving of Medial and Lateral Collateral Ligaments, as shown in the figure.

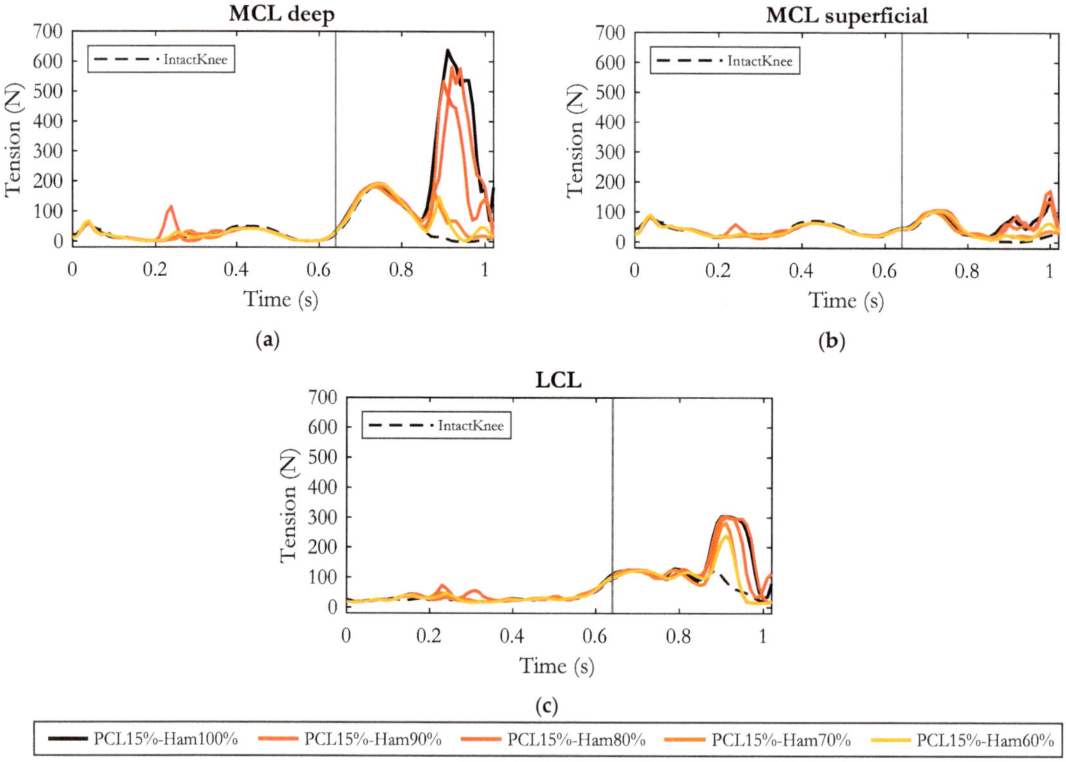

Figure 10. Tension of deep Medial Collateral Ligament (**a**), superficial Medial Collateral Ligament (**b**), Lateral Collateral Ligament (**c**) along the gait cycle in the five simulated conditions. Dashed curve represents the intact knee condition. The gray vertical line refers to the toe-off event, marking the end of the stance phase and the beginning of the swing phase.

The MCL superficial gradually relaxed similarly to the other collateral ligaments, but compared to them, its variations were not remarkable. With respect to the reference condition ("PCL15%-Ham100%"), when decreasing the hamstring's force to 90%, the deep bundles of MCL reduced their tension by 100 N on average, while in the LCL no appreciable effect was recorded. Assuming a force generated by the hamstrings equal to 80% of the physiological condition, MCL-deep experienced a further relieving with a reduction in tension of 120 N on average.

A significant tension drop (reduction of 248 N on average with respect to "PCL15%-Ham100%") was highlighted for MCL-deep when hamstring force was decreased to 70%. In the latter case, a more noticeable effect was observed on the LCL which experienced a tension reduction of 68 N on average.

Overall, considering all the simulated conditions, the strongest effects were recorded in the case of 60% hamstring force, in which the LCL and MCL-deep were relieved by an average of 82 N and 280 N, respectively.

3.2.3. Effects on Tibial–Femoral Contact Force

As regards the tibial–femoral contact force, in the late swing phase, a gradual decrease was recorded as the hamstring's force was progressively reduced, as depicted in Figure 11. Compared to the "PCL15%-Ham100%" condition, in the second half of the swing phase, an average force reduction of 190 N and 280 N was recorded when the hamstring's residual force was assumed to be 90% and 80%, respectively.

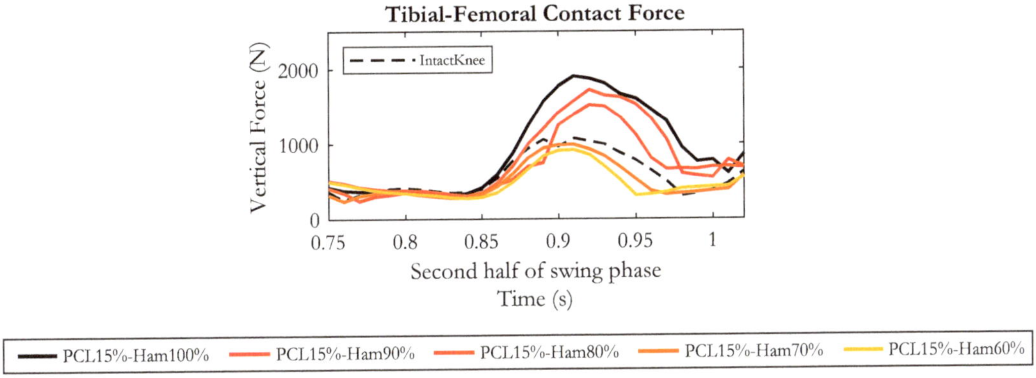

Figure 11. Tibial–femoral contact force in the second half of the swing phase, resulting from the five simulated conditions. Dashed curve represents the intact knee condition.

Further, an additional 10% drop in the muscular force decreased the contact force by 430 N on average, and again a decrease in the hamstring's force to 60% lead to a mean decrease of 520 N.

Interestingly, the curves obtained in these two last conditions were below the curve of the intact knee, demonstrating that PCL injury compensated by a reduced activity of the hamstrings can reduce the total force exchanged between femur and tibia.

4. Discussion

In the physiological knee, the posterior traction transmitted to the tibia by the hamstrings during the second half of the swing phase is efficiently sustained by the recruitment of the PCL (as shown in the previous study [25]). Consistently, this study showed that in this time window the gradual reduction in PCL stiffness resulted in a progressive increment of the posterior tibial displacement with respect to the intact knee. Specifically, in correspondence with a PCL injury such that the residual ligament stiffness is reduced to the 50% or 40% of the healthy state, the tibial posterior displacement reached an average increase of 1.3 mm and 1.5 mm, respectively, with respect to the intact knee. The average posterior sliding of the tibia gradually augmented as the PCL residual stiffness was reduced to the most critical simulated condition (residual stiffness of 15%), in which the lesion led to an averaged increase of 10 mm in the posterior tibial displacement.

As reported by [18,54,55], one of the major complications in PCL rupture is caused by the posterior displacement of the tibia, which leads to the degeneration of both the tibial–femoral and patellofemoral joints. Consistently, the outcomes of our study highlighted that, in the late-swing phase, the contact force between femoral condyles and tibial plateaus had a mild gradual increase as the PCL residual stiffness was reduced from 50% to 30%. In the most critical conditions (PCL residual stiffness of 20% and 15%), the average increment in contact force reached up to approximately 500 N compared to the healthy condition. As reported in our findings, this effect could be related to the enhancement in the remaining ligaments' tension (mainly in deep MCL and LCL) compensating for the PCL lesion, which resulted in increased longitudinal traction force between femur and tibia. Indeed, the contact force, as well as the tension of the remaining ligaments, was unaffected by the PCL lesion until the 80% point of the gait cycle, while considerable changes were evident from the 80% point to the end of the gait cycle, when the PCL should express its maximum force.

A typical cartilage degeneration pattern for a PCL-deficient knee presents anterior wear on the medial tibial plateau [56]. Our findings are consistent with anterior tibial cartilage wear, given the posterior translation of the tibia and the consequent anterior translation of the tibial–femoral contact point [57–59].

On the other hand, no significant changes were recorded on patellar–femoral contact force (not depicted in this study) and this result was in agreement with the findings of Kang et al. [24], who simulated the gait loading conditions by a subject-specific knee model. The cited authors outlined that no differences were found on patellar–femoral joint contact force between the intact and PCL-injured conditions at 0° and 60° of knee flexion during gait, while the contact force consistently increased during high flexion in the squat-loading condition.

It is worth mentioning that the patellar–femoral contact force we predicted presented two peaks in correspondence with the quadriceps contractions occurring approximately at 15% and 50% of the stride cycle [25]. Their values were 0.88 BW for the first peak and 0.74 BW for the second one. These values are within the range estimated by [60], who applied computational modeling of experimental data collected from six healthy subjects during walking. Since these contact forces occurred in correspondence with a negligible PCL tension, the change in the PCL stiffness did not affect the patellar–femoral contact force.

Actually, in our opinion, the increase in the contact force between patella and femur would occur if the backward displacement of the tibia with respect to the femur was so large that the angle of the patellar tendon with respect to the tibial plateau would change significantly. Moreover, if such a displacement occurred in a phase in which no quadriceps force is present, this effect would be not completely evident. This could explain the reason why, during the second half of the swing phase (no quadriceps activation), corresponding to the interval of the gait cycle in which the PCL would be mostly recruited under physiological conditions, no change in the patellar–femoral contact force can be appreciated. Then, for the rest of the gait cycle, the recruitment of the PCL is very limited and, consequently, as consistently outlined in this study, no appreciable changes in the kinematics, and therefore not even in the patellar–femoral force, would be expected.

Moreover, this study showed that a reduction of the contraction force of the hamstring muscles could compensate for PCL injury. Indeed, in an intact knee the PCL would be recruited during the second half of the swing phase, and at the same time the knee flexors would be activated to brake the forward movement of the shank. The required knee flexion moment is thus produced by the backward-oriented hamstring force and the forward-oriented PCL tension. In the absence of PCL, or in the case of reduced PCL stiffness, the point of attachment of the hamstrings on the tibia constitutes a sort of pivot point, so that the distal leg moves forward while the proximal tibia slides backwards, and this is depicted well in our results. Our outcomes showed that a gradual reduction of hamstring activity in a PCL-injured knee (PCL residual stiffness of 15%) resulted in a progressive reduction in the posterior tibial excursion. These findings are consistent with the well-known posterior traction effect of the hamstrings on the proximal tibia [61–65]. The reduction in the hamstring's force limited the posterior sliding of the tibia and, consequently, the tension of the ligaments compensating for PCL injury (LCL and mainly deep MCL) decreased. Accordingly, thanks to the reduction in ligament tension, the tibial–femoral contact force appeared as gradually reduced.

As previously reported, a chronic anteriorization of the tibial–femoral contact point is an issue in the PCL-deficient knee, especially for the medial compartment. This could occur if the posterior sliding of the tibia is increased, and our results show that the hamstring's force associated with PCL deficiency can be responsible for that. In this context, our results highlight the importance of targeting specific muscle groups in the rehabilitation setting during conservative treatment of PCL injury. The focus should be on promoting quadriceps recruitment, while hamstrings are best addressed with stretching and relaxation in order to avoid fostering their negative effect on tibial translation during activities of daily living [66,67].

Moreover, this study reports evidence of the compensation role of the MCL in restraining the posterior tibial translation in the case of PCL injury, and this is coherent with previous literature [68]. This observation raises awareness regarding the inherent difference between isolated PCL injury and PCL injury combined with lesions of the medial ligamentous structures. The latter condition should probably be approached with more caution, particularly during the initial phase, since one of the most important compensation structures is lacking. For example, even if at present there is limited evidence about the importance of knee bracing in the initial phase, we can assume that this indication should be stronger if the PCL lesion is combined with MCL injury.

In our model, the main limitations are the lack of cartilage and menisci between the articulating surfaces and the simplification of the patellofemoral joint. Furthermore, this study is based on a single, exemplary subject, and does not pretend to represent any specific population. Many studies showed that there is a great variability among subjects in terms of both attachment points and mechanical properties [31,32,69,70]. The realization of a subject-specific model is a challenging objective to be pursued. However, at present, the repositioning of the ligaments and changing the geometry of the joint surfaces in our model is a very delicate procedure, and this makes it difficult to adapt the model to different sizes.

However, the implemented musculoskeletal model has shown its potential in allowing the non-invasive estimation of loads on ligaments and articular surfaces and of the redistribution of loads in case of PCL injury. These aspects could hardly be evaluated in vivo.

5. Conclusions

The implemented musculoskeletal model appeared as a valuable tool for (i) non-invasive estimation of tension redistribution on remaining ligaments in the case of PCL injury, (ii) investigation on knee biomechanics modification in response to PCL injury and (iii) analysis of a conceivable compensatory muscular mechanism on a PCL-injured knee. The obtained results have to be considered as generic and not representative of a specific population, since the implemented musculoskeletal model refers to a single exemplary subject. However, this approach allowed us to understand and analyze the effects of a single-element modification through dynamic simulations. Moreover, it made it possible to investigate conditions hardly verifiable or reproducible in experiments, such as a gradual reduction of the hamstring's force, as well as aspects difficult to evaluate in vivo due to ethical and technical issues, such as the ligaments' tension or the anterior–posterior displacement of the tibia during dynamic tasks. Specifically, this computational study simulated different degrees of PCL injury and outlined their effects on the knee biomechanics: both the posterior tibial sliding and the ligaments' load redistribution were quantified and analyzed. In the implemented model, in the case of PCL lesion, the Medial Collateral Ligament adopted the role of the main passive stabilizer of the tibial posterior translation, with a synergistic recruitment of the Lateral Collateral Ligament. However, this coping strategy resulted in an enhancement of the tibial–femoral contact force with respect to the intact knee condition and did not prevent the posterior displacement of the tibia and consequent anteriorization of the tibial–femoral contact point. This can explain the well-known conditions of tibial cartilage degeneration in a PCL-injured knee. Moreover, this study showed that reducing the hamstring activity would allow for relieving all ligaments further recruited for compensating the PCL injury and, as a consequence, limiting the contact force between the distal femur and the proximal tibia.

Author Contributions: Conceptualization, L.D., V.S. and C.A.F.; methodology, L.D. and C.A.F.; software, L.D. and C.A.F.; validation, L.D., A.G., V.P., V.S. and C.A.F.; formal analysis, L.D. and C.A.F.; investigation, L.D., A.G., V.P., V.S. and C.A.F.; data curation, L.D.; writing—original draft preparation, L.D. and C.A.F.; writing—review and editing, L.D., A.G., V.P., V.S. and C.A.F.; visualization, L.D.; supervision, V.S. and C.A.F. All authors have read and agreed to the published version of the manuscript.

Funding: This research received no external funding.

Institutional Review Board Statement: Not applicable.

Informed Consent Statement: Not applicable.

Data Availability Statement: The data presented in this study are available on request from the corresponding author.

Acknowledgments: This work is part of a PhD project (Lucia Donno), supported by Ministry of University and Research, Politecnico di Milano, 37th cycle.

Conflicts of Interest: The authors declare no conflict of interest.

References

1. Majewski, M.; Susanne, H.; Klaus, S. Epidemiology of Athletic Knee Injuries: A 10-Year Study. *Knee* **2006**, *13*, 184–188. [CrossRef]
2. Hassebrock, J.D.; Gulbrandsen, M.T.; Asprey, W.L.; Makovicka, J.L.; Chhabra, A. Knee Ligament Anatomy and Biomechanics. *Sports Med. Arthrosc. Rev.* **2020**, *28*, 80–86. [CrossRef]
3. Grood, E.S.; Stowers, S.F.; Noyes, F.R. Limits of Movement in the Human Knee. Effect of Sectioning the Posterior Cruciate Ligament and Posterolateral Structures. *J. Bone Jt. Surg.* **1988**, *70*, 88–97. [CrossRef]
4. Logterman, S.L.; Wydra, F.B.; Frank, R.M. Posterior Cruciate Ligament: Anatomy and Biomechanics. *Curr. Rev. Musculoskelet. Med.* **2018**, *11*, 510–514. [CrossRef]
5. Gollehon, D.L.; Torzilli, P.A.; Warren, R.F. The Role of the Posterolateral and Cruciate Ligaments in the Stability of the Human Knee. A Biomechanical Study. *J. Bone Jt. Surg.* **1987**, *69*, 233–242. [CrossRef]
6. LaPrade, C.M.; Civitarese, D.M.; Rasmussen, M.T.; LaPrade, R.F. Emerging Updates on the Posterior Cruciate Ligament: A Review of the Current Literature. *Am. J. Sports Med.* **2015**, *43*, 3077–3092. [CrossRef]
7. Victor, J.; Banks, S.; Bellemans, J. Kinematics of Posterior Cruciate Ligament-Retaining and -Substituting Total Knee Arthroplasty: A prospective randomised outcome study. *J. Bone Jt. Surgery. Br. Vol.* **2005**, *87 Pt B*, 646–655. [CrossRef]
8. Ewald, F.C.; Jacobs, M.A.; Miegel, R.E.; Walker, P.S.; Poss, R.; Sledge, C.B. Kinematic Total Knee Replacement. *J. Bone Jt. Surg. Am* **1984**, *66*, 1032–1040. [CrossRef]
9. Schultz, R.A.; Miller, D.C.; Kerr, C.S.; Micheli, L. Mechanoreceptors in Human Cruciate Ligaments. A Histological Study. *J Bone Jt. Surg. Am.* **1984**, *66*, 1072–1076. [CrossRef]
10. Laskin, R.S.; Maruyama, Y.; Villaneuva, M.; Bourne, R. Deep-Dish Congruent Tibial Component Use in Total Knee Arthroplasty: A Randomized Prospective Study. *Clin. Orthop. Relat. Res.* **2000**, *380*, 36–44. [CrossRef]
11. Innocenti, B. High Congruency MB Insert Design: Stabilizing Knee Joint Even with PCL Deficiency. *Knee Surg. Sports Traumatol. Arthrosc.* **2020**, *28*, 3040–3047. [CrossRef]
12. Mazzucchelli, L.; Deledda, D.; Rosso, F.; Ratto, N.; Bruzzone, M.; Bonasia, D.E.; Rossi, R. Cruciate Retaining and Cruciate Substituting Ultra-Congruent Insert. *Ann. Transl. Med.* **2016**, *4*, 2. [CrossRef]
13. Peters, C.L.; Mulkey, P.; Erickson, J.; Anderson, M.B.; Pelt, C.E. Comparison of Total Knee Arthroplasty with Highly Congruent Anterior-Stabilized Bearings versus a Cruciate-Retaining Design. *Clin. Orthop. Relat. Res.* **2014**, *472*, 175–180. [CrossRef]
14. Oehme, S.; Moewis, P.; Boeth, H.; Bartek, B.; Lippert, A.; von Tycowicz, C.; Ehrig, R.; Duda, G.N.; Jung, T. PCL Insufficient Patients with Increased Translational and Rotational Passive Knee Joint Laxity Have No Increased Range of Anterior–Posterior and Rotational Tibiofemoral Motion during Level Walking. *Sci. Rep.* **2022**, *12*, 13232. [CrossRef]
15. Shelbourne, K.D.; Benner, R.; Ringenberg, J.; Gray, T. Optimal Management of Posterior Cruciate Ligament Injuries: Current Perspectives. *ORR* **2017**, *9*, 13–22. [CrossRef]
16. Okazaki, K.; Takayama, Y.; Osaki, K.; Matsuo, Y.; Mizu-Uchi, H.; Hamai, S.; Honda, H.; Iwamoto, Y. Subclinical Cartilage Degeneration in Young Athletes with Posterior Cruciate Ligament Injuries Detected with T1ρ Magnetic Resonance Imaging Mapping. *Knee Surg. Sports Traumatol. Arthrosc.* **2015**, *23*, 3094–3100. [CrossRef]
17. Pearsall, A.W.; Hollis, J.M. The Effect of Posterior Cruciate Ligament Injury and Reconstruction on Meniscal Strain. *Am. J. Sports Med.* **2004**, *32*, 1675–1680. [CrossRef]
18. Boynton, M.D.; Tietjens, B.R. Long-Term Followup of the Untreated Isolated Posterior Cruciate Ligament-Deficient Knee. *Am. J. Sports Med.* **1996**, *24*, 306–310. [CrossRef]
19. Grassmayr, M.J.; Parker, D.A.; Coolican, M.R.J.; Vanwanseele, B. Posterior Cruciate Ligament Deficiency: Biomechanical and Biological Consequences and the Outcomes of Conservative Treatment. *J. Sci. Med. Sport* **2008**, *11*, 433–443. [CrossRef]

20. Fanelli, G.C.; Edson, C.J. Posterior Cruciate Ligament Injuries in Trauma Patients: Part II. *Arthrosc. J. Arthrosc. Relat. Surg.* **1995**, *11*, 526–529. [CrossRef]
21. Fregly, B.J. A Conceptual Blueprint for Making Neuromusculoskeletal Models Clinically Useful. *Appl. Sci.* **2021**, *11*, 2037. [CrossRef]
22. Killen, B.A.; Falisse, A.; De Groote, F.; Jonkers, I. In Silico-Enhanced Treatment and Rehabilitation Planning for Patients with Musculoskeletal Disorders: Can Musculoskeletal Modelling and Dynamic Simulations Really Impact Current Clinical Practice? *Appl. Sci.* **2020**, *10*, 7255. [CrossRef]
23. Donno, L.; Sansone, V.; Galluzzo, A.; Frigo, C.A. Walking in the Absence of Anterior Cruciate Ligament: The Role of the Quadriceps and Hamstrings. *Appl. Sci.* **2022**, *12*, 8667. [CrossRef]
24. Kang, K.-T.; Koh, Y.-G.; Jung, M.; Nam, J.-H.; Son, J.; Lee, Y.H.; Kim, S.-J.; Kim, S.-H. The Effects of Posterior Cruciate Ligament Deficiency on Posterolateral Corner Structures under Gait- and Squat-Loading Conditions: A Computational Knee Model. *Bone Jt. Res.* **2017**, *6*, 31–42. [CrossRef]
25. Frigo, C.A.; Donno, L. The Effects of External Loads and Muscle Forces on the Knee Joint Ligaments during Walking: A Musculoskeletal Model Study. *Appl. Sci.* **2021**, *11*, 2356. [CrossRef]
26. Frigo, C.; Rabuffetti, M.; Kerrigan, D.C.; Deming, L.C.; Pedotti, A. Functionally Oriented and Clinically Feasible Quantitative Gait Analysis Method. *Med. Biol. Eng. Comput.* **1998**, *36*, 179–185. [CrossRef]
27. Chandler, R.F.; Clauser, C.E.; McConville, J.T.; Reynolds, H.M.; Young, J.W. *Investigation of Inertial Properties of the Human Body*; Technical Report AMRL-TR-74-137; National Technical Information Service: Alexandria, VA, USA, 1975.
28. Merkher, Y.; Sivan, S.; Etsion, I.; Maroudas, A.; Halperin, G.; Yosef, A. A Rational Human Joint Friction Test Using a Human Cartilage-on-Cartilage Arrangement. *Tribol. Lett.* **2006**, *22*, 29–36. [CrossRef]
29. Grood, E.S.; Suntay, W.J. A Joint Coordinate System for the Clinical Description of Three-Dimensional Motions: Application to the Knee. *J. Biomech. Eng.* **1983**, *105*, 136–144. [CrossRef]
30. Rasmussen, J.; Damsgaard, M.; Voigt, M. Muscle Recruitment by the Min/Max Criterion—A Comparative Numerical Study. *J. Biomech.* **2001**, *34*, 409–415. [CrossRef]
31. Blankevoort, L.; Huiskes, R.; de Lange, A. Recruitment of Knee Joint Ligaments. *J. Biomech. Eng.* **1991**, *113*, 94–103. [CrossRef]
32. Blankevoort, L.; Huiskes, R. Ligament-Bone Interaction in a Three-Dimensional Model of the Knee. *J. Biomech. Eng.* **1991**, *113*, 263–269. [CrossRef]
33. Shelburne, K.B.; Pandy, M.G. A Musculoskeletal Model of the Knee for Evaluating Ligament Forces during Isometric Contractions. *J. Biomech.* **1997**, *30*, 163–176. [CrossRef]
34. Bersini, S.; Sansone, V.; Frigo, C.A. A Dynamic Multibody Model of the Physiological Knee to Predict Internal Loads during Movement in Gravitational Field. *Comput. Methods Biomech. Biomed. Eng.* **2016**, *19*, 571–579. [CrossRef]
35. Amiri, S.; Cooke, D.; Kim, I.Y.; Wyss, U. Mechanics of the Passive Knee Joint. Part 2: Interaction between the Ligaments and the Articular Surfaces in Guiding the Joint Motion. *Proc. Inst. Mech. Eng. Part H* **2007**, *221*, 821–832. [CrossRef]
36. Kapandji, A.I. *Physiologie Articulaire*; Maloine: Paris, France, 1970; Volume 2.
37. Marra, M.A.; Vanheule, V.; Fluit, R.; Koopman, B.H.F.J.M.; Rasmussen, J.; Verdonschot, N.; Andersen, M.S. A Subject-Specific Musculoskeletal Modeling Framework to Predict In Vivo Mechanics of Total Knee Arthroplasty. *J. Biomech. Eng.* **2015**, *137*, 020904. [CrossRef]
38. Smith, C.R.; Vignos, M.F.; Lenhart, R.L.; Kaiser, J.; Thelen, D.G. The Influence of Component Alignment and Ligament Properties on Tibiofemoral Contact Forces in Total Knee Replacement. *J. Biomech. Eng.* **2016**, *138*, 021017. [CrossRef]
39. Scanlan, S.F.; D'Lima, D.D.; Colwell, C.W.; Andriacchi, T.P. Step-to-Step Variability in Tibiofemoral Contact Force During Walking Measured by an Instrumented Knee Implant. In Proceedings of the ASME 2009 Summer Bioengineering Conference, Parts A and B, Lake Tahoe, CA, USA, 17–21 June 2009; American Society of Mechanical Engineers: Lake Tahoe, CA, USA, 2009; pp. 907–908.
40. Taylor, S.J.G.; Walker, P.S.; Perry, J.S.; Cannon, S.R.; Woledge, R. The Forces in the Distal Femur and the Knee during Walking and Other Activities Measured by Telemetry. *J. Arthroplast.* **1998**, *13*, 428–437. [CrossRef] [PubMed]
41. Kinney, A.L.; Besier, T.F.; D'Lima, D.D.; Fregly, B.J. Update on Grand Challenge Competition to Predict in Vivo Knee Loads. *J. Biomech. Eng.* **2013**, *135*, 021012. [CrossRef]
42. Damm, P.; Kutzner, I.; Bergmann, G.; Rohlmann, A.; Schmidt, H. Comparison of in Vivo Measured Loads in Knee, Hip and Spinal Implants during Level Walking. *J. Biomech.* **2017**, *51*, 128–132. [CrossRef]
43. Hu, J.; Chen, Z.; Xin, H.; Zhang, Q.; Jin, Z. Musculoskeletal Multibody Dynamics Simulation of the Contact Mechanics and Kinematics of a Natural Knee Joint during a Walking Cycle. *Proc. Inst. Mech. Eng. Part H* **2018**, *232*, 508–519. [CrossRef]
44. Winby, C.R.; Lloyd, D.G.; Besier, T.F.; Kirk, T.B. Muscle and External Load Contribution to Knee Joint Contact Loads during Normal Gait. *J. Biomech.* **2009**, *42*, 2294–2300. [CrossRef] [PubMed]
45. Meireles, S.; De Groote, F.; Reeves, N.D.; Verschueren, S.; Maganaris, C.; Luyten, F.; Jonkers, I. Knee Contact Forces Are Not Altered in Early Knee Osteoarthritis. *Gait Posture* **2016**, *45*, 115–120. [CrossRef] [PubMed]
46. Morrison, J. The mechanics of the knee joint in relation to normal walking. *J. Biomech.* **1970**, *3*, 51–61. [CrossRef] [PubMed]
47. Markolf, K.L.; Mensch, J.S.; Amstutz, H.C. Stiffness and Laxity of the Knee—The Contributions of the Supporting Structures. *J. Bone Jt. Surg.* **1975**, *58 Pt A*, 12.
48. Hof, A.L.; Elzinga, H.; Grimmius, W.; Halbertsma, J.P.K. Detection of Non-Standard EMG Profiles in Walking. *Gait Posture* **2005**, *21*, 171–177. [CrossRef]

49. Pedotti, A.; Krishnan, V.V.; Stark, L. Optimization of Muscle-Force Sequencing in Human Locomotion. *Math. Biosci.* **1978**, *38*, 57–76. [CrossRef]
50. Crowninshield, R.D.; Brand, R.A. A Physiologically Based Criterion of Muscle Force Prediction in Locomotion. *J. Biomech.* **1981**, *14*, 793–801. [CrossRef]
51. Anderson, F.C.; Pandy, M.G. Static and Dynamic Optimization Solutions for Gait Are Practically Equivalent. *J. Biomech.* **2001**, *34*, 153–161. [CrossRef]
52. Martelli, S.; Valente, G.; Viceconti, M.; Taddei, F. Sensitivity of a Subject-Specific Musculoskeletal Model to the Uncertainties on the Joint Axes Location. *Comput. Methods Biomech. Biomed. Eng.* **2015**, *18*, 1555–1563. [CrossRef]
53. Martelli, S.; Taddei, F.; Cappello, A.; Van Sint Jan, S.; Leardini, A.; Viceconti, M. Effect of Sub-Optimal Neuromotor Control on the Hip Joint Load during Level Walking. *J. Biomech.* **2011**, *44*, 1716–1721. [CrossRef]
54. MacDonald, P.; Miniaci, A.; Fowler, P.; Marks, P.; Finlay, B. A Biomechanical Analysis of Joint Contact Forces in the Posterior Cruciate Deficient Knee. *Knee Surg. Sports Traumatol. Arthrosc.* **1996**, *3*, 252–255. [CrossRef] [PubMed]
55. Skyhar, M.J.; Warren, R.F.; Ortiz, G.J.; Schwartz, E.; Otis, J.C. The Effects of Sectioning of the Posterior Cruciate Ligament and the Posterolateral Complex on the Articular Contact Pressures within the Knee. *J. Bone Jt. Surg.* **1993**, *75*, 694–699. [CrossRef]
56. Van de Velde, S.K.; Bingham, J.T.; Gill, T.J.; Li, G. Analysis of Tibiofemoral Cartilage Deformation in the Posterior Cruciate Ligament-Deficient Knee. *J. Bone Jt. Surg.-Am. Vol.* **2009**, *91*, 167–175. [CrossRef] [PubMed]
57. Logan, M.; Williams, A.; Lavelle, J.; Gedroyc, W.; Freeman, M. The Effect of Posterior Cruciate Ligament Deficiency on Knee Kinematics. *Am. J. Sports Med.* **2004**, *32*, 1915–1992. [CrossRef] [PubMed]
58. Goyal, K.; Tashman, S.; Wang, J.H.; Li, K.; Zhang, X.; Harner, C. In Vivo Analysis of the Isolated Posterior Cruciate Ligament–Deficient Knee During Functional Activities. *Am. J. Sports Med.* **2012**, *40*, 777–785. [CrossRef]
59. Li, G.; Papannagari, R.; Li, M.; Bingham, J.; Nha, K.W.; Allred, D.; Gill, T. Effect of Posterior Cruciate Ligament Deficiency on in Vivo Translation and Rotation of the Knee during Weightbearing Flexion. *Am. J. Sports Med.* **2008**, *36*, 474–479. [CrossRef]
60. Thomeer, L.T.; Lin, Y.-C.; Pandy, M.G. Load Distribution at the Patellofemoral Joint During Walking. *Ann. Biomed. Eng.* **2020**, *48*, 2821–2835. [CrossRef]
61. Withrow, T.J.; Huston, L.J.; Wojtys, E.M.; Ashton-Miller, J.A. Effect of Varying Hamstring Tension on Anterior Cruciate Ligament Strain During In Vitro Impulsive Knee Flexion and Compression Loading. *J. Bone Jt. Surg.-Am. Vol.* **2008**, *90*, 815–823. [CrossRef]
62. Solomonow, M.; Baratta, R.; Zhou, B.H.; Shoji, H.; Bose, W.; Beck, C.; D'Ambrosia, R. The Synergistic Action of the Anterior Cruciate Ligament and Thigh Muscles in Maintaining Joint Stability. *Am. J. Sports Med.* **1987**, *15*, 207–213. [CrossRef]
63. Draganich, L.F.; Reider, B.; Ling, M.; Samuelson, M. An in Vitro Study of an Intraarticular and Extraarticular Reconstruction in the Anterior Cruciate Ligament Deficient Knee. *Am. J. Sports Med.* **1990**, *18*, 262–266. [CrossRef]
64. Baratta, R.; Solomonow, M.; Zhou, B.H.; Letson, D.; Chuinard, R.; D'Ambrosia, R. Muscular Coactivation: The Role of the Antagonist Musculature in Maintaining Knee Stability. *Am. J. Sports Med.* **1988**, *16*, 113–122. [CrossRef] [PubMed]
65. Aagaard, P.; Simonsen, E.B.; Andersen, J.L.; Magnusson, S.P.; Halkjær-Kristensen, J.; Dyhre-Poulsen, P. Neural Inhibition during Maximal Eccentric and Concentric Quadriceps Contraction: Effects of Resistance Training. *J. Appl. Physiol.* **2000**, *89*, 2249–2257. [CrossRef] [PubMed]
66. Pierce, C.M.; O'Brien, L.; Griffin, L.W.; LaPrade, R.F. Posterior Cruciate Ligament Tears: Functional and Postoperative Rehabilitation. *Knee Surg. Sports Traumatol. Arthrosc.* **2013**, *21*, 1071–1084. [CrossRef] [PubMed]
67. Veltri, D.M.; Warren, R.F. Isolated and Combined Posterior Cruciate Ligament Injuries. *J. Am. Acad. Orthop. Surg.* **1993**, *1*, 67–75. [CrossRef]
68. Wang, D.; Graziano, J.; Williams, R.J.; Jones, K.J. Nonoperative Treatment of PCL Injuries: Goals of Rehabilitation and the Natural History of Conservative Care. *Curr. Rev. Musculoskelet. Med.* **2018**, *11*, 290–297. [CrossRef]
69. Shelburne, K.B.; Pandy, M.G.; Anderson, F.C.; Torry, M.R. Pattern of Anterior Cruciate Ligament Force in Normal Walking. *J. Biomech.* **2004**, *37*, 797–805. [CrossRef]
70. Pandy, M.G.; Shelburne, K.B. Dependence of Cruciate-Ligament Loading on Muscle Forces and External Load. *J. Biomech.* **1997**, *30*, 1015–1024. [CrossRef]

Disclaimer/Publisher's Note: The statements, opinions and data contained in all publications are solely those of the individual author(s) and contributor(s) and not of MDPI and/or the editor(s). MDPI and/or the editor(s) disclaim responsibility for any injury to people or property resulting from any ideas, methods, instructions or products referred to in the content.

bioengineering

MDPI

Review

The Psychological Nature of Female Gait Attractiveness

Hiroko Tanabe [1,*] and Kota Yamamoto [2]

1 Graduate School of Humanities and Human Sciences, Hokkaido University, Kita 10, Nishi 7, Kita-ku, Sapporo 060-0810, Japan
2 School of Humanities, Hokusei Gakuen University, 2-3-1, Ohyachi-Nishi, Atsubetsu-ku, Sapporo 004-8631, Japan; ko-yamamoto@hokusei.ac.jp
* Correspondence: h.tanabe@let.hokudai.ac.jp; Tel.: +81-(0)11-706-4163

Abstract: Walking, a basic physical movement of the human body, is a resource for observers in forming interpersonal impressions. We have previously investigated the expression and perception of the attractiveness of female gaits. In this paper, drawing on our previous research, additional analysis, and reviewing previous studies, we seek to deepen our understanding of the function of gait attractiveness. First, we review previous research on gait as nonverbal information. Then, we show that fashion models' gaits reflect sociocultural genderlessness, while nonmodels express reproductive-related biological attractiveness. Next, we discuss the functions of gait attractiveness based on statistical models that link gait parameters and attractiveness scores. Finally, we focus on observers' perception of attractiveness, constructing a model of the visual information processing with respect to gait attractiveness. Overall, our results suggest that there are not only biological but also sociocultural criteria for gait attractiveness, and men and women place greater importance on the former and latter criteria, respectively, when assessing female gait attractiveness. This paper forms a major step forward in neuroaesthetics to understand the beauty of the human body and the generation of biological motions.

Keywords: gait; attractiveness; self-expression; interpersonal cognition; nonverbal communication; structural equation modeling; gait observation; gaze behavior

Citation: Tanabe, H.; Yamamoto, K. The Psychological Nature of Female Gait Attractiveness. *Bioengineering* **2024**, *11*, 1037. https://doi.org/10.3390/bioengineering11101037

Academic Editor: Massimiliano Pau

Received: 19 September 2024
Revised: 4 October 2024
Accepted: 16 October 2024
Published: 17 October 2024

Copyright: © 2024 by the authors. Licensee MDPI, Basel, Switzerland. This article is an open access article distributed under the terms and conditions of the Creative Commons Attribution (CC BY) license (https://creativecommons.org/licenses/by/4.0/).

1. Introduction

The movements of the human body interact with the social environment. In other words, human movements can be modulated by social stimuli and can convey nonverbal messages, taking the form of social stimuli, such as the actor's interpersonal attitude, intention, gender, and emotional state (e.g., inclusive or noninclusive attitudes toward a person that are implicitly conveyed by the physical positioning of two people) [1–5]. Charles Darwin systematically described the connection between emotions and behavioral expression, clarifying the role of animals' bodily expression in their social environment [6]. Darwin's theory was developed into the Laban movement analysis, which describes emotional expressions, such as the characteristics and quality of movement (effort) and geometrical characteristics (shape) [7,8]. In addition, Norton classified linguistic and paralinguistic information (including body motion) into nine communicative styles (i.e., dramatic, dominant, animated, relaxed, attentive, open, friendly, argumentative, and impression-leaving) in relation to how observers interpret and give meaning to it [9]. These styles have three dimensions, assertiveness, responsiveness, and versatility, relating to the ability to attract or control the observer's interest and attention, his or her emotional, flexible, and approachable nature, and the ability to change behavior in response to others, respectively [10]. In this way, human body movements have an important role to play in interpersonal communication, conveying the psychological state of the person who is performing the movement.

Gait, the most basic human movement, functions as a social stimulus, conveying the attributes and psychological state of the walker as nonverbal information. Among other features, there are gender differences in gait kinematics, reflected in hip flexion angle, knee flexion moment, and pelvic tilt angle [11–13]. Using these differences, it is possible to generate motions of gait that are characteristic of each gender [14]. Age also alters gait, including increases in step width and double support period and decreases in stride length, cadence, gait symmetry, toe-off angle, and walking speed [15,16]. In addition, gait reflects the walker's emotions, and observers can estimate a walker's emotions from walking speed, posture, and dynamic cues [17–21]. Even where interpersonal communication is active, gait patterns affect the evaluations made by others. For example, in job interviews, the gait patterns of applicants convey an impression of their temperament and personality [22]. In this way, walking motion embodies the physiological and psychological states of the walker. However, it remains unclear what specific dynamic cues in gait affect interpersonal communication and what functions they have.

Immediacy, defined as a certain level of physical and psychological intimacy that is recognized between two people [10], is important for smooth interpersonal communication. People who behave in a way characterized by high immediacy are more likely to be liked by others, highly evaluated, and selected [5,23]. In nonverbal immediacy, physical attractiveness in appearance and behavior plays an important role in making others perceive one as attractive, thus increasing the closeness between two people [10]. Many previous studies have identified the characteristics of attractive physical appearance. For example, it has been reported that men who have a muscular body shape, called mesomorphic [24–26], and male bodies with healthy amounts of fat and muscle mass [27,28] are considered attractive by women. In addition, waist-to-hip ratio [29] and lumbar curvature, when viewed from the side [30,31], are cues for the evaluation of attractiveness in female bodies. Because these physical characteristics are related to health and reproductive function, physical attractiveness may function as material for detecting individuals' health [32,33]. This is the dominant evolutionary psychological interpretation of the function of physical attractiveness. However, in some cases, physical features deviating from a healthy state are evaluated as attractive, especially in female bodies (e.g., a BMI [28] and fat mass [34] lower than the healthy level are considered attractive), which may be influenced by the cultural environment reflected by mass media [35]. To the best of our knowledge, there is no consistent view of physical attractiveness, including in the context of the sociocultural environment. In addition, physical attractiveness is not only seen in static representations of the face and body but also in physical movements (dance, walking, gestures, etc. [36,37]), and Morris et al. suggested that stride length and hip rotation are factors in the assessment in the attractiveness of the female gait [38]. However, detailed whole-body features related to attractiveness have not been elucidated, and the causal relationship between the parameters of the gait and impression evaluation, as well as how observers perceive the attractiveness of gait, remain unclear. Against this background, we have conducted research into the expression of attractiveness in terms of walking movements and the perception and evaluation of gait attractiveness [39–41].

In psychology, "Attractiveness refers to a person's perceived qualities, characteristics, or traits that evoke positive feelings, interest, or desire for social interaction. The multifaceted concept of attractiveness encompasses several components, including physical appearance, personality, intelligence, social skills, and other traits." [42]. In this study, we focus on attractiveness as perceived by observers in terms of physical appearance, including human movement. Following our previous research on the expression and perception of gait attractiveness and using additional data analysis, we seek to deepen our understanding of the function of gait attractiveness, as well as construct a model of the visual information processing process in gait attractiveness perception. Along with reviews of various previous studies on nonverbal communication, evolutionary psychology, and visual information processing, we structured this paper in the following three sections.

- Section 2: Expressing the attractiveness of female walkers. What gait patterns do women use to express attractiveness and increase their immediacy to others? From our previous work [39] and additional analyses of comparisons between fashion models and nonmodels, as well as impression evaluation by observers, we comprehensively identify the expression of physical attractiveness in walking motion.
- Section 3: Attractiveness evaluation model of female gait. We outline statistical estimation models for female gait attractiveness using the parameters of gait as explanatory variables [40]. The models we constructed partially show elements that may be inconsistent with evolutionary adaptations. Thus, we discuss the influence of the biological and sociocultural environment on the attractiveness of physical movements in relation to a review of previous research.
- Section 4: Perception of gait attractiveness. From our research on gaze behavior and its gender differences when evaluating gait attractiveness [41], we discuss how the visual gait information related to attractiveness is processed in observers, conducting a review of previous research. Finally, we develop a neuroaesthetic model for visual information processing for gait attractiveness.

In this paper, we use the terms "gender" and "sex" differently depending on how previous research is described. Moreover, in our data analysis and interpretation, we use the term "sex" instead of "gender" because the latter includes the meaning of self-representation of one's own sexuality, which is not covered in this study.

2. Expressing the Attractiveness of Female Walkers

2.1. Biomechanical Strategies to Maximize Gait Attractiveness

Nonverbal behavior has the function of increasing immediacy in interpersonal relationships. Gait kinematics may be a cue for female attractiveness, which is a factor in increasing immediacy [43]. Gait can also exhibit the intentional expression of beauty, such as the gait of a runway model [44], although the strategy for expressing attractiveness in gait remains unclear. Thus, we investigated the biomechanical basis of how women express their attractiveness in walking [39].

We recruited professional fashion models (N = 7; 42.4 ± 7.0 years; 170.6 ± 3.7 cm; 55.6 ± 3.4 kg) and nonmodels (N = 10; 34.0 ± 7.2 years; 162.0 ± 5.4 cm; 54.7 ± 7.7 kg) and had them walk barefoot or in heels on a treadmill at a fixed speed of 1.0 m/s. In each footwear condition, the participants walked under two types of instructions: (1) walking as casually as possible, in the normal condition, and (2) walking as attractively as possible, in the attractive-conscious condition. Using a motion capture system, the 3D positions of the anatomical feature points of the entire body of the walker were recorded, and then we used these in subsequent offline analyses to calculate the 3D joint angles for 13 joints throughout the body. The most challenging part of this study (i.e., [39]) was selecting gait parameters to capture the expression of gait attractiveness. From the Laban movement analysis [7,8], the performer's intention is expressed by the speed and energy of the movement, the use of space exhibited, and the movement's geometric alignment (as shown in the silhouette). Because a gait pattern in a straight line has a limited degree of spatial freedom, this study focused on the kinetic energy and silhouette of each joint and compared them between conditions.

The results suggest that female gait attractiveness is expressed through the following five strategies (Figure 1): increased hip joint energy immediately following heel contact, increased trunk curvature in the side view, head tilted and facing frontward, knee extension during push-off (i.e., before and after toe-off), and upper arm take-back. Following previous research on the characteristics of physical attractiveness, these strategies may have the function of expressing femininity and the health of walkers. For example, trunk curvature could embody the silhouette of the body in relation to reproductive function [30,31,45]. In addition, increased hip joint energy and drawing back of the upper arm have the effect of emphasizing the chest and buttocks [46]. The knee extension during push-off is associated with walking capacity, related to the muscle strength of the knee extension [47,48], which expresses the health of the walker. It should be noted that this study revealed a relationship between head silhouette and female gait attractiveness. The possible interpretations of this result are as follows: the change in head silhouette is a byproduct of the change in the alignment of the whole body, or, alternatively, the head silhouette conveys the defensive attitude of the walker [22] or emotion [19,21], which then leads observers to perceive attractiveness. The function of the head silhouette in attractiveness requires further investigation and is discussed in Section 3.2. This study showed that movements expressing health and reproductive function may be nonverbal elements inducing the perception of attractiveness (at least, performers adopt these elements as a strategy for the expression of physical attractiveness). The walking on a constant-speed treadmill captured in this study (differing in part from the gait kinematics of overground walking [49–51]) may affect the attractiveness ratings (which are influenced by walking speed [52,53]), so further research is needed to capture the attractiveness of gaits in more everyday environments.

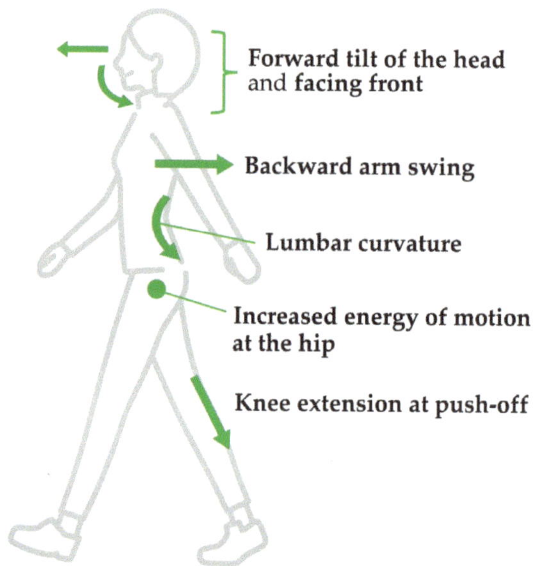

Figure 1. Five strategies for women to have an attractive gait.

2.2. Differences Between Fashion Models and Nonmodels: What Is the Attractive-Conscious Gait Generated by Nonmodels?

How did the nonmodels generate attractive gait in our study [39]? If seeking to reproduce the gait of fashion models in a neural representation, the expression strategies of attractiveness for both models and nonmodels should be similar. The focus of our study was the strategy for expressing attractiveness, and we did not examine statistical differences between the groups. Therefore, here, we additionally compared the models and nonmodels using statistical parametric mapping (SPM) [54,55]. Figures 2 and 3 present time series data (% gait cycle) for the gait parameters (silhouette-related and energy-related parameters, respectively) for each group (models: blue and dark blue plots; nonmodels: yellow and orange plots) and condition (the left and right rows represent the normal and attractive-conscious conditions, respectively). The SPM{t} function using alpha = 0.05 was calculated with the SPM1D MATLAB package (http://www.spm1d.org (accessed on 15 October 2024)), and the black bars on the horizontal axis show time points with statistically significant group differences. As a result, the group differences in body silhouette in walking were observed in head tilt and rotation, as well as thoracolumbar curvature in the frontal plane (Figure 2). In other words, relative to nonmodels, models tilted their heads more forward (in both the normal and attractive-conscious conditions: 0–100% of the gait cycle), rotated their heads horizontally while pulling back their shoulders (normal condition: 31–37.5% and 82–88% of the gait cycle, with a similar tendency in the attractive-conscious condition), and with increased hip twisting when seen from the front (attractive-conscious condition: 9.5–15% of the gait cycle, around 60–65%, and a normal condition with a similar tendency). In addition, while no statistically significant difference was seen, the models tended to take their upper arms back more than the nonmodels. No statistically significant group differences were found for energy-related gait parameters (Figure 3), but the models tended to exhibit smaller hip and thoracolumbar joint energies in the double-leg support phase (around 15–35% and 65–75% for the hip and around 20–40% and 70–90% of the gait cycle).

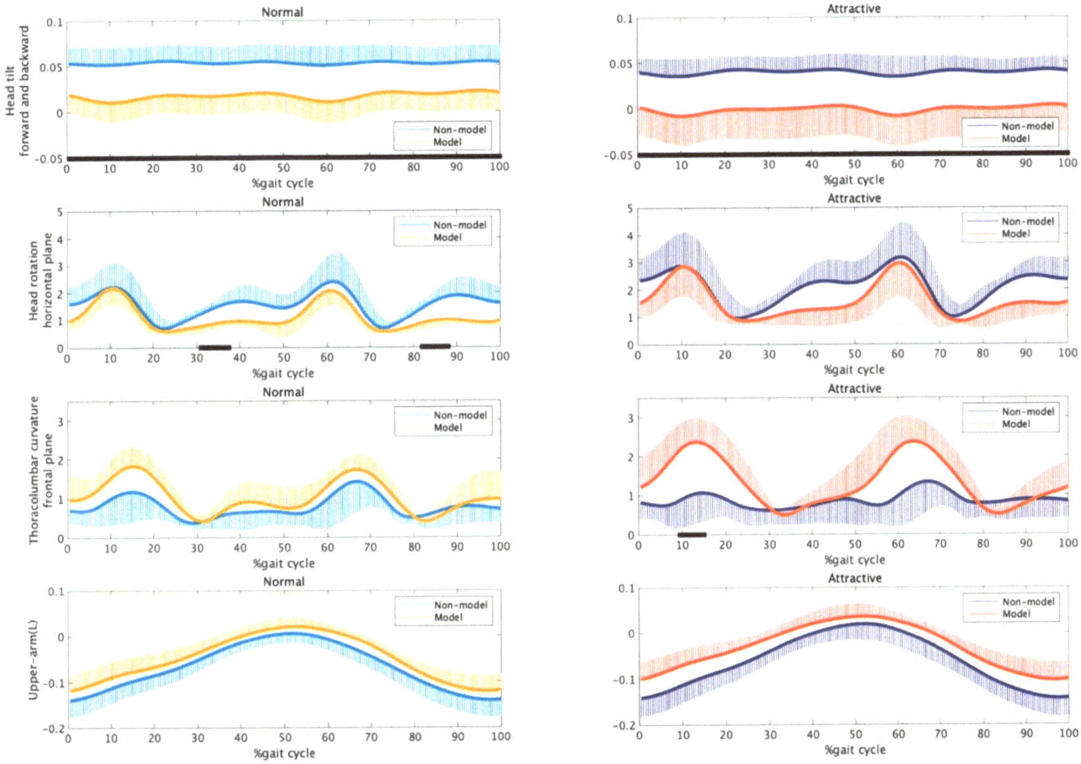

Figure 2. A between-group comparison in silhouette-related gait parameters: The values in the first (**left**) and second (**right**) rows show a comparison between groups in the normal and attractive-conscious conditions, respectively. The yellow and orange plots represent the model, and the blue and dark blue plots represent the nonmodel. The black bars on the horizontal axis indicate time points showing statistically significant group differences. The horizontal axis shows one stride beginning with a right-heel contact, followed by a left-toe-off (approximately 10.3%), a left-heel contact (approximately 50.7%), and a right-toe-off (approximately 60.3%), ending with a subsequent right-heel contact. Vertical axis: A larger value of each parameter (i.e., first to the fourth rows) represents a further backward head tilt, smaller head rotation in the horizontal plane (that is, facing the front), larger hip twist from the front view, and more backward arm swing, respectively.

The results above suggest that certain strategies that women adopt when expressing attractiveness go against the characteristics of the gaits of the fashion models. For example, the kinetic energy of the hip and thoracolumbar joints increased in both groups when they expressed attractiveness, but when the groups were compared, little kinetic energy was a characteristic of the model's gait (Figure 3). In addition, the horizontal rotation of the head is smaller in the attractive-conscious condition, i.e., the head gazes straight ahead, but the horizontal rotation of the models' heads is greater than that of the nonmodels' heads. Of course, other attractiveness-expression strategies were consistent with the model's gait. In the attractive-conscious condition, the woman tilts her head forward, forming a gait characteristic of fashion models. Likewise, the strategy of twisting the thoracolumbar joints on the frontal plane and taking back the upper arms is also emphasized in the models' walks. What features were seen in the attractive-conscious gait of the nonmodels? If they were seeking to reproduce fashion models' gait as a neural representation, the attractiveness-oriented gait of the nonmodel should go in the direction of closing the gap with the models' gait. However, both models and nonmodels expressed patterns that were partially opposite to the strategies that are learned and embodied by models in their gait.

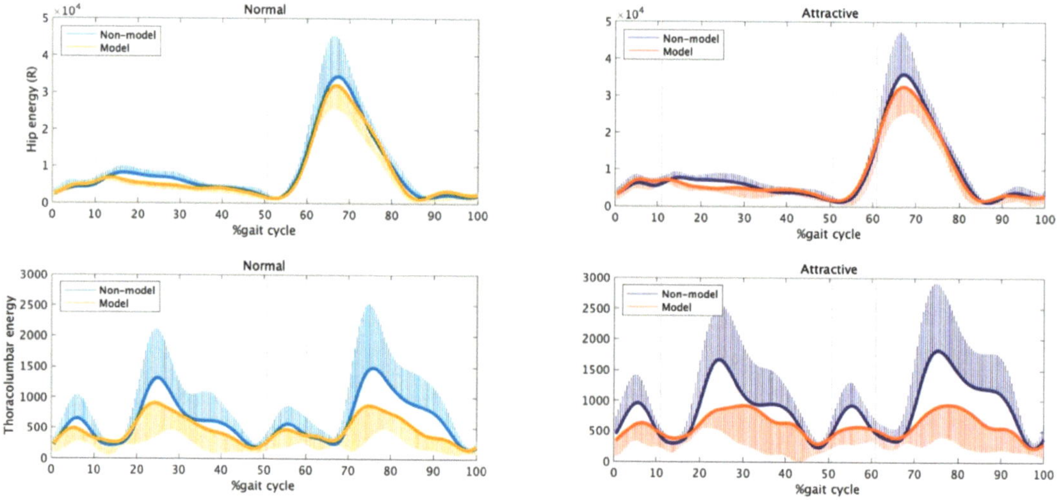

Figure 3. Between-group comparison of energy-related gait parameters: The values for the first (**left**) and second (**right**) rows present a comparison between groups in the normal and attractive-conscious conditions, respectively. The yellow and orange lines represent the models, and the blue and dark blue lines represent the nonmodels. The black bars on the horizontal axis present the time points using statistically significant group differences.

There may be a qualitative difference here between the expression of biological attractiveness and sociocultural attractiveness. The gait of a fashion model has as its purpose and function the showcasing of the attractiveness of clothes in the fashion industry, and the garments and styles change with the times [56]. In recent years, the tendency for models to cross their legs (employing the so-called cat walk) in their fashion display walks has decreased, and this change in gait appears to be due to the less gendered nature of clothing in a more gender-neutral or gender-fluid environment [56]. In addition, fashion model gaits are characterized by a relaxation of the upper body and a lack of excessive movement [57]. Our results can be interpreted as follows: A model's gait is influenced by a sociocultural environment exhibiting genderlessness, with less crossing of the legs, resulting in less hip joint energy. In addition, the kinetic energy of the thoracolumbar joints is also reduced in the relaxation of the upper body, with the prevention of excessive chest movement. The increase in horizontal head rotation is a matter of further investigation, but it reflects a strategy of attracting the audience's attention through increasing head movement [56]. Conversely, when increasing the attractiveness of their own gait, without reference to the fashion industry, women tend to increase their biological attractiveness by increasing the energy expressed in the hip and thoracolumbar joints, which could go against genderlessness. The function of the attractiveness expressed by head movements, such as facing front and having a forward tilt, needs to be clarified, but taking into account that the head silhouette can be a nonverbal message, reflecting a defensive posture [22] or emotion [19,21], a strategy to induce the perception of attractiveness could involve expressions of psychological states in one's gait. Thus, it may be that an attractive-conscious gait is not simply a reproduction of a model's gait but could be a biological adaptation emphasizing gender.

2.3. Was the Strategy of Expressing Attractiveness Successful?

Do observers perceive the attractiveness that is expressed by pedestrians? A total of 60 subjects (30 women, aged 38.50 ± 13.26 years; and 30 men, aged 40.70 ± 10.59 years) rated the attractiveness of all gaits on a 7-point scale (1: very ugly, 7: very attractive). In the barefoot condition, the attractiveness scores for the nonmodel gaits were 3.09 ± 0.53 in the normal condition and 3.50 ± 0.43 in the attractive-conscious condition, while the attractiveness scores of the models' gaits were 3.77 ± 0.44 in the normal condition and 4.29 ± 0.64 in the attractive-conscious condition. After the normality and homogeneity of variance in the data were checked, a 2-way ANOVA was performed, and the results show a main effect of condition ($F(1,30) = 6.8$, $p = 0.014$, $\eta^2 = 0.13$) and group ($F(1,30) = 17.2$, $p < 0.001$, $\eta^2 = 0.32$), with no interaction. The same analysis was performed for the heel condition: the attractiveness scores for the nonmodels' gaits were 3.49 ± 0.75 in the normal condition and 4.20 ± 0.64 in the attractive-conscious condition, while the attractiveness scores of the models' gaits were 4.85 ± 0.69 in the normal condition and 5.41 ± 0.48 in the attractive-conscious condition. A 2-way ANOVA showed a main effect of condition ($F(1,30) = 7.74$, $p = 0.0093$, $\eta^2 = 0.11$) and group ($F(1,30) = 31.3$, $p < 0.001$, $\eta^2 = 0.45$), showing no interaction. These results indicate that, in both footwear conditions, the attractive-conscious gait was rated as significantly more attractive than the normal condition, where the models' gaits were rated as more attractive than the nonmodels' gait. The great social value of this result is that others' attractiveness ratings are not only influenced by body shape and appearance, which are innate and unchanging or require significant time to change, but also by physical movements that can be instantly adjusted at will by the performer. Saito et al. [58] found that the tilted pelvic posture in gait affects the attractiveness evaluation, regardless of one's figure. These facts may help steer individuals away from the ideal of thinness as promulgated in the mass media and social media [35], with its resulting unhealthy eating behaviors [59].

3. Attractiveness Evaluation Model of Female Gait

3.1. Statistical Models for Estimating Gait Attractiveness

Section 2 shows that women can change their gait to express attractiveness, and this makes observers perceive them to be more attractive. What causal relationship is there between gait parameters and attractiveness evaluation? D'Argenio et al. [60] investigated the causal relationship between body dynamics and femininity and attractiveness, revealing that body poses that are less dynamic are preferred for the female body, and that such poses are linked to attractiveness through femininity. However, they did not note the movement elements that affect the evaluation of the visual impression or their function in terms of nonverbal messages. We conducted a study to examine the function of gait parameters in attractiveness evaluation by clarifying the causal relationship between a range of gait parameters and impression scores (i.e., attractiveness and femininity) [40]. We created statistical models to explain gait attractiveness and femininity with the use of gait parameters. To construct the models, we performed structural equation modeling (SEM), which has the advantage of combining measurement models and structural (i.e., regression) models into one overarching model that can optimally handle measurement errors in predictors [61,62].

A 30 s gait animation was created using motion capture data obtained in Section 2.1. In all of the animations, the angle of presentation was rotated 180 degrees at a constant speed for 30 s to observe the walkers from the front and from the side. We recruited

60 subjects (30 women, aged 38.50 ± 13.26 years, and 30 men, aged 40.70 ± 10.59 years), and each subject watched all of the animations, rating the attractiveness and femininity of each animation on a 7-point Likert scale (as above). To obtain explanatory variables for SEM, we first performed a correlation analysis between the impression scores and the walkers' parameters, including gait parameters and physical characteristics. The walkers' parameters were as follows: height, weight, BMI, waist-to-hip ratio, lumbar curvature, upper arm pull-back, forward head tilt, horizontal rotation of the head, stride CV, cadence, clearance, symmetry, knee extension, and toe-off angle. Next, we used the parameters of the walkers with a moderate or higher correlation with the impression score as explanatory variables to construct statistical models using SEM. Because it is possible for observers to distinguish between barefoot and heel walking, SEM was constructed for the footwear conditions separately.

From the correlation analysis, BMI, lumbar curve (barefoot only), upper arm take-back, head forward tilt, horizontal rotation of the head (heels only), cadence, clearance (heels only), knee extension, and toe-off angle (barefoot only) were moderately correlated with impression scores. Using SEM with these parameters as explanatory variables, two models were constructed for each footwear condition: one model consisting of silhouette elements of the head and trunk (Model 1), and a model consisting of body shape, trunk silhouette, and health elements (Model 2). We attempted to merge Models 1 and 2, but they diverged, which is likely because they were independent. Figure 4 forms an overview of the walkers' elements, which affected the observers' perception of attractiveness, based on the SEM models constructed in [40]. The functions of Models 1 and 2 are discussed in Section 3.2. Because it is possible to predict target values based on SEM models [63], this study likewise made it possible to estimate gait attractiveness using gait parameters and artificially tune gait attractiveness in the generation of biological motion.

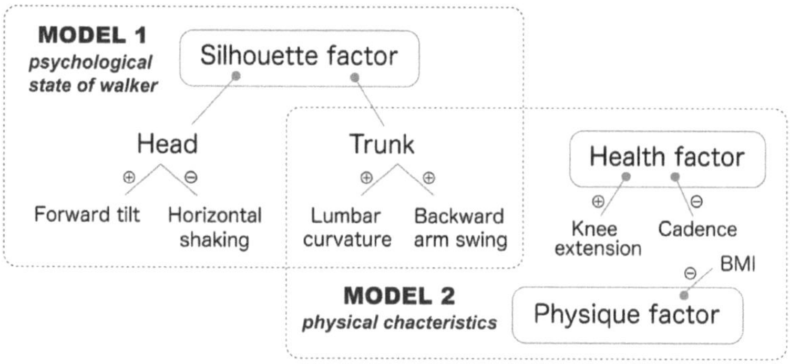

Figure 4. An overview of the walkers' elements affecting the observers' attractiveness perception. Models 1 and 2 indicate the evaluation criteria for attractiveness (silhouette-related and health–physique-related standards, respectively) in the two models that were constructed in our previous study [40]. The positive and negative signs indicate whether larger or smaller values lead to higher attractiveness ratings.

3.2. Function of Gait Attractiveness in Evolutionary Psychological and Sociocultural Contexts

Model 2 (Figure 4) is composed of the elements of health-related gait parameters, body shape, and trunk silhouette in relation to femininity, which are the criteria reflecting the walker's physical characteristics. In support of previous studies [30,31,58], it was found that the lumbar curvature (i.e., the anterior pelvic tilt) promotes the observer's perception of attractiveness in walking movements. The criterion of the knee extension, which decreases with age, is also consistent with the explanation of attractiveness according to evolutionary psychology, which states that the perception of attractiveness is a psychological process that detects an individual's health (e.g., [32,33]). Our findings suggest that observers sense information related to evolutionary adaptability using visual information of the movements of the body, linking it to the perception of attractiveness. However, other evaluation criteria went against the perception of health. BMI is an index of weight, where a normal weight is between 20 and 24.9 kg/m^2 [64]. The BMI of the walkers in this study was 20.01 $\pm$ 1.89 (minimum of 17.02 and maximum of 24.07), within the range of underweight to normal weight. A negative correlation was observed between the BMI and attractiveness score. This is consistent with previous research, which shows that women having BMIs lower than the healthy standard are preferred [28], which may reflect a feature of physical appearance that is shaped by the sociocultural environments [65], such as in the tendency of the mass media to promote underweight body appearance as part of the standard of beauty as a gender role norm for girls [35,66,67]. In addition, the negative correlation we found between cadence and attractiveness scores [40] is inconsistent with an association between health and attractiveness, in that lower cadences reflect a reduced ability to modulate the gait cycle duration, which can be linked to age-related diseases, such as knee osteoarthritis [68]. Why does attractiveness perception based on the physical characteristics of walkers both coincide with and contradict their health? One reason for this may be that the evaluations conducted by men and women are mixed. From the perspective of evolutionary psychology, males' evaluations of female attractiveness should have qualitative differences from females' evaluations of female attractiveness. We discuss this point in Section 4.

Model 1 (Figure 4) is based on the silhouette of the trunk and head. To the best of our knowledge, the relationship and function between head alignment and attractiveness is unclear. Head alignment is linked to emotional expression and interpersonal attitudes. For example, sadness [19,21] and a defensive posture [22] are expressed in a tilt of the head. In addition, the relationship between the horizontal rotation of the head is a negative correlation with the attractiveness score, which suggests that a silhouette with the head tilted forward and facing front in walking may convey a message of low openness and conservatism toward others, and this psychological state may be perceived as favorable. This is consistent with a preference for a static pose in female bodies [60]. Thus, Model 1 could form a criterion to judge the walker's emotions and interpersonal attitudes. Where first impressions are being formed in romantic relationships, women tend to communicate their approachability in nonverbal behaviors, such as tilting their head, shrugging their shoulders, looking down, looking to the side, frowning, holding hands, hugging, patting, having their hands in front of their bodies, and smoothing their hair [69]. Therefore, head movements are important for conveying women's nonverbal immediacy. The function and meaning of head movements in women's nonverbal immediacy must be further examined. In summary, it was suggested that the attractiveness of the female gait is determined by two criteria (Figure 4): the physical state of the walker, i.e., both evolutionarily adaptive and sociocultural criteria (Model 2), and the walker's interpersonal attitude (Model 1).

4. Perception of Gait Attractiveness

4.1. Gaze Behavior Associated with Gait Attractiveness Evaluation and Its Sex Differences

Section 3 identifies the criteria according to which observers evaluate the female gait attractiveness. What gaze behavior gathers information on gait and leads to the perception of attractiveness and positive evaluation? Gender differences appear in the perception of physical attractiveness. For example, in the observation of both men's and women's body images, male observers pay attention sooner and for longer to women's chests, while female observers pay attention sooner to men's legs [70]. Moreover, men pay more attention to female breasts and heads than women do [71]. Pazhoohi et al. [72] also show that there are sex differences in terms of the prominence of the shoulder-to-hip ratio (SHR), where men paid more attention to SHR. Here, two forces of sexual selection exist: intersexual (i.e., mate choice) and intrasexual (i.e., the sexual competition between individuals of the same sex) selection [73]. In human sexual selection, physical attractiveness is an important criterion in mate selection [74], and women compete with the same sex through attracting men, while men tend to compete with the same sex through physical strength [75]. We thus investigated what bodily cues are used to assess female gait attractiveness, focusing not only on men's evaluation of women, which could affect mate selection, but also on women's evaluation of their potential competitors [41]. In addition, when an observer perceives attractiveness, their attention interacts with their preference of bodily appearance [76]. We thus investigated the relationship between gait attractiveness and gaze behavior, in addition to sex differences.

We analyzed the gaze behavior of 33 (17 women; 38.06 ± 14.21 years, and 16 men; 42.25 ± 10.32 years) of the 60 participants in the impression evaluation experiment described in Section 3. We analyzed their gaze-tracking data using eye-tracking software (Tobii Pro Lab, Screen Edition, version 1.181) and a Tobii Pro Nano screen-based eye-tracking camera (Tobii, Danderyd, Sweden) attached to the bottom of the computer monitor. We set five AOIs (head, trunk, hip, leg, and others) for each walking animation in the software, and we calculated the fixation rate for each AOI and each participant. The observers' gaze area could have been influenced by their sex and preferences (i.e., for gait attractiveness). Therefore, we compared fixation rates according to the four factors of sex, preference (i.e., top vs. bottom 5 animations in attractiveness ranking among 68 animations), phase (phase 0 is the first 1.0 s from the start of the animation, and phases 1–4 are each a period of 7.5 s, totaling 30 s), and AOI (i.e., head, trunk, hip, leg, and others). We also constructed models that link gait parameters and the attractiveness scores for each observer's sex by using SEM and qualitatively compared standardized estimates.

A sex difference was found in visual attention in the fixation rates of the trunk and leg. Male observers were highly fixed on the trunk, while female observers tended to shift their attention from the trunk to the legs, especially in the observation of high-preference gait animations. In addition, the models constructed by SEM for each sex showed a tendency for male observers to place a greater weight on the walkers' trunk silhouette, while female observers prioritized parameters requiring whole-body observation. In human sexual selection, women's physical attractiveness is an important criterion for men's mate selection [74,77–79], and women compete with the same sex by attracting men [75]. We found that in the evaluation of female walkers' attractiveness, men place importance on the gait silhouette of the female trunk, which supports the idea that the reproductive regions of the female body are important for men in mate choices. Meanwhile, female observers tended to highlight BMI, knee extension, cadence, and head silhouette, requiring full-body observation when evaluating female gait attractiveness. Considering that the preferred women's BMI and fat mass values are below the healthy standard [28] and that mass media uses underweight bodies as icons of beauty and gender norms for girls [35,66,67], women may consider other women's attractiveness in relation to the dominant beauty standards of their sociocultural environment [65]. Further, taking into account that head silhouettes can function as nonverbal messages that reflect a walker's interpersonal attitude [22] and emotions [19,21], the walker's psychological state is a potential factor used in evaluating the attractiveness of potential competitors.

4.2. Visual Information Processing of Gait Attractiveness

The central nervous system processes visual information in a hierarchical and parallel manner [80–82]. Sequential hierarchy is divided into three stages [83]: an early vision that extracts simple elements such as color, shape, motion, and position [84,85]; intermediate vision, which groups elements into coherent entities [86–89]; and late vision, which selects areas of coherence for the direction of attention and recalling memories that allow us to recognize and attach meaning to objects [80,90]. According to neuroaesthetics [91,92], an observer's perception of an aesthetic object includes visual information processing, impression evaluation, and emotional reaction to the object. Each of these processes can be further deconstructed as follows: Once the basic visual elements of the object have been received, this information is collated in higher-level cognitive and attentional processes to form an impression and a final determination of aesthetic evaluation. This principle applies not only to visual aesthetics but also to music, dance, literature, and aesthetic judgments of the body [92]. In the evaluation of gait attractiveness, the motion of walking as a stimulus was initially deconstructed into visual features and sorted and processed. Then, higher-level cognitive and attentional processes are used to evaluate walkers' gaits attractiveness.

Drawing on information processing in neuroaesthetics [92], we created a model that showed the visual information processing of the attractiveness of female gaits (Figure 5). When walking motion is input as visual information, gait parameters such as lumbar curvature and head tilt are first extracted in early vision, and then they are recognized as a whole in intermediate vision. Here, male observers emphasize the trunk silhouette, while female observers emphasize the silhouette and movement of the head and whole body, and attention is induced in such sex-dependent vision. Based on this information, men and women attach meanings that are related to reproductive function, sociocultural standards of beauty, and the psychological state of the walker, respectively. In this way, the extraction of visual elements, attention, and meaning (i.e., context) interact to evoke the emotional responses of attractiveness and the final decision making of aesthetic judgment. Because attention and emotional response interact (i.e., they induce each other) [76], a reciprocal path appears between them. To the best of our knowledge, this is the first attempt to build an information processing model for gait attractiveness. It not only deepens our understanding of the central perceptual processes of physical attractiveness but also provides a deeper descriptive texture to bodily aesthetics (i.e., the psychological nature of the reward of the attractive experience). However, our study features limitations, such as the limitation of the visual stimuli to the female gait and the observers being limited to individuals of Japanese culture in a small sample. Thus, further verification is needed.

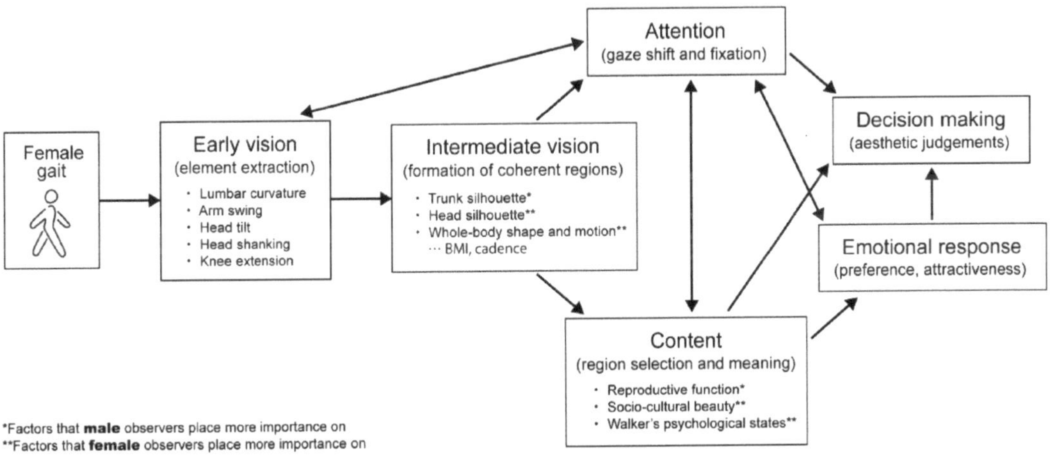

Figure 5. Visual information processing for the attractiveness perception and evaluation of female gaits. The arrows represent the flow of visual processing.

5. Conclusions

In this paper, we deepened our understanding of the meaning and function of walking in terms of nonverbal information regarding our previous study of the expression and perception of gait attractiveness in a comprehensive literature review. In Section 2, we focused on gait patterns used by women to express attractiveness and introduced five strategies of increase in hip joint energy immediately after heel contact, increase in trunk curvature in the side view, tilt and front facing of the head, knee extension during push-off (i.e., before and after toe-off), and upper arm drawback. Additional analysis of the differences between models and nonmodels showed the following qualitative differences between the expression of biological attractiveness and cultural attractiveness: In the models' gait, influenced by the sociocultural environment, leg crossing is reduced, which reduces the energy of the hip joint. In addition, the kinetic energy of the thoracolumbar joint is reduced to prevent the excessive movement of the upper body. These features are thought to be due to the trend toward genderlessness in the fashion industry. However, when trying to increase the attractiveness of their gait, women, including fashion models, increase the energy of their hip and thoracolumbar joints, which is in tension with genderlessness. Thus, the generation of attractive gait in nonmodels may not be a reproduction of the gait of fashion models but a biological adaptation to emphasize gender.

In Sections 3 and 4, we discussed the link between the gait parameters and evaluation of attractiveness, the visual information processing of the gait in observers, and gait attractiveness in relation to the evolutionary psychological and sociocultural perspectives. There are not only criteria for gait attractiveness consistent with the reproductive function and health of walkers but also sociocultural criteria that contradict them. Male observers tend to place more importance on the former criteria when evaluating female gait attractiveness as intersexual selection, while female observers tend to assign more importance to the latter when evaluating the attractiveness of their competitors in intrasexual selection. From these results, we proposed a visual information processing model for the attractiveness perception and evaluation of female gaits. This forms a major step forward in the neuroaesthetics of the beauty of the human body. The generation of biological motion for walking is widely used in the robot and AI industries, and the review in this paper provides a useful psychological basis for industrial implementation. In further studies, we plan to investigate the function of attractiveness of human movements and behavior, including social factors like clothing and context.

Author Contributions: H.T. designed this study, analyzed and interpreted data, and drafted and revised the manuscript. K.Y. designed this study, interpreted data, and revised the manuscript. All authors have read and agreed to the published version of the manuscript.

Funding: This work was supported by JSPS KAKENHI Grant Number JP23K16747.

Acknowledgments: The authors would like to thank the member of Center for Experimental Research in Social Sciences (CERSS) of Hokkaido University for useful comments and discussions, when the first author gave a talk at CERSS workshop.

Conflicts of Interest: The authors declare no conflicts of interest.

References

1. Mehrabian, A. Measures of achieving tendency. *Educ. Psychol. Meas.* **1969**, *29*, 445–451. [CrossRef]
2. Mehrabian, A. Significance of posture and position in the communication of attitude and status relationships. *Psychol. Bull.* **1969**, *71*, 359–372. [CrossRef]
3. Mehrabian, A. Verbal and nonverbal interaction of strangers in a waiting situation. *J. Exp. Res. Pers.* **1971**, *5*, 127–138.
4. Mehrabian, A. *Nonverbal Communication*, 1st ed.; Routledge: New York, NY, USA, 1972.
5. Mehrabian, A. *Silent Messages: Implicit Communication of Emotions and Attitudes*, 2nd ed.; Wadsworth Pub. Co.: Belmont, CA, USA, 1981.
6. Darwin, C.R. *The Expression of the Emotions in Man and Animals*, 1st ed.; John Murray: London, UK, 1872.
7. Bartenieff, I.; Lewis, D. *Body Movement: Coping with the Environment*, 1st ed.; Routledge: London, UK, 1980.
8. Amighi, J.K.; Lewis, P.; Loman, S.; Sossin, K.M. *The Meaning of Movement: Embodied Developmental, Clinical, and Cultural Perspectives of the Kestenberg Movement Profile*, 2nd ed.; Gordon & Breach: Amsterdam, The Netherlands, 2018.
9. Norton, R. *Communicator Style: Theory, Applications, and Measures*, 1st ed.; SAGE: Riverside County, CA, USA, 1983.
10. McCroskey, J.C.; Richmond, V.P. *Fundamentals of Human Communication: An Interpersonal Perspective*, 1st ed.; Waveland Press: Long Grove, IL, USA, 1996.
11. Kerrigan, D.C.; Todd, M.K.; Della Croce, U. Gender differences in joint biomechanics during walking: Normative study in young adults. *Am. J. Phys. Med. Rehabil.* **1998**, *77*, 2–7. [CrossRef]
12. Cho, S.H.; Park, J.M.; Kwon, O.Y. Gender differences in three dimensional gait analysis data from 98 healthy Korean adults. *Clin. Biomech.* **2004**, *19*, 145–152. [CrossRef]
13. Chumanov, E.S.; Wall-Scheffler, C.; Heiderscheit, B.C. Gender differences in walking and running on level and inclined surfaces. *Clin. Biomech.* **2008**, *23*, 1260–1268. [CrossRef]
14. Troje, N.F. The little difference: Fourier based gender classification from biological motion. In *Dynamic Perception*, 1st ed.; Würtz, R.P., Lappe, M., Eds.; Aka Press: Berlin, Germany, 2002; pp. 115–120.
15. Patterson, K.K.; Nadkarni, N.K.; Black, S.E.; McIlroy, W.E. Gait symmetry and velocity differ in their relationship to age. *Gait Posture* **2012**, *35*, 590–594. [CrossRef]

16. Aboutorabi, A.; Arazpour, M.; Bahramizadeh, M.; Hutchins, S.W.; Fadayevatan, R. The effect of aging on gait parameters in able-bodied older subjects: A literature review. *Aging Clin. Exp. Res.* **2016**, *28*, 393–405. [CrossRef]
17. Roether, C.L.; Omlor, L.; Christensen, A.; Giese, M.A. Critical features for the perception of emotion from gait. *J. Vis.* **2009**, *9*, 15. [CrossRef]
18. Naugle, K.M.; Hass, C.J.; Joyner, J.; Coombes, S.A.; Janelle, C.M. Emotional state affects the initiation of forward gait. *Emotion* **2011**, *11*, 267–277. [CrossRef]
19. Gross, M.M.; Crane, E.A.; Fredrickson, B.L. Effort-shape and kinematic assessment of bodily expression of emotion during gait. *Hum. Mov. Sci.* **2012**, *31*, 202–221. [CrossRef]
20. Barliya, A.; Omlor, L.; Giese, M.A.; Berthoz, A.; Flash, T. Expression of emotion in the kinematics of locomotion. *Exp. Brain Res.* **2013**, *225*, 159–176. [CrossRef]
21. Venture, G.; Kadone, H.; Zhang, T.; Grèzes, J.; Berthoz, A.; Hicheur, H. Recognizing emotions conveyed by human gait. *Int. J. Soc. Robot.* **2014**, *6*, 621–632. [CrossRef]
22. Bernhardt, C. *Nonverbal Communication in Recruiting: How to Identify Suitable Applicants and Attract Them to Your Company*, 1st ed.; Springer Nature: Wiesbaden, Germany, 2022; pp. 163–182.
23. Mehrabian, A. Immediacy: An indicator of attitudes in linguistic communication. *J. Pers.* **1966**, *34*, 26–34. [CrossRef]
24. Sheldon, W.H.; Stevens, S.S.; Tucker, W.B. *The Varieties of Human Physique: An Introduction to Constitutional Psychology*, 1st ed.; Harper & Brothers Publishers: New York, NY, USA, 1940.
25. Sheldon, W.H.; Stevens, S.S. *The Varieties of Temperament: A Psychology of Constitutional Differences*, 1st ed.; Harper: New York, NY, USA, 1942.
26. Sheldon, W.A. *Atlas of Men: A Guide for Somatotyping the Adult Male at All Ages*, 1st ed.; Harper: New York, NY, USA, 1954.
27. Dixson, B.J.; Grimshaw, G.M.; Ormsby, D.K.; Dixson, A.F. Eye-tracking women's preferences for men's somatotypes. *Evol. Hum. Behav.* **2014**, *35*, 73–79. [CrossRef]
28. Brierley, M.E.; Brooks, K.R.; Mond, J.; Stevenson, R.J.; Stephen, I.D. The body and the beautiful: Health, attractiveness and body composition in men's and women's bodies. *PLoS ONE* **2016**, *11*, e0156722. [CrossRef]
29. Singh, D.; Singh, D. Shape and significance of feminine beauty: An evolutionary perspective. *Sex Roles* **2011**, *64*, 723–731. [CrossRef]
30. Lewis, D.M.; Russell, E.M.; Al-Shawaf, L.; Buss, D.M. Lumbar curvature: A previously undiscovered standard of attractiveness. *Evol. Hum. Behav.* **2015**, *36*, 345–350. [CrossRef]
31. Lewis, D.M.G.; Russell, E.M.; Al-Shawaf, L.; Ta, V.; Senveli, Z.; Ickes, W.; Buss, D.M. Why women wear high heels: Evolution, lumbar curvature, and attractiveness. *Front. Psychol.* **2017**, *8*, 1875. [CrossRef]
32. Symons, D. Beauty is in the adaptations of the beholder: The evolutionary psychology of human female sexual attractiveness. In *Sexual Nature, Sexual Culture*, 1st ed.; Abramson, P.R., Pinkerton, S.D., Eds.; University of Chicago Press: Chicago, IL, USA, 1995; pp. 80–119.
33. Gangestad, S.W.; Scheyd, G.J. The evolution of human physical attractiveness. *Annu. Rev. Anthropol.* **2005**, *34*, 523–548. [CrossRef]
34. Stephen, I.D.; Perera, A.T.-M. Judging the difference between attractiveness and health: Does exposure to model images influence the judgments made by men and women? *PLoS ONE* **2014**, *9*, e86302. [CrossRef]
35. Groesz, L.M.; Levine, M.P.; Murnen, S.K. The effect of experimental presentation of thin media images on body satisfaction: A meta-analytic review. *Int. J. Eat. Disord.* **2002**, *31*, 1–16. [CrossRef]
36. Fink, B.; Weege, B.; Neave, N.; Pham, M.N.; Shackelford, T.K. Integrating body movement into attractiveness research. *Front. Psychol.* **2015**, *6*, 220. [CrossRef]
37. Rodero, E. Effectiveness, attractiveness, and emotional response to voice pitch and hand gestures in public speaking. *Front. Commun.* **2022**, *7*, 869084. [CrossRef]
38. Morris, P.H.; White, J.; Morrison, E.R.; Fisher, K. High heels as supernormal stimuli: How wearing high heels affects judgements of female attractiveness. *Evol. Hum. Behav.* **2013**, *34*, 176–181. [CrossRef]
39. Tanabe, H.; Fujii, K.; Kaneko, N.; Yokoyama, H.; Nakazawa, K. Biomechanical strategies to maximize gait attractiveness among women. *Front. Sports Act. Living* **2023**, *5*, 1091470. [CrossRef]
40. Tanabe, H.; Yamamoto, K. Structural equation modeling of female gait attractiveness using gait kinematics. *Sci. Rep.* **2023**, *13*, 17823. [CrossRef]
41. Tanabe, H.; Yamamoto, K. Gender differences in attentional processes and attractiveness evaluation models during gait observation. *Front. Psychol.* **2024**. *submitted*.
42. Agthe, M.; Klümper, L.; Schwarz, S. Attractiveness. In *Encyclopedia of Sexual Psychology and Behavior*; Shackelford, T.K., Ed.; Springer: Cham, Switzerland, 2023.
43. Fink, B.; Apalkova, Y.; Butovskaya, M.L.; Shackelford, T.K. Are there differences in experts' and lay assessors' attractiveness judgments of non-professional Men's dance/gait movements? *Percept. Mot. Ski.* **2021**, *128*, 492–506. [CrossRef]
44. Or, J. Computer simulations of a humanoid robot capable of walking like fashion models. *IEEE Trans. Syst. Man Cybern. C (Appl. Rev.)* **2011**, *42*, 241–248. [CrossRef]
45. Whitcome, K.K.; Shapiro, L.J.; Lieberman, D.E. Fetal load and the evolution of lumbar lordosis in bipedal hominins. *Nature* **2007**, *450*, 1075–1078. [CrossRef]
46. Furnham, A.; Swami, V. Perception of female buttocks and breast size in profile. *Soc. Behav. Pers. Int. J.* **2007**, *35*, 1–8. [CrossRef]

47. Hamrin, E.; Eklund, G.; Hillgren, A.K.; Borges, O.; Hall, J.; Hellström, O. Muscle strength and balance in post-stroke patients. *Ups. J. Med. Sci.* **1982**, *87*, 11–26. [CrossRef]
48. Nakamura, R.; Hosokawa, T.; Tsuji, I. Relationship of muscle strength for knee extension to walking capacity in patients with spastic hemiparesis. *Tohoku J. Exp. Med.* **1985**, *145*, 335–340. [CrossRef]
49. Strathy, G.M.; Chao, E.Y.; Laughman, R.K. Changes in knee function associated with treadmill ambulation. *J. Biomech.* **1983**, *16*, 517–522. [CrossRef]
50. Alton, F.; Baldey, L.; Caplan, S.; Morrissey, M.C. A kinematic comparison of overground and treadmill walking. *Clin. Biomech.* **1998**, *13*, 434–440. [CrossRef]
51. Hollman, J.H.; Watkins, M.K.; Imhoff, A.C.; Braun, C.E.; Akervik, K.A.; Ness, D.K. A comparison of variability in spatiotemporal gait parameters between treadmill and overground walking conditions. *Gait Posture* **2016**, *43*, 204–209. [CrossRef]
52. Schmitt, A.; Atzwanger, K. Walking fast-ranking high: A sociobiological perspective on pace. *Ethol. Sociobiol.* **1995**, *16*, 451–462. [CrossRef]
53. Fink, B.; Wübker, M.; Ostner, J.; Butovskaya, M.L.; Mezentseva, A.; Muñoz-Reyes, J.A.; Sela, Y.; Shackelford, T.K. Cross-cultural investigation of male gait perception in relation to physical strength and speed. *Front. Psychol.* **2017**, *8*, 1427. [CrossRef]
54. Pataky, T.C.; Robinson, M.A.; Vanrenterghem, J. Vector field statistical analysis of kinematic and force trajectories. *J. Biomech.* **2013**, *46*, 2394–2401. [CrossRef]
55. Bańkosz, Z.; Winiarski, S. Statistical parametric mapping reveals subtle gender differences in angular movements in table tennis topspin backhand. *Int. J. Environ. Res. Public Health* **2020**, *17*, 6996. [CrossRef]
56. Kobayashi, Y.; Saito, S.; Murahori, T. Classification of fashion models' walking styles using publicly available data, pose detection technology, and multivariate analysis: From past to current trendy walking styles. *Sensors* **2024**, *24*, 3865. [CrossRef]
57. Guo, R.L.; Liu, B.; Huang, H.Y. The study of female fashion Model's basic walking posture. *Adv. Mater. Res.* **2011**, *332*, 1272–1275. [CrossRef]
58. Saito, S.; Saito, M.; Kondo, M.; Kobayashi, Y. Gait pattern can alter aesthetic visual impression from a third-person perspective. *Sci. Rep.* **2024**, *14*, 6602. [CrossRef]
59. Xiang, K.; Kong, F. Passive social networking sites use and disordered eating behaviors in adolescents: The roles of upward social comparison and body dissatisfaction and its sex differences. *Appetite* **2024**, *198*, 107360. [CrossRef]
60. D'Argenio, G.; Finisguerra, A.; Urgesi, C. Motion and gender-typing features interact in the perception of human bodies. *Front. Neurosci.* **2020**, *14*, 277. [CrossRef]
61. Ullman, J.B.; Bentler, P.M. Structural equation modeling. In *Handbook of Psychology: Research Methods in Psychology*, 2nd ed.; Schinka, J.A., Velicer, W.F., Weiner, I.B., Eds.; John Wiley & Sons: Hoboken, NJ, USA, 2013; pp. 661–690.
62. Gholami Fesharaki, M. Structural equation modeling and its application in psychological studies: A review study. *Clin. Psychol. Pers.* **2020**, *16*, 253–265.
63. de Rooij, M.; Karch, J.D.; Fokkema, M.; Bakk, Z.; Pratiwi, B.C.; Kelderman, H. SEM-based out-of-sample predictions. *Struct. Equ. Model. Multidiscip. J.* **2023**, *30*, 132–148. [CrossRef]
64. Nuttall, F.Q. Body mass index: Obesity, BMI, and health: A critical review. *Nutr. Today* **2015**, *50*, 117–128. [CrossRef]
65. Nicholson, L. Interpreting gender. *Signs J. Women Cult. Soc.* **1994**, *20*, 79–105. [CrossRef]
66. Hawkins, N.; Richards, P.S.; Granley, H.M.; Stein, D.M. The impact of exposure to the thin-ideal media image on women. *Eat. Disord.* **2004**, *12*, 35–50. [CrossRef]
67. Grabe, S.; Ward, L.M.; Hyde, J.S. The role of the media in body image concerns among women: A meta-analysis of experimental and correlational studies. *Psychol. Bull.* **2008**, *134*, 460–476. [CrossRef]
68. Boekesteijn, R.J.; Van Gerven, J.; Geurts, A.C.H.; Smulders, K. Objective gait assessment in individuals with knee osteoarthritis using inertial sensors: A systematic review and meta-analysis. *Gait Posture* **2022**, *98*, 109–120. [CrossRef]
69. Romaniuk, O.; Teran, L. First impression sexual scripts of romantic encounters: Effect of gender on verbal and non-verbal immediacy behaviors in American media dating culture. *J. Soc. Pers. Relat.* **2022**, *39*, 107–131. [CrossRef]
70. Hewig, J.; Trippe, R.H.; Hecht, H.; Straube, T.; Miltner, W.H. Gender differences for specific body regions when looking at men and women. *J. Nonverbal Behav.* **2008**, *32*, 67–78. [CrossRef]
71. Garza, R.; Heredia, R.R.; Cieslicka, A.B. Male and female perception of physical attractiveness: An eye movement study. *Evol. Psychol.* **2016**, *14*, 1474704916631614. [CrossRef]
72. Pazhoohi, F.; Garza, R.; Doyle, J.F.; Macedo, A.F.; Arantes, J. Sex differences for preferences of shoulder to hip ratio in men and women: An eye tracking study. *Evol. Psychol. Sci.* **2019**, *5*, 405–415. [CrossRef]
73. Darwin, C. *The Decent of Man and Selection in Relation to Sex*, 1st ed.; Murray: London, UK, 1871.
74. Puts, D.A. Mechanisms of sexual selection in humans Beauty and the Beast. *Evol. Hum. Behav.* **2010**, *31*, 157–175. [CrossRef]
75. Puts, D. Human sexual selection. *Curr. Opin. Psychol.* **2016**, *7*, 28–32. [CrossRef]
76. Shimojo, S.; Simion, C.; Shimojo, E.; Scheier, C. Gaze bias both reflects and influences preference. *Nat. Neurosci.* **2003**, *6*, 1317–1322. [CrossRef]
77. Elder, G.H., Jr. Appearance and education in marriage mobility. *Am. Sociol. Rev.* **1969**, *34*, 519–533. [CrossRef]
78. Buss, D.M. Human mate selection: Opposites are sometimes said to attract, but in fact we are likely to marry someone who is similar to us in almost every variable. *Am. Sci.* **1985**, *73*, 47–51.
79. Buss, D.M.; Schmitt, D.P. Mate preferences and their behavioral manifestations. *Annu. Rev. Psychol.* **2019**, *70*, 77–110. [CrossRef]

80. Farah, M.J. *The Cognitive Neuroscience of Vision*, 1st ed.; Wiley-Blackwell: Hoboken, NJ, USA, 2000.
81. Zeki, S. *A Vision of the Brain*, 1st ed.; Blackwell Scientific Publications: Hoboken, NJ, USA, 1993.
82. Van Essen, D.C.; Felleman, D.J.; DeYoe, E.A.; Olavarria, J.; Knierim, J. Modular and hierarchical organization of extrastriate visual cortex in the macaque monkey. *Cold Spring Harb. Symp. Quant. Biol.* **1990**, *55*, 679–696. [CrossRef]
83. Marr, D. *Vision. A Computational Investigation into the Human Representation and Processing of Visual Information*, 1st ed.; W. H. Freeman and Company: New York, NY, USA, 1982.
84. Livingstone, M.S.; Hubel, D.H. Psychophysical evidence for separate channels for the perception of form, color, movement, and depth. *J. Neurosci.* **1987**, *7*, 3416–3468. [CrossRef]
85. Livingstone, M.; Hubel, D. Segregation of form, color, movement, and depth: Anatomy, physiology, and perception. *Science* **1988**, *240*, 740–749. [CrossRef]
86. Biederman, I.; Cooper, E.E. Priming contour-deleted images: Evidence for intermediate representations in visual object recognition. *Cogn. Psychol.* **1991**, *23*, 393–419. [CrossRef]
87. Grossberg, S.; Mingolla, E.; Ross, W.D. Visual brain and visual perception: How does the cortex do perceptual grouping? *Trends Neurosci.* **1997**, *20*, 106–111. [CrossRef]
88. Vecera, S.P.; Behrmann, M. Spatial attention does not require preattentive grouping. *Neuropsychology* **1997**, *11*, 30–43. [CrossRef]
89. Ricci, R.; Vaishnavi, S.; Chatterjee, A. A deficit of intermediate vision: Experimental observations and theoretical implications. *Neurocase* **1999**, *5*, 1–12. [CrossRef]
90. Chatterjee, A. Neglect. A disorder of spatial attention. In *Neurological Foundations of Cognitive Neuroscience*, 1st ed.; D'esposito, M., Ed.; MIT Press: Cambridge, MA, USA, 2003; pp. 1–26.
91. Chatterjee, A. The neuropsychology of visual artistic production. *Neuropsychologia* **2004**, *42*, 1568–1583. [CrossRef]
92. Chatterjee, A. Neuroaesthetics: A coming of age story. *J. Cogn. Neurosci.* **2011**, *23*, 53–62. [CrossRef]

Disclaimer/Publisher's Note: The statements, opinions and data contained in all publications are solely those of the individual author(s) and contributor(s) and not of MDPI and/or the editor(s). MDPI and/or the editor(s) disclaim responsibility for any injury to people or property resulting from any ideas, methods, instructions or products referred to in the content.

bioengineering

MDPI

Review

Muscle Synergy Analysis as a Tool for Assessing the Effectiveness of Gait Rehabilitation Therapies: A Methodological Review and Perspective

Daniele Borzelli [1,2,*,†], Cristiano De Marchis [3,†], Angelica Quercia [1], Paolo De Pasquale [4], Antonino Casile [1], Angelo Quartarone [4], Rocco Salvatore Calabrò [4] and Andrea d'Avella [2,5]

1 Department of Biomedical and Dental Sciences and Morphofunctional Imaging, University of Messina, 98125 Messina, Italy; angelica.quercia@unime.it (A.Q.); acasile@unime.it (A.C.)
2 Laboratory of Neuromotor Physiology, IRCCS Fondazione Santa Lucia, 00179 Rome, Italy; a.davella@hsantalucia.it
3 Engineering Department, University of Messina, Messina 98166, Italy; cristiano.demarchis@unime.it
4 IRCCS Centro Neurolesi "Bonino Pulejo", 98124 Messina, Italy; paolo.depasquale@irccsme.it (P.D.P.); angelo.quartarone@irccsme.it (A.Q.); roccos.calabro@irccsme.it (R.S.C.)
5 Department of Biology, University of Rome Tor Vergata, 00133 Rome, Italy
* Correspondence: dborzelli@unime.it
† These authors contributed equally to this work.

Citation: Borzelli, D.; De Marchis, C.; Quercia, A.; De Pasquale, P.; Casile, A.; Quartarone, A.; Calabrò, R.S.; d'Avella, A. Muscle Synergy Analysis as a Tool for Assessing the Effectiveness of Gait Rehabilitation Therapies: A Methodological Review and Perspective. *Bioengineering* **2024**, *11*, 793. https://doi.org/10.3390/bioengineering11080793

Academic Editor: Carlo Albino Frigo

Received: 25 June 2024
Revised: 19 July 2024
Accepted: 29 July 2024
Published: 5 August 2024

Copyright: © 2024 by the authors. Licensee MDPI, Basel, Switzerland. This article is an open access article distributed under the terms and conditions of the Creative Commons Attribution (CC BY) license (https://creativecommons.org/licenses/by/4.0/).

Abstract: According to the modular hypothesis for the control of movement, muscles are recruited in synergies, which capture muscle coordination in space, time, or both. In the last two decades, muscle synergy analysis has become a well-established framework in the motor control field and for the characterization of motor impairments in neurological patients. Altered modular control during a locomotion task has been often proposed as a potential quantitative metric for characterizing pathological conditions. Therefore, the purpose of this systematic review is to analyze the recent literature that used a muscle synergy analysis of neurological patients' locomotion as an indicator of motor rehabilitation therapy effectiveness, encompassing the key methodological elements to date. Searches for the relevant literature were made in Web of Science, PubMed, and Scopus. Most of the 15 full-text articles which were retrieved and included in this review identified an effect of the rehabilitation intervention on muscle synergies. However, the used experimental and methodological approaches varied across studies. Despite the scarcity of studies that investigated the effect of rehabilitation on muscle synergies, this review supports the utility of muscle synergies as a marker of the effectiveness of rehabilitative therapy and highlights the challenges and open issues that future works need to address to introduce the muscle synergies in the clinical practice and decisional process.

Keywords: motor coordination; musculo-skeletal redundancy; motor modules; neural disease; stroke; Parkinson's disease; cerebral palsy; multiple sclerosis; myelopathy; brain tumor

1. Introduction

Gait is a complex task that requires the coordination of multiple muscles, driven by different cortical and spinal structures [1–4]. Neurological diseases can lead to impairments in the neuromuscular control of gait, with an increased risk of accidental falls [5] and the consequent reduced independence in performing activities of daily living, due to alterations in coordination strategies and in spatiotemporal gait parameters [6]. In this scenario, gait rehabilitation plays a crucial role, and recovering a functional gait is a fundamental step for independent living. Gait rehabilitation may require the action of physiotherapists alone, or it could additionally involve the use of rehabilitation devices. Gait assistive equipment comes in a variety of forms that can be roughly classified into two major categories that can be identified as follows [7,8]: alternative devices, which are used by patients with no movements and do not involve exercises for injured extremities, and augmentative devices, employed by people with limited mobility to generate movements or workouts with a rehabilitative purpose. Augmentative gait devices can be used on a treadmill, on footplates, overground through the use of mobile robotic bases [9], or they can be stationary [10]. The quantitative assessment of gait is of great importance to characterize the stage of a neurological pathology [11–14], to validate the beneficial effects of a therapy [15–18], or for the early detection of conditions such as risk of fall [19], dementia [20], and Parkinson's disease [11]. The quantification of the motion of body segments in two or three dimensions is frequently exploited to estimate the spatiotemporal and kinematic parameters of gait [21]. This is usually carried out through marker-based [22] or marker-less [23,24] motion capture systems, even though the use of Inertial Measurement Units (IMUs) have started to gain popularity [25–28]. This kind of analysis is often complemented by the measurement of contact forces during the stance phase of gait to quantify impact forces, loading rates, propulsive and breaking forces, and to track variations in the center of pressure [29,30].

However, the main window on the physiological mechanisms underlying the neural control of gait is provided by electromyography (EMG), which allows to gain quantitative information regarding muscle coordination. Earlier EMG systems used cables to transmit recorded signals. This additional wiring represented a major limitation as it potentially reduced the subject's range of motion. However, over the last two decades, EMG systems have been improved by the incorporation of technology that enables the data to be delivered wirelessly or kept in a data logger worn by the subject to detect relevant pathological features [31–34] or to drive robotic devices [35–43].

Concurrently, data analysis techniques have been proposed that allow us to investigate multi-muscle activity in order to quantitatively characterize the coordination among muscles acting across different joints. Muscle synergy analysis, i.e., the analysis of low-dimensional structures through the factorization of multi-muscle sEMG recordings, has gained popularity as it has been shown to provide relevant information regarding the neuromechanics of movement for a variety of motor tasks and motor impairment in neurological conditions [44–57]. Indeed, muscle synergies are suggested to derive from neural inputs that drive multiple muscles [58–60]. This idea is supported by the presence of synchronized oscillations within the firings of motor unit pools across different muscles, potentially stemming from cortical oscillations [61–63]. The sEMG signal represents the sum of action potentials induced in multiple muscle fibers [64,65]. Motor unit action potentials have been modeled as electromagnetic resonant modes [66], and their spectral characteristics [67] are known to change during fatiguing tasks [36,68,69], with aging [70], and at different muscle temperatures [71]. The synchronous modulation of motor unit pools is thought to reduce kinematic noise [72] and the existence of a common drive has been suggested to respect both the size principle [73–76] and the onion skin principle [77].

Many previous studies have shown that muscle coordination during human gait can be well described by the combination of a small number of muscle synergies and that each synergy is associated with specific biomechanical functions of gait in various conditions [53,78–82]. There is relatively vast literature analyzing muscle synergies in gait in physiological and pathological conditions, indicating how changes in the muscle synergy dimensionality and spatiotemporal structures can account for several behavioral correlates and clinical scales as well. This body of literature collectively suggests that muscle synergy analysis might be a good quantitative tool to investigate the neural correlates of gait performance and functional gait recovery.

Following the acknowledgment of motor synergies as a potential tool to compactly index motor functions [83–85], reviews have been published that examine the alteration of synergies after neurological diseases such as Parkinson's disease [86], stroke [87,88], or neurodevelopmental disorders [89]. However, to the best of our knowledge, no review to date has comprehensively examined the alteration of muscle synergies identified during locomotion tasks after rehabilitation therapy. If muscle synergies capture the neurophysiological mechanisms underlying the control of gait and provide a reliable indicator of gait performance, they should also highlight specific changes in motor coordination during motor recovery. Therefore, this work aims to systematically review the literature to assess the state of the art in research on the use of muscle synergies during a locomotion task, as a quantitative tool to analyze the efficacy of rehabilitation therapy.

2. Materials and Methods

2.1. Search Strategy

A systematic search strategy was conducted using the electronic databases of PubMed, Scopus, and Web of Science and performed on the 1 April 2024. The time of publication was restricted to the interval between January 2011 and March 2024. The lower bound of January 2011 was that of the publication date of the first study that suggested muscle synergies as a tool for clinicians to assess healthy and pathological muscle activity [83]. As shown in Table 1, the searches involved a combination of the following words only in the title or abstract: "Gait", or any word beginning with "walk" or "locomot"; "Therapy", or "training", or any word beginning with "rehabilit" or "neurorehabilit"; "synergy", or "synergies", or "muscle coordination", or "motor module", or "motor modules", or "primitive", or "primitives".

Table 1. Search query used in all systematic searches (PubMed version shown in table).

Search Query (PubMed, Scopus, WoS)	("gait"[Title/Abstract] OR "walk*"[Title/Abstract]) AND ("therapy"[Title/Abstract] OR "rehabilit*"[Title/Abstract] OR "neurorehabilit*"[Title/Abstract] OR "training"[Title/Abstract]) AND ("muscle synerg*"[Title/Abstract] OR "synergies"[Title/Abstract] OR "muscle coordination"[Title/Abstract] OR "motor module*"[Title/Abstract] OR "primitive*"[Title/Abstract])

2.2. Study Eligibility: Inclusion and Exclusion Criteria

Studies were retained if the following eligibility criteria were met: (1) a pathological condition was investigated; (2) a therapy, which was finalized to recover gait motor function, was administered; (3) the muscle synergies of the patients were assessed before and after the therapy; (4) studies followed clear and reproducible methodological stages [90], and studies were excluded when the therapy was not adequately specified, the task was not clear and reproducible, or the assessment of muscle synergies did not include the lower limb; (5) the study was published in English in peer-reviewed journals. Reviews, meta-analysis articles, perspective/position papers, editorials, commentaries, and conference papers were excluded, but case reports were retained.

2.3. Study Selection and Data Extraction

In compliance with the PRISMA statement [91,92], the eligibility of potentially relevant studies was based on title and abstract adherence to inclusion/exclusion criteria, and the screening was conducted by three authors independently (D.B., C.D., and Angelica Quercia). Full texts were then retrieved and evaluated thoroughly to confirm eligibility based on the described inclusion and exclusion criteria. Conflicts were resolved by discussions among the authors. Reference lists of the included studies were manually screened to identify additional relevant studies. From the retained studies, we extracted the following information and imported them into an Excel spreadsheet: First author, year of publication, the number of patients, the assessed pathology, the collected muscle set (for details, see Table 2), the type of task during which muscle synergies were extracted, the type of training and its duration, the algorithm used to extract the muscle synergies and define their number, the metrics adopted to compare synergies extracted before and after the training were tabulated for each study. The number of patients and the number of control participants enrolled in each study were both recorded in the spreadsheet. Additionally, a check on the inclusion of healthy control participants, the acquired clinical scales, and the presence of other kinematic measures was conducted. Figure 1 shows the PRISMA flowchart for study inclusion/exclusion.

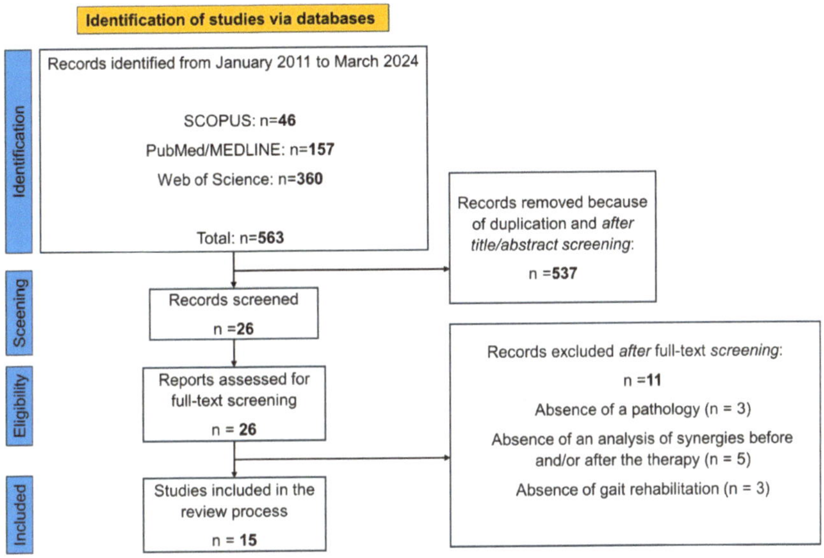

Figure 1. PRISMA flowchart for study inclusion/exclusion.

3. Results

3.1. Selected Studies

Our search query returned a total of 563 studies, of which 26 full-text articles were retained and further screened based on their title and abstract and the inclusion/exclusion criteria. Manual reference list screening did not result in any additional studies. A total of 15 articles passed this further screening and were thus included in this review.

3.2. Study Characteristics

3.2.1. Patients

Most of the retained studies (9 out of 15) investigated the effect of physical gait rehabilitation on the muscle synergies of stroke survivors [93–101]. The other studies investigated Parkinson's disease [102,103], cerebral palsy [104], multiple sclerosis [105], myelopathy [106], and brain tumor [107].

Four studies enrolled a large population of patients ($\geq$20 patients) [96,99,100,103], while two studies were case reports [106,107], and two other studies only enrolled two patients each [94,95].

Three studies included as control groups patients who underwent conventional therapy [99], cognitive training [100], or treadmill training with body weight support [97].

Seven studies also enrolled healthy control participants to define a set of representative healthy muscle synergies sets and activation profiles for reference data [94–97,101,103,105], five of which selected age-matched healthy controls [95–97,103,105].

3.2.2. Task

During walking, the changes in the muscle synergies were investigated overground in 10 studies [94,95,97–100,102,103,105,106] but also on a treadmill [96,104,107]. Both overground and treadmill conditions were investigated in one study [101]. In addition to gait, in two studies, muscle synergies were also explored during a reactive balance task [102] and a recumbent cycling task [93].

3.2.3. Training and Clinical Evaluation

In the 15 retained studies, patients underwent rehabilitative training for an average of 4 weeks (range: 3–12 weeks). Rehabilitative sessions (14 on average; range: 9–36) were administered 2 to 5 times a week, and each session lasted 33 min on average (range: 5–90 min).

In 6 out of our 15 selected studies [97–99,101,105,107], gait rehabilitative training of the lower limbs of patients was based on robotic exoskeletons [108,109]. Lower-limb exoskeleton-assisted training was mainly used for post-stroke patients [97–99,101], in children with cerebral palsy [104] and in cases of thoracic myelopathy [106]. In 4 of the selected studies, multichannel Functional Electrical Stimulation (FES) was applied as a rehabilitation method combined with walking [94,95,107] and cycling [93]. FES is considered an effective intervention for lower-limb rehabilitation, particularly suitable in stroke patients, as the combination of FES with walking or cycling can enhance motor learning and plasticity, improve locomotion ability, and strengthen lower limb muscles and motor coordination [110,111]. In the absence of severe musculoskeletal and neurologic pathologies that could influence gait, treadmill walking sessions can also be used for locomotor rehabilitation [112,113]. In the retained studies, treadmill training sessions were administered both in stroke patients [96] and patients with multiple sclerosis [105]. Finally, in three out of the 15 studies [101,103,104], the rehabilitation protocol did not involve an assistive device. One study [100] investigated muscle activation patterns after trunk training in stroke patients to provide new insights in gait recovery. Another study [102], instead, investigated changes in the neuromuscular control of gait and balance after dance-based rehabilitation in Parkinson's patients, specifically, an Adapted Tango (AT) dance. Finally, a third study [103] investigated the alteration in muscle synergies after a bilateral deep brain stimulation of the subthalamic nucleus.

Across the retained studies, participants were evaluated at baseline (T0) and after an intervention period (T1) with the following validated clinical scales used to measure patient motor functions: Functional Independence Measure-Motor General (FIM-M) [114]; Functional Independence Measure-Locomotion (FIM-Locomotion) [114]; Functional Ambulation Categories (FACs) [115]; The Motricity Index (MI) [116]; Fugl-Meyer Assessment, Lower Extremity (FMA-LE) [117]; Barthel Index (BI) [118]; Functional gait assessment (FGA) [119]; the Berg Balance scale (BBS) for standing balance [120]; Mini Best test (MBT), which evaluated the dynamic balance [121]; Fullerton Advanced Balance scale (FAB) [122]; Dynamic Gait Index (DGI) [123]; the 2 Minute Walking Test (2MWT) for gait endurance [124]; the 10 m timed walk (10MTW) for gait speed [125]; time up and go (TUG) [126]; the 6 min walk test (6MWT) [124]; Tinetti Performance Oriented Mobility Assessment (POMA) [127]; the Brunnstrom recovery stage (BRS), which evaluated impairment of the lower limb [128]; Trunk Control Test (TCT) [129]; and Trunk Impairment Scale (TIS) [130]. The motor function of children with cerebral palsy in [104] was evaluated following the Gross Motor Function Classification System (GMFCS) [131] to establish the ability to walk and which lower limb was most affected.

3.2.4. Muscles

Across the selected studies, the EMG activity was collected either unilaterally or bilaterally. Given a set of available EMG sensors, a unilateral arrangement allows to collect the activity of more muscles in a single leg, and therefore, it provides a finer characterization of the locomotor patterns which may allow for a better assessment of intra-subject variability [82]. On the contrary, a bilateral sensor arrangement allows to investigate gait asymmetries occurring as a consequence of several diseases [132]. Six out of the 15 identified studies collected muscle activity unilaterally [94,95,97,102,104,105], and seven studies collected muscle activity bilaterally [93,96,98,99,101,106,107]. One study [100] collected unilaterally the activity of six leg muscles and bilaterally the activity of one trunk muscle (i.e., the erector spinae), while another study [103] collected unilaterally the activity of eleven leg muscles and bilaterally the activity of one trunk muscle (i.e., the longissimus dorsi).

Muscle selection varied across studies. In Table 2, the number and the identity of the muscles included in each study are reported, while in Figure 2, the percentage of studies which included each muscle is reported. Due to the redundancy of the musculoskeletal system [133] or its abundance [134], different studies selected different muscles with similar function, e.g., while nine studies collected the activity only from Vastus Medialis muscle (VM), three studies collected the activity from the Vastus Lateralis muscle (VL) whose action is the same as the VM. Moreover, one study [103] collected the activity from the lateral and medial hamstring without specifying whether the activity was from the semitendinosus or the semimembranosus muscles (medial hamstring) or the short or long head of the biceps femoris (lateral hamstring).

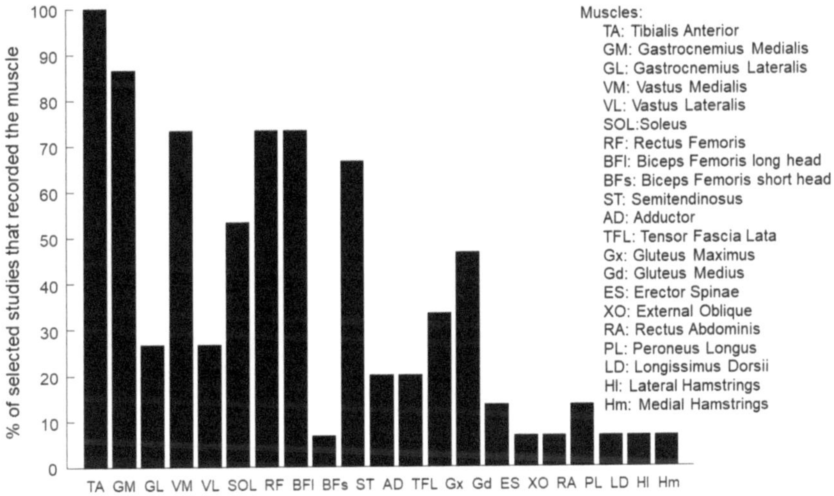

Figure 2. Percentage of the selected studies in which the activity of a muscle is collected.

Table 2. Muscles that were collected from the selected studies. An 'X' in correspondence to the column related to a muscle indicates that the muscle was acquired. The muscles were Tibialis Anterior (TA), Gastrocnemius Medialis (GM), Gastrocnemius Lateralis (GL), Vastus Medialis (VM), Vastus Lateralis (VL), Soleus (SOL), Rectus Femoris (RF), Biceps Femoris long head (BFl), Biceps Femoris short head (BFs), Semitendinosus (ST), Adductor (AD), Tensor Fascia Lata (TFL), Gluteus Maximus (Gx), Gluteus Medius (Gd), Erector Spinae (ES), External Oblique (XO), Rectus Abdominis (RA), Peroneus Longus (PL), Longissimus Dorsii (LD), Lateral Hamstrings (Hl), and Medial Hamstrings (Hm). The 'number of muscles' column reports how many muscles were collected on a single side (unilaterally, U) and on both sides (bilaterally, B).

	Muscles																					
Reference	TA	GM	GL	VM	VL	SOL	RF	BFl	BFs	ST	AD	TFL	Gx	Gd	ES	XO	RA	PL	LD	Hl	Hm	Number
Allen et al., 2017 [102]	X	X	X	X		X	X	X				X		X	X	X	X	X				13 U
Ambrosini et al., 2020 [93]	X	X			X	X	X	X	X			X	X									9 B
Conner et al., 2021 [104]	X				X	X				X												4 U
Ferrante et al., 2016 [94]	X	X		X			X	X		X			X									8 U
Ghislieri et al., 2023 [103]	X		X	X		X	X			X		X		X				X	X	X	X	12U + 1B
Jonsdottir et al., 2020 [105]	X	X	X	X		X	X			X				X								8 U
Kadone et al., 2020 [106]	X	X		X						X			X									5 B
Kinugawa et al., 2022 [107]	X	X					X	X														4 B
Lim et al., 2021 [95]	X	X		X			X	X		X	X			X								8 U
Routson et al., 2013 [96]	X	X		X		X	X	X		X				X								8 B
Srivastava et al., 2016 [97]	X	X	X	X	X	X	X	X		X				X								10 U
Tan et al., 2020 [99]	X	X		X				X			X		X									6 B
Tan et al., 2018 [98]	X	X		X				X			X		X									6 B
Van Criekinge et al., 2021 [100]	X	X			X		X	X		X					X							6 U + 1 B
Zhu et al., 2021 [101]	X	X		X		X	X	X		X				X								8 B

3.2.5. Muscle Synergies Extraction

In all the retained studies, muscle synergies were analyzed within the framework of the spatial (or synchronous) model, in which the muscle activity $\boldsymbol{m}$ of M muscles can be represented as the linear combination of a set of $N < M$ time-invariant modules $\boldsymbol{W}$ modulated by time-varying activation profiles $\boldsymbol{C}$, as follows:

$$\boldsymbol{m}(t) = \sum_{i=1}^{N} \boldsymbol{w}_i \, c_i(t) + \boldsymbol{\varepsilon}(t)$$

In which $\boldsymbol{m}(t)$ is a vector of EMG data samples of all the recorded muscles at time t, $\boldsymbol{w}_i$ is the i-th synergy vector, $c_i(t)$ is the synergy activation coefficient of the i-th synergy at time t, and $\boldsymbol{\varepsilon}(t)$ is additive noise vector at time t. In all studies but one, muscle synergies were extracted through a Non-negative Matrix Factorization (NMF) algorithm [135], which decomposes the muscle activation data matrix $\boldsymbol{M} = [\boldsymbol{m}(t_1) \ldots \boldsymbol{m}(t_T)]$, whose dimensions are $[M \times T]$ where M is the number of muscles and T the number of time samples, into two matrices such that $\boldsymbol{M} = \boldsymbol{WC} + \varepsilon$, where $\boldsymbol{W}$ is an $[M \times N]$ matrix with N muscle synergies, and $\boldsymbol{C}$ is an $[N \times T]$ matrix of synergy activation coefficients. One study [93] implemented an algorithm, the Weighted Non-negative Matrix Factorization (WNMF) [136], that differs from the traditional NMF as it assigns each data sample a weight (1 = EMG present, 0 = EMG absent), to accommodate clinical data that contain poor or missing EMG channels.

The number of muscle synergies, which is a free parameter of the factorization algorithm, was defined in all studies but one, according to the uncentered Variance Accounted For (*VAF*), which measures the quality of the experimental EMG data with the extracted synergies:

$$VAF = 100 \cdot \left(1 - \frac{\sum_t (\boldsymbol{m}_t - (\boldsymbol{W}\boldsymbol{c}_t))^2}{\sum_t (\boldsymbol{m}_t)^2}\right)$$

One study [103] used a centered *VAF*, otherwise called the coefficient of determination (R^2), defined as follows:

$$R^2 = 100 \cdot \left(1 - \frac{\sum_t (\boldsymbol{m}_t - (\boldsymbol{W}\boldsymbol{c}_t))^2}{\sum_t (\boldsymbol{m}_t - \overline{\boldsymbol{m}})^2}\right)$$

where $\overline{\boldsymbol{m}}$ is the mean of the observed EMG data and retained the number of synergies at which the R^2 vs. number of synergies curve achieves the highest curvature [137].

Seven studies identified the number of synergies as the set whose *VAF* was higher than 90% [93,96,97,100,102,105,106]. Three studies [94,98,99] added to the condition of *VAF* > 90% a second condition, by which a new synergy, added to the identified set, did not increase the *VAF* more than 5%. One study [101] exploited a 'three-way *VAF* > 90%', which imposed that the overall *VAF*, calculated with all muscles through the entire gait cycle, each of the *VAF* values separately calculated for each muscle throughout the entire gait cycle, and each of the *VAF* values calculated with all muscles within six separate gait phases were higher than 90%. In contrast, two studies [104,107] calculated the total variance accounted for by one synergy from the EMG data (*VAF1*), and one study [95] reconstructed the EMG signals collected from patients with four synergies extracted from healthy participants.

3.2.6. Muscle Synergy Analysis and Improvement-Related Metrics

To assess the effectiveness of the adopted rehabilitation therapy, the retained studies relied on a set of metrics related to the spatiotemporal structure of the extracted muscle synergies and compared their evolution along the rehabilitation process, mostly comparing the T0–T1 modifications in these metrics before (T0) and after (T1) the therapy and sometimes comparing them with the metrics of healthy participants, when available.

- Clinical evaluation

As shown in Table 3, overall, patients improved in all clinical scales after the rehabilitation protocols. The most sensitive scales to clinical improvement in motor function resulted in the Functional Independence Measure-Locomotion (FIM-Locomotion) [114], the 10 m timed walk (10MTW) for gait speed [125], and 6 min walk test (6MWT) [124]. However, only [93] investigated correlations between muscle synergies and clinical scales. The authors found from moderate to high correlations between BBS, TCT, and motor subscale of FIM and VAF1 of the affected and unaffected leg.

Table 3. Effect of the rehabilitation therapy on measured clinical scales.

	Clinical Scales		
	Improved	**Not Improved**	**Not Altered**
Allen et al., 2017 [102]	UPDRS-III, BBS, FAB, DGI, TUG, 6MWT		
Ambrosini et al., 2020 [93]	MI, TCT, BBS, FIMM		
Conner et al., 2021 [104]			
Ferrante et al., 2016 [94]	mini best test, Fugl-Meyer motor		
Ghislieri et al., 2023 [103]	UPDRS-III, FAB		MMSE
Jonsdottir et al., 2020 [105]	2MWT, 10MWT		BBS
Kadone et al., 2020 [106]	FIM motor, Barthel, FAC	10MWT	
Kinugawa et al., 2022 [107]	10MWT	FMA, BRS	
Lim et al., 2021 [95]	10MWT		BBS
Routson et al., 2013 [96]	FMA		DGI
Srivastava et al., 2016 [97]	FMA, FGA, 6MWT, TUG		
Tan et al., 2020 [99]	FIM motor, FIM locomotion, FMA lower ex		
Tan et al., 2018 [98]	FIM motor, FIM locomotion, FMA lower ex		
Van Criekinge et al., 2021 [100]	FAC, TIS, POMA Tinetti,		Barthel
Zhu et al., 2021 [101]	10MWT, 6MWT	TUG	

- Number of synergies

Among the analyzed parameters, the number of extracted synergies was the most commonly adopted metric (see Table 4), as 10 out of the 15 retained studies used the number of extracted synergies as a marker of the complexity of the modular organization [79] to assess rehabilitation effectiveness [93,94,96,98–103,105].

Table 4. Type of rehabilitation therapy (R = robotic based, F = FES-based, O = other techniques) and its effect of the rehabilitation therapy on muscle synergies characteristics (+: improvement post-rehabilitation, −: no improvement post-rehabilitation, n/a: comparison not performed).

	Type of Therapy	Number of Synergies	Spatial Synergies	Temporal Activations	Coordination Symmetry
Allen et al., 2017 [102]	O	−	+	n/a	n/a
Ambrosini et al., 2020 [93]	F	−	+	+	−
Conner et al., 2021 [104]	R	+	n/a	n/a	n/a
Ferrante et al., 2016 [94]	F	+	n/a	n/a	n/a

Table 4. *Cont.*

	Type of Therapy	Number of Synergies	Spatial Synergies	Temporal Activations	Coordination Symmetry
Ghislieri et al., 2023 [103]	O	+	+	+	n/a
Jonsdottir et al., 2020 [105]	O	–	–	+	n/a
Kadone et al., 2020 [106]	R	–	n/a	n/a	n/a
Kinugawa et al., 2022 [107]	F	+	n/a	n/a	n/a
Lim et al., 2021 [95]	F	n/a	+	+	n/a
Routson et al., 2013 [96]	O	+	+	+	n/a
Srivastava et al., 2016 [97]	R	n/a	+	–	n/a
Tan et al., 2020 [99]	R	+	n/a	n/a	+
Tan et al., 2018 [98]	R	–	n/a	n/a	+
Van Criekinge et al., 2021 [100]	O	–	+	n/a	n/a
Zhu et al., 2021 [101]	R	–	–	+	n/a

No accordance was identified among studies in terms of changes in the number of muscle synergies after the intervention. Some of the studies on stroke patients reported that an increase in the number of muscle synergies reflected an increase in performance [94,96,100] but other studies on stroke patients [93] and on patients with multiple sclerosis [105] or with Parkinson's disease [103] did not show such change after the therapy. One study even reported a reduction in the number of muscle synergies for a subset of participants [102].

- Spatial and temporal organization

A total of 10 out of the 15 studies made quantitative assessments of the spatial composition of the extracted muscle synergies, either by comparing their spatial structure [93–101,103,105] or their degree of bilateral symmetry [98,99] to that of healthy individuals. The same studies also used the temporal activation coefficients to define quantitative metrics, either by calculating the temporal symmetry between sides or by measuring the similarity with the synergy activation pattern of healthy controls. A subset of studies (4 out of 15) defined VAF-based metrics to compactly indicate the complexity of muscle coordination with the VAF explained by one synergy [93,104,106,107] or two and three synergies [106]. One study constructed specific metrics on the synergy vectors, assessing generalizability, sparsity, and variability [102].

An increase in the complexity after rehabilitation was identified in patients with brain tumor [107] or cerebral palsy [104] but not in stroke patients [93] or patients with myelopathy [106]. Finally, an increase in the synergy coefficients within gait subphases was correlated with performance increase in patients with multiple sclerosis [105].

Some neurological diseases, such as unilateral stroke, led to a kinematical asymmetry in the lower limbs during locomotion [138], which is reflected in an asymmetry in muscle synergies [139]. Therefore, an improvement in the synergy symmetry, which was accompanied by an improvement in performance, was demonstrated by [98], or a synergy timing symmetry, not accompanied by improvements in clinical scores, was identified by [99].

4. Discussion

The goal of the present review was to assess the state of the art of the investigation of the effect of gait rehabilitation in patients with neurological diseases in terms of changes in the organization and recruitment of muscle synergies. Even though muscle synergies as a tool to investigate motor coordination was introduced over two decades ago [140–142] and despite several studies demonstrating that neurological patients showed altered muscle synergies with respect to healthy participants [51,86,143–146] and that synergies were proposed as a potential candidate marker for the quantitative assessment of neurological pathologies [143,147], only a few studies have specifically investigated the alteration of muscle synergies after rehabilitation. Specifically, we found only 15 studies in the last 13 years that investigated the effect of gait rehabilitation therapies on muscle synergies in neurological patients and to what extent the changes in muscle synergies are related to clinical improvements.

Fourteen out of the 15 selected studies reported a modification of the muscle synergies after gait rehabilitation, and only one study [106] identified no clear effect on muscle synergies. Most of the studies that involved a control group demonstrated that rehabilitation makes the muscle synergies more similar to those of healthy participants in terms of structure. In contrast, the temporal patterns of activation of the muscle synergies identified in patients with multiple sclerosis after rehabilitation differed from those of healthy controls when walking at comparable speeds [105]. An improvement in the synergy symmetry [98], or timing synergy symmetry [99], were also identified.

No accordance could be found across studies, in terms of changes in the number of muscle synergies. While some studies on stroke patients demonstrated an increase in the number of muscle synergies, also related to a performance increase [94,96,100], other studies that enrolled patients with multiple sclerosis [105], Parkinson's disease [102], or stroke patients [93] either found no change or even a reduction in the number of muscle synergies [102]. Moreover, an increase in the number of synergies after rehabilitation was identified in patients with brain tumor [107] or cerebral palsy [104] but not in stroke patients [93] or patients with myelopathy [106]. Finally, an increase in the synergy coefficients within gait subphases was correlated with performance increase in patients with multiple sclerosis [105].

Overall, this review supports the hypothesis that modifications in the muscle synergies can index the progression of the rehabilitation process in an interpretable and quantitatively measurable manner.

Discrepancies were found among the studies. These discrepancies may be a consequence of the investigated neurological pathologies, which differ among the studies, and therefore may differently influence motor control and muscle synergies. Moreover, these discrepancies may reflect the lack of standardized protocols for processing and investigating muscle synergies and to compare them with those extracted from healthy participants.

When characterizing the modular control of locomotion in healthy subjects, four-to--five muscle synergies are typically extracted, and they are usually related to four specific biomechanical sub-functions along the gait cycle [79]:

- M1—a knee-hip extensor module, activated during early stance, serving as body support and weight acceptance
- M2—a calf plantar-flexor muscle module, activated during late stance, with forward propulsion, body support, and swing preparation function
- M3—an ankle dorsiflexion module, activated during early swing, contributing to the ground clearance of the foot
- M4—a knee flexor module, activated during late swing, to decelerate the leg an prepare heel strike

Despite the previously discussed large variability both in adopted methodologies and pathological conditions across studies, few invariances could be identified regarding the effect of rehabilitation therapy on muscle synergies. Indeed, regardless of the adopted metric to characterize the change, all those studies showing modifications in the spatial organization or temporal recruitment of muscle synergies after the therapy [70–73,77,78,82] reported a modification related to the plantarflexion module M2 and/or the knee flexor module M4, likely indicating an effect on forward propulsion and balance.

A recommendation for future research is to define standardized protocols and algorithms for the characterization of synergy structures and patterns in neurological patients, and for their comparison with healthy participants [148]. This would allow for a systematic characterization and comparison of the muscle synergies of patients with different pathologies and different levels of motor impairment.

Moreover, this review underlines the need for further research on muscle synergy analysis in the assessment of neurological patients' rehabilitation, before it can be fully transferred to the clinical practice as a marker of the progression of the rehabilitation and as a support to clinical decision making [85], which would provide clinicians and therapists with a novel instrument to assess the efficacy of a therapy and whether and how it should be changed, not only by analyzing movement kinematics and kinetics but also through the lens of the underlying neural control strategies.

5. Conclusions

The present review strongly suggests that muscle synergy analysis has a very high potential as a tool to quantitatively assess the efficacy of rehabilitation therapies in neurological patients. However, it also highlights the significant lack of studies specifically investigating the effect of physical rehabilitation on muscle synergies. In the last 13 years, only 15 studies have examined the alteration of muscle synergies after physical rehabilitation during locomotion tasks. This indicates that the potential of muscle synergy analysis remains largely untapped as it has been scarcely used in clinical studies so far. For this promising approach to move to the clinical practice, the scientific community working on these topics should spend a greater effort in defining a methodological standardization of the assessment protocols and algorithms for the extraction and description of muscle synergies, together with the creation of an extensive publicly available database of the synergies identified in patients with different neurological pathologies and different levels of impairment. This would facilitate the standardization of the adopted procedures, together with the related metrics to quantify the effect of motor rehabilitation on the muscle synergies of a patient. This would provide clinicians and physiotherapists with a novel tool to be used as a marker of the effectiveness of therapy, as well as a source of information to develop innovative therapies.

Author Contributions: Conceptualization, D.B., C.D.M. and A.d.; methodology, D.B., C.D.M. and A.Q. (Angelica Quercia); formal analysis, D.B., C.D.M. and A.Q. (Angelica Quercia); data curation, D.B., C.D.M. and A.Q. (Angelica Quercia); writing—original draft preparation, D.B., C.D.M. and A.Q. (Angelica Quercia); writing—review and editing, D.B., C.D.M., A.Q. (Angelica Quercia), P.D.P., A.C., A.Q. (Angelo Quartarone), R.S.C. and A.d. All authors have read and agreed to the published version of the manuscript.

Funding: This research was funded by the American Ministry of Defense (DoD Grant), grant number W81XWH-19-1-0810.

Institutional Review Board Statement: Not applicable.

Informed Consent Statement: Not applicable.

Data Availability Statement: Data can be made available on request by the authors.

Conflicts of Interest: The authors declare no conflicts of interest.

References

1. Knutsson, E.; Richards, C. Different Types of Disturbed Motor Control in Gait of Hemiparetic Patients. *Brain* **1979**, *102*, 405–430. [CrossRef] [PubMed]
2. Petersen, T.H.; Willerslev-Olsen, M.; Conway, B.A.; Nielsen, J.B. The Motor Cortex Drives the Muscles during Walking in Human Subjects. *J. Physiol.* **2012**, *590*, 2443–2452. [CrossRef] [PubMed]
3. Schubert, M.; Curt, A.; Jensen, L.; Dietz, V. Corticospinal Input in Human Gait: Modulation of Magnetically Evoked Motor Responses. *Exp. Brain Res.* **1997**, *115*, 234–246. [CrossRef]
4. Winter, D.A.; MacKinnon, C.D.; Ruder, G.K.; Wieman, C. Chapter 32 An Integrated EMG/Biomechanical Model of Upper Body Balance and Posture during Human Gait. In *Progress in Brain Research*; Natural and Artificial Control of Hearing and Balance, Allum, J.H.J., Allum-Mecklenburg, D.J., Harris, F.P., Probst, R., Eds., Elsevier: Amsterdam, The Netherlands, 1993; Volume 97, pp. 359–367.
5. Stolze, H.; Klebe, S.; Zechlin, C.; Baecker, C.; Friege, L.; Deuschl, G. Falls in Frequent Neurologicaldiseases. *J. Neurol.* **2004**, *251*, 79–84. [CrossRef] [PubMed]

6. Ebersbach, G.; Sojer, M.; Valldeoriola, F.; Wissel, J.; Müller, J.; Tolosa, E.; Poewe, W. Comparative Analysis of Gait in Parkinson's Disease, Cerebellar Ataxia and Subcortical Arteriosclerotic Encephalopathy. *Brain* **1999**, *122*, 1349–1355. [CrossRef] [PubMed]
7. Chaparro-Cárdenas, S.L.; Lozano-Guzmán, A.A.; Ramirez-Bautista, J.A.; Hernández-Zavala, A. A Review in Gait Rehabilitation Devices and Applied Control Techniques. *Disabil. Rehabil. Assist. Technol.* **2018**, *13*, 819–834. [CrossRef]
8. Vitiello, N.; Oddo, C.M.; Lenzi, T.; Roccella, S.; Beccai, L.; Vecchi, F.; Carrozza, M.C.; Dario, P. Neuro-Robotics Paradigm for Intelligent Assistive Technologies. In *Intelligent Assistive Robots: Recent Advances in Assistive Robotics for Everyday Activities*; Springer Tracts in Advanced Robotics; Mohammed, S., Moreno, J.C., Kong, K., Amirat, Y., Eds.; Springer International Publishing: Cham, Switzerland, 2015; pp. 1–40. ISBN 978-3-319-12922-8.
9. Senanayake, C.; Senanayake, S.M.N.A. Emerging Robotics Devices for Therapeutic Rehabilitation of the Lower Extremity. In Proceedings of the 2009 IEEE/ASME International Conference on Advanced Intelligent Mechatronics, Singapore, 14–17 July 2009; pp. 1142–1147.
10. Díaz, I.; Gil, J.J.; Sánchez, E. Lower-Limb Robotic Rehabilitation: Literature Review and Challenges. *J. Robot.* **2011**, *2011*, e759764. [CrossRef]
11. Balaji, E.; Brindha, D.; Balakrishnan, R. Supervised Machine Learning Based Gait Classification System for Early Detection and Stage Classification of Parkinson's Disease. *Appl. Soft Comput.* **2020**, *94*, 106494. [CrossRef]
12. Kalron, A.; Achiron, A.; Dvir, Z. Muscular and Gait Abnormalities in Persons With Early Onset Multiple Sclerosis. *J. Neurol. Phys. Ther.* **2011**, *35*, 164–169. [CrossRef]
13. Pistacchi, M.; Gioulis, M.; Sanson, F.; De Giovannini, E.; Filippi, G.; Rossetto, F.; Marsala, S.Z. Gait Analysis and Clinical Correlations in Early Parkinson's Disease. *Funct. Neurol.* **2017**, *32*, 28–34. [CrossRef]
14. Alito, A.; Fenga, D.; Portaro, S.; Leonardi, G.; Borzelli, D.; Sanzarello, I.; Calabrò, R.; Milone, D.; Tisano, A.; Leonetti, D. Early Hip Fracture Surgery and Rehabilitation. How to Improve Functional Quality Outcomes. A Retrospective Study. *Folia Med.* **2023**, *65*, 879–884. [CrossRef]
15. Borggraefe, I.; Meyer-Heim, A.; Kumar, A.; Schaefer, J.S.; Berweck, S.; Heinen, F. Improved Gait Parameters after Robotic-Assisted Locomotor Treadmill Therapy in a 6-Year-Old Child with Cerebral Palsy. *Mov. Disord.* **2008**, *23*, 280–283. [CrossRef]
16. Gunning, E.; Uszynski, M.K. Effectiveness of the Proprioceptive Neuromuscular Facilitation Method on Gait Parameters in Patients With Stroke: A Systematic Review. *Arch. Phys. Med. Rehabil.* **2019**, *100*, 980–986. [CrossRef]
17. Rocha, P.A.; Porfírio, G.M.; Ferraz, H.B.; Trevisani, V.F.M. Effects of External Cues on Gait Parameters of Parkinson's Disease Patients: A Systematic Review. *Clin. Neurol. Neurosurg.* **2014**, *124*, 127–134. [CrossRef]
18. Maggio, M.G.; Cezar, R.P.; Milardi, D.; Borzelli, D.; De Marchis, C.; D'avella, A.; Quartarone, A.; Calabrò, R.S. Do Patients with Neurological Disorders Benefit from Immersive Virtual Reality? A Scoping Review on the Emerging Use of the Computer-Assisted Rehabilitation Environment. *Eur. J. Phys. Rehabil. Med.* **2024**, *60*, 37–43. [CrossRef]
19. Ruiz-Ruiz, L.; Jimenez, A.R.; Garcia-Villamil, G.; Seco, F. Detecting Fall Risk and Frailty in Elders with Inertial Motion Sensors: A Survey of Significant Gait Parameters. *Sensors* **2021**, *21*, 6918. [CrossRef]
20. Kondragunta, J.; Hirtz, G. Gait Parameter Estimation of Elderly People Using 3D Human Pose Estimation in Early Detection of Dementia. In Proceedings of the 2020 42nd Annual International Conference of the IEEE Engineering in Medicine & Biology Society (EMBC), Montreal, QC, Canada, 20–24 July 2020; pp. 5798–5801.
21. Pfister, A.; West, A.M.; Bronner, S.; Noah, J.A. Comparative Abilities of Microsoft Kinect and Vicon 3D Motion Capture for Gait Analysis. *J. Med. Eng. Technol.* **2014**, *38*, 274–280. [CrossRef]
22. Sigal, L.; Balan, A.O.; Black, M.J. HumanEva: Synchronized Video and Motion Capture Dataset and Baseline Algorithm for Evaluation of Articulated Human Motion. *Int. J. Comput. Vis.* **2010**, *87*, 4–27. [CrossRef]
23. Kanko, R.M.; Laende, E.K.; Strutzenberger, G.; Brown, M.; Selbie, W.S.; DePaul, V.; Scott, S.H.; Deluzio, K.J. Assessment of Spatiotemporal Gait Parameters Using a Deep Learning Algorithm-Based Markerless Motion Capture System. *J. Biomech.* **2021**, *122*, 110414. [CrossRef]
24. Mathis, M.W.; Mathis, A. Deep Learning Tools for the Measurement of Animal Behavior in Neuroscience. *Curr. Opin. Neurobiol.* **2020**, *60*, 1–11. [CrossRef]
25. Gu, C.; Lin, W.; He, X.; Zhang, L.; Zhang, M. IMU-Based Motion Capture System for Rehabilitation Applications: A Systematic Review. *Biomim. Intell. Robot.* **2023**, *3*, 100097. [CrossRef]
26. Gujarathi, T.; Bhole, K. Gait Analysis Using Imu Sensor. In Proceedings of the 2019 10th International Conference on Computing, Communication and Networking Technologies (ICCCNT), Kanpur, India, 6–8 July 2019; pp. 1–5.
27. Panero, E.; Digo, E.; Agostini, V.; Gastaldi, L. Comparison of Different Motion Capture Setups for Gait Analysis: Validation of Spatio-Temporal Parameters Estimation. In Proceedings of the 2018 IEEE International Symposium on Medical Measurements and Applications (MeMeA), Rome, Italy, 11–13 June 2018; pp. 1–6.
28. Washabaugh, E.P.; Kalyanaraman, T.; Adamczyk, P.G.; Claflin, E.S.; Krishnan, C. Validity and Repeatability of Inertial Measurement Units for Measuring Gait Parameters. *Gait Posture* **2017**, *55*, 87–93. [CrossRef]
29. Belli, A.; Bui, P.; Berger, A.; Geyssant, A.; Lacour, J.-R. A Treadmill Ergometer for Three-Dimensional Ground Reaction Forces Measurement during Walking. *J. Biomech.* **2001**, *34*, 105–112. [CrossRef]
30. Dierick, F.; Penta, M.; Renaut, D.; Detrembleur, C. A Force Measuring Treadmill in Clinical Gait Analysis. *Gait Posture* **2004**, *20*, 299–303. [CrossRef]

31. Campanini, I.; Disselhorst-Klug, C.; Rymer, W.Z.; Merletti, R. Surface EMG in Clinical Assessment and Neurorehabilitation: Barriers Limiting Its Use. *Front. Neurol.* **2020**, *11*, 934. [CrossRef]
32. Stlberg, E.; Falck, B. The Role of Electromyography in Neurology. *Electroencephalogr. Clin. Neurophysiol.* **1997**, *103*, 579–598. [CrossRef]
33. Zwarts, M.J.; Drost, G.; Stegeman, D.F. Recent Progress in the Diagnostic Use of Surface EMG for Neurological Diseases. *J. Electromyogr. Kinesiol.* **2000**, *10*, 287–291. [CrossRef]
34. Borzelli, D.; Cesqui, B.; Berger, D.J.; Burdet, E.; d'Avella, A. Muscle Patterns Underlying Voluntary Modulation of Co-Contraction. *PLoS ONE* **2018**, *13*, e0205911. [CrossRef]
35. Borzelli, D.; Pastorelli, S.; Gastaldi, L. Determination of the Human Arm Stiffness Efficiency with a Two Antagonist Muscles Model. In *Advances in Italian Mechanism Science: Proceedings of the First International Conference of IFToMM Italy, Vicenza, Italy, 1–2 December 2016*; Boschetti, G., Gasparetto, A., Eds.; Springer International Publishing: Cham, Switzerland, 2017; pp. 71–78.
36. Borzelli, D.; Burdet, E.; Pastorelli, S.; d'Avella, A.; Gastaldi, L. Identification of the Best Strategy to Command Variable Stiffness Using Electromyographic Signals. *J. Neural Eng.* **2020**, *17*, 016058. [CrossRef]
37. Cimolato, A.; Driessen, J.J.M.; Mattos, L.S.; De Momi, E.; Laffranchi, M.; De Michieli, L. EMG-Driven Control in Lower Limb Prostheses: A Topic-Based Systematic Review. *J. Neuroeng. Rehabil.* **2022**, *19*, 43. [CrossRef]
38. Manal, K.; Gonzalez, R.V.; Lloyd, D.G.; Buchanan, T.S. A Real-Time EMG-Driven Virtual Arm. *Comput. Biol. Med.* **2002**, *32*, 25–36. [CrossRef]
39. Yang, H.; Wan, J.; Jin, Y.; Yu, X.; Fang, Y. EEG- and EMG-Driven Poststroke Rehabilitation: A Review. *IEEE Sens. J.* **2022**, *22*, 23649–23660. [CrossRef]
40. Borzelli, D.; Pastorelli, S.; Gastaldi, L. Model of the Human Arm Stiffness Exerted by Two Antagonist Muscles. In *Advances in Robot Design and Intelligent Control: Proceedings of the 25th Conference on Robotics in Alpe-Adria-Danube Region (RAAD16), Belgrade, Serbia, 30 June–2 July 2016*; Rodić, A., Borangiu, T., Eds.; Springer International Publishing: Cham, Switzerland, 2017; pp. 285–292.
41. Borzelli, D.; Pastorelli, S.; d'Avella, A.; Gastaldi, L. Virtual Stiffness: A Novel Biomechanical Approach to Estimate Limb Stiffness of a Multi-Muscle and Multi-Joint System. *Sensors* **2023**, *23*, 673. [CrossRef]
42. Borzelli, D.; Gurgone, S.; De Pasquale, P.; Lotti, N.; d'Avella, A.; Gastaldi, L. Use of Surface Electromyography to Estimate End-Point Force in Redundant Systems: Comparison between Linear Approaches. *Bioengineering* **2023**, *10*, 234. [CrossRef]
43. Borzelli, D.; Pastorelli, S.; Gastaldi, L. Elbow Musculoskeletal Model for Industrial Exoskeleton with Modulated Impedance Based on Operator's Arm Stiffness. *Int. J. Autom. Technol.* **2017**, *11*, 442–449. [CrossRef]
44. Berger, D.J.; Gentner, R.; Edmunds, T.; Pai, D.K.; d'Avella, A. Differences in Adaptation Rates after Virtual Surgeries Provide Direct Evidence for Modularity. *J. Neurosci.* **2013**, *33*, 12384–12394. [CrossRef]
45. Berger, D.J.; Masciullo, M.; Molinari, M.; Lacquaniti, F.; d'Avella, A. Does the Cerebellum Shape the Spatiotemporal Organization of Muscle Patterns? Insights from Subjects with Cerebellar Ataxias. *J. Neurophysiol.* **2020**, *123*, 1691–1710. [CrossRef]
46. Berger, D.J.; Borzelli, D.; d'Avella, A. Task Space Exploration Improves Adaptation after Incompatible Virtual Surgeries. *J. Neurophysiol.* **2022**, *127*, 1127–1146. [CrossRef]
47. Borzelli, D.; Berger, D.; Pai, D.; D'avella, A. Effort Minimization and Synergistic Muscle Recruitment for Three-Dimensional Force Generation. *Front. Comput. Neurosci.* **2013**, *7*, 186. [CrossRef]
48. Cheung, V.C.K.; Piron, L.; Agostini, M.; Silvoni, S.; Turolla, A.; Bizzi, E. Stability of Muscle Synergies for Voluntary Actions after Cortical Stroke in Humans. *Proc. Natl. Acad. Sci. USA* **2009**, *106*, 19563–19568. [CrossRef]
49. d'Avella, A.; Portone, A.; Fernandez, L.; Lacquaniti, F. Control of Fast-Reaching Movements by Muscle Synergy Combinations. *J. Neurosci.* **2006**, *26*, 7791–7810. [CrossRef]
50. De Marchis, C.; Schmid, M.; Bibbo, D.; Bernabucci, I.; Conforto, S. Inter-Individual Variability of Forces and Modular Muscle Coordination in Cycling: A Study on Untrained Subjects. *Hum. Mov. Sci.* **2013**, *32*, 1480–1494. [CrossRef]
51. Dipietro, L.; Krebs, H.I.; Fasoli, S.E.; Volpe, B.T.; Stein, J.; Bever, C.; Hogan, N. Changing Motor Synergies in Chronic Stroke. *J. Neurophysiol.* **2007**, *98*, 757–768. [CrossRef] [PubMed]
52. Gentner, R.; Edmunds, T.; Pai, D.; D'avella, A. Robustness of Muscle Synergies during Visuomotor Adaptation. *Front. Comput. Neurosci.* **2013**, *7*, 120. [CrossRef]
53. Ivanenko, Y.P.; Poppele, R.E.; Lacquaniti, F. Five Basic Muscle Activation Patterns Account for Muscle Activity during Human Locomotion. *J. Physiol.* **2004**, *556*, 267–282. [CrossRef] [PubMed]
54. Roh, J.; Rymer, W.Z.; Perreault, E.J.; Yoo, S.B.; Beer, R.F. Alterations in Upper Limb Muscle Synergy Structure in Chronic Stroke Survivors. *J. Neurophysiol.* **2013**, *109*, 768–781. [CrossRef]
55. Torres-Oviedo, G.; Ting, L.H. Subject-Specific Muscle Synergies in Human Balance Control Are Consistent Across Different Biomechanical Contexts. *J. Neurophysiol.* **2010**, *103*, 3084–3098. [CrossRef]
56. Borzelli, D.; Gurgone, S.; Mezzetti, M.; De Pasquale, P.; Berger, D.J.; Milardi, D.; Acri, G.; D'Avella, A. Adaptation to Virtual Surgeries Across Multiple Practice Sessions. In *Converging Clinical and Engineering Research on Neurorehabilitation IV: Proceedings of the 5th International Conference on Neurorehabilitation (ICNR2020), Online, 13–16 October 2020*; Torricelli, D., Akay, M., Pons, J.L., Eds.; Springer International Publishing: Cham, Switzerland, 2022; pp. 563–568.

57. Borzelli, D.; Gurgone, S.; De Pasquale, P.; Berger, D.J.; d'Avella, A. Consistency of Myoelectric Control Across Multiple Sessions. In *Converging Clinical and Engineering Research on Neurorehabilitation III: Proceedings of the 4th International Conference on NeuroRehabilitation (ICNR2018), Pisa, Italy, 16–20 October 2018*; Masia, L., Micera, S., Akay, M., Pons, J.L., Eds.; Springer International Publishing: Cham, Switzerland, 2019; pp. 1166–1170.
58. Borzelli, D.; Vieira, T.M.M.; Botter, A.; Gazzoni, M.; Lacquaniti, F.; d'Avella, A. Synaptic Inputs to Motor Neurons Underlying Muscle Co-Activation for Functionally Different Tasks Have Different Spectral Characteristics. *J. Neurophysiol.* **2024**, *131*, 1126–1142. [CrossRef]
59. Laine, C.M.; Martinez-Valdes, E.; Falla, D.; Mayer, F.; Farina, D. Motor Neuron Pools of Synergistic Thigh Muscles Share Most of Their Synaptic Input. *J. Neurosci.* **2015**, *35*, 12207–12216. [CrossRef]
60. Levine, J.; Avrillon, S.; Farina, D.; Hug, F.; Pons, J.L. Two Motor Neuron Synergies, Invariant across Ankle Joint Angles, Activate the Triceps Surae during Plantarflexion. *J. Physiol.* **2023**, *601*, 4337–4354. [CrossRef]
61. David, F.J.; Munoz, M.J.; Corcos, D.M. The Effect of STN DBS on Modulating Brain Oscillations: Consequences for Motor and Cognitive Behavior. *Exp. Brain Res.* **2020**, *238*, 1659–1676. [CrossRef]
62. Leonardi, G.; Ciurleo, R.; Cucinotta, F.; Fonti, B.; Borzelli, D.; Costa, L.; Tisano, A.; Portaro, S.; Alito, A. The Role of Brain Oscillations in Post-Stroke Motor Recovery: An Overview. *Front. Syst. Neurosci.* **2022**, *16*, 947421. [CrossRef]
63. MacKay, W.A. Synchronized Neuronal Oscillations and Their Role in Motor Processes. *Trends Cogn. Sci.* **1997**, *1*, 176–183. [CrossRef] [PubMed]
64. Basmajian, J.V. Muscles Alive. Their Functions Revealed by Electromyography. *Acad. Med.* **1962**, *37*, 802.
65. De Luca, C.J. Physiology and Mathematics of Myoelectric Signals. *IEEE Trans. Biomed. Eng.* **1979**, *BME-26*, 313–325. [CrossRef]
66. Lugo, J.E.; Doti, R.; Faubert, J. Planckian Power Spectral Densities from Human Calves during Posture Maintenance and Controlled Isometric Contractions. *PLoS ONE* **2015**, *10*, e0131798. [CrossRef]
67. Lindstrom, L.H.; Magnusson, R.I. Interpretation of Myoelectric Power Spectra: A Model and Its Applications. *Proc. IEEE* **1977**, *65*, 653–662. [CrossRef]
68. Merletti, R.; Parker, P.J. *Electromyography: Physiology, Engineering, and Non-Invasive Applications*; John Wiley & Sons: Hoboken, NJ, USA, 2004; ISBN 978-0-471-67580-8.
69. Castronovo, A.M.; De Marchis, C.; Schmid, M.; Conforto, S.; Severini, G. Effect of Task Failure on Intermuscular Coherence Measures in Synergistic Muscles. *Appl. Bionics Biomech.* **2018**, *2018*, e4759232. [CrossRef]
70. Yamada, H.; Okada, M.; Oda, T.; Nemoto, S.; Shiozaki, T.; Kizuka, T.; Kuno, S.; Masuda, T. Effects of Aging on Emg Variables During Fatiguing Isometric Contractions. *J. Hum. Ergol.* **2000**, *29*, 7–14. [CrossRef]
71. Petrofsky, J.; Laymon, M. Muscle Temperature and EMG Amplitude and Frequency during Isometric Exercise. *Aviat. Space Environ. Med.* **2005**, *76*, 1024–1030.
72. Farina, D.; Negro, F.; Dideriksen, J.L. The Effective Neural Drive to Muscles Is the Common Synaptic Input to Motor Neurons: Effective Neural Drive to Muscles. *J. Physiol.* **2014**, *592*, 3427–3441. [CrossRef]
73. Henneman, E.; Somjen, G.; Carpenter, D.O. Functional Significance of Cell Size in Spinal Motoneurons. *J. Neurophysiol.* **1965**, *28*, 560–580. [CrossRef]
74. Henneman, E. Relation between Size of Neurons and Their Susceptibility to Discharge. *Science* **1957**, *126*, 1345–1347. [CrossRef] [PubMed]
75. Monster, A.W.; Chan, H. Isometric Force Production by Motor Units of Extensor Digitorum Communis Muscle in Man. *J. Neurophysiol.* **1977**, *40*, 1432–1443. [CrossRef]
76. Hug, F.; Avrillon, S.; Ibáñez, J.; Farina, D. Common Synaptic Input, Synergies and Size Principle: Control of Spinal Motor Neurons for Movement Generation. *J. Physiol.* **2023**, *601*, 11–20. [CrossRef] [PubMed]
77. de Luca, C.J.; LeFever, R.S.; McCue, M.P.; Xenakis, A.P. Behaviour of Human Motor Units in Different Muscles during Linearly Varying Contractions. *J. Physiol.* **1982**, *329*, 113–128. [CrossRef]
78. Chvatal, S.A.; Ting, L.H. Voluntary and Reactive Recruitment of Locomotor Muscle Synergies during Perturbed Walking. *J. Neurosci.* **2012**, *32*, 12237–12250. [CrossRef]
79. Clark, D.J.; Ting, L.H.; Zajac, F.E.; Neptune, R.R.; Kautz, S.A. Merging of Healthy Motor Modules Predicts Reduced Locomotor Performance and Muscle Coordination Complexity Post-Stroke. *J. Neurophysiol.* **2010**, *103*, 844–857. [CrossRef]
80. De Marchis, C.; Ranaldi, S.; Serrao, M.; Ranavolo, A.; Draicchio, F.; Lacquaniti, F.; Conforto, S. Modular Motor Control of the Sound Limb in Gait of People with Trans-Femoral Amputation. *J. NeuroEng. Rehabil.* **2019**, *16*, 132. [CrossRef] [PubMed]
81. Neptune, R.R.; Clark, D.J.; Kautz, S.A. Modular Control of Human Walking: A Simulation Study. *J. Biomech.* **2009**, *42*, 1282–1287. [CrossRef] [PubMed]
82. Rimini, D.; Agostini, V.; Knaflitz, M. Intra-Subject Consistency during Locomotion: Similarity in Shared and Subject-Specific Muscle Synergies. *Front. Hum. Neurosci.* **2017**, *11*, 292485. [CrossRef]
83. Safavynia, S.; Torres-Oviedo, G.; Ting, L. Muscle Synergies: Implications for Clinical Evaluation and Rehabilitation of Movement. *Top. Spinal Cord Inj. Rehabil.* **2011**, *17*, 16–24. [CrossRef]
84. Zhao, K.; Zhang, Z.; Wen, H.; Liu, B.; Li, J.; d'Avella, A.; Scano, A. Muscle Synergies for Evaluating Upper Limb in Clinical Applications: A Systematic Review. *Heliyon* **2023**, *9*, e16202. [CrossRef]
85. Scano, A.; Lanzani, V.; Brambilla, C.; d'Avella, A. Transferring Sensor-Based Assessments to Clinical Practice: The Case of Muscle Synergies. *Sensors* **2024**, *24*, 3934. [CrossRef]

86. Mileti, I.; Zampogna, A.; Santuz, A.; Asci, F.; Del Prete, Z.; Arampatzis, A.; Palermo, E.; Suppa, A. Muscle Synergies in Parkinson's Disease. *Sensors* **2020**, *20*, 3209. [CrossRef] [PubMed]
87. Hong, Y.N.G.; Ballekere, A.N.; Fregly, B.J.; Roh, J. Are Muscle Synergies Useful for Stroke Rehabilitation? *Curr. Opin. Biomed. Eng.* **2021**, *19*, 100315. [CrossRef]
88. Van Criekinge, T.; Vermeulen, J.; Wagemans, K.; Schröder, J.; Embrechts, E.; Truijen, S.; Hallemans, A.; Saeys, W. Lower Limb Muscle Synergies during Walking after Stroke: A Systematic Review. *Disabil. Rehabil.* **2020**, *42*, 2836–2845. [CrossRef]
89. Beltrame, G.; Scano, A.; Marino, G.; Peccati, A.; Molinari Tosatti, L.; Portinaro, N. Recent Developments in Muscle Synergy Analysis in Young People with Neurodevelopmental Diseases: A Systematic Review. *Front. Bioeng. Biotechnol.* **2023**, *11*, 1145937. [CrossRef] [PubMed]
90. Khan, M.H.; Farid, M.S.; Grzegorzek, M. Vision-Based Approaches towards Person Identification Using Gait. *Comput. Sci. Rev.* **2021**, *42*, 100432. [CrossRef]
91. Liberati, A.; Altman, D.G.; Tetzlaff, J.; Mulrow, C.; Gøtzsche, P.C.; Ioannidis, J.P.A.; Clarke, M.; Devereaux, P.J.; Kleijnen, J.; Moher, D. The PRISMA Statement for Reporting Systematic Reviews and Meta-Analyses of Studies That Evaluate Health Care Interventions: Explanation and Elaboration. *Ann. Intern. Med.* **2009**, *151*, W-65–W-94. [CrossRef]
92. Page, M.J.; McKenzie, J.E.; Bossuyt, P.M.; Boutron, I.; Hoffmann, T.C.; Mulrow, C.D.; Shamseer, L.; Tetzlaff, J.M.; Akl, E.A.; Brennan, S.E.; et al. The PRISMA 2020 Statement: An Updated Guideline for Reporting Systematic Reviews. *Int. J. Surg.* **2021**, *88*, 105906. [CrossRef] [PubMed]
93. Ambrosini, E.; Parati, M.; Peri, E.; De Marchis, C.; Nava, C.; Pedrocchi, A.; Ferriero, G.; Ferrante, S. Changes in Leg Cycling Muscle Synergies after Training Augmented by Functional Electrical Stimulation in Subacute Stroke Survivors: A Pilot Study. *J. NeuroEng. Rehabil.* **2020**, *17*, 35. [CrossRef]
94. Ferrante, S.; Chia Bejarano, N.; Ambrosini, E.; Nardone, A.; Turcato, A.M.; Monticone, M.; Ferrigno, G.; Pedrocchi, A. A Personalized Multi-Channel FES Controller Based on Muscle Synergies to Support Gait Rehabilitation after Stroke. *Front. Neurosci.* **2016**, *10*, 425. [CrossRef]
95. Lim, J.; Lim, T.; Lee, J.; Sim, J.; Chang, H.; Yoon, B.; Jung, H. Patient-Specific Functional Electrical Stimulation Strategy Based on Muscle Synergy and Walking Posture Analysis for Gait Rehabilitation of Stroke Patients. *J. Int. Med. Res.* **2021**, *49*, 03000605211016782. [CrossRef]
96. Routson, R.L.; Clark, D.J.; Bowden, M.G.; Kautz, S.A.; Neptune, R.R. The Influence of Locomotor Rehabilitation on Module Quality and Post-Stroke Hemiparetic Walking Performance. *Gait Posture* **2013**, *38*, 511–517. [CrossRef] [PubMed]
97. Srivastava, S.; Kao, P.C.; Reisman, D.S.; Scholz, J.P.; Agrawal, S.K.; Higginson, J.S. Robotic Assist-As-Needed as an Alternative to Therapist-Assisted Gait Rehabilitation. *Int. J. Phys. Med. Rehabil.* **2016**, *4*, 370. [CrossRef] [PubMed]
98. Tan, C.K.; Kadone, H.; Watanabe, H.; Marushima, A.; Yamazaki, M.; Sankai, Y.; Suzuki, K. Lateral Symmetry of Synergies in Lower Limb Muscles of Acute Post-Stroke Patients After Robotic Intervention. *Front. Neurosci.* **2018**, *12*, 276. [CrossRef]
99. Tan, C.K.; Kadone, H.; Watanabe, H.; Marushima, A.; Hada, Y.; Yamazaki, M.; Sankai, Y.; Matsumura, A.; Suzuki, K. Differences in Muscle Synergy Symmetry between Subacute Post-Stroke Patients with Bioelectrically-Controlled Exoskeleton Gait Training and Conventional Gait Training. *Front. Bioeng. Biotechnol.* **2020**, *8*, 770. [CrossRef]
100. Van Criekinge, T.; Hallemans, A.; Herssens, N.; Lafosse, C.; Claes, D.; De Hertogh, W.; Truijen, S.; Saeys, W. SWEAT2 Study: Effectiveness of Trunk Training on Gait and Trunk Kinematics After Stroke: A Randomized Controlled Trial. *Phys. Ther.* **2020**, *100*, 1568–1581. [CrossRef]
101. Zhu, F.; Kern, M.; Fowkes, E.; Afzal, T.; Contreras-Vidal, J.-L.; Francisco, G.E.; Chang, S.-H. Effects of an Exoskeleton-Assisted Gait Training on Post-Stroke Lower-Limb Muscle Coordination. *J. Neural Eng.* **2021**, *18*, 046039. [CrossRef]
102. Allen, J.L.; McKay, J.L.; Sawers, A.; Hackney, M.E.; Ting, L.H. Increased Neuromuscular Consistency in Gait and Balance after Partnered, Dance-Based Rehabilitation in Parkinson's Disease. *APSselect* **2017**, *4*, 363–373. [CrossRef] [PubMed]
103. Ghislieri, M.; Lanotte, M.; Knaflitz, M.; Rizzi, L.; Agostini, V. Muscle Synergies in Parkinson's Disease before and after the Deep Brain Stimulation of the Bilateral Subthalamic Nucleus. *Sci. Rep.* **2023**, *13*, 6997. [CrossRef] [PubMed]
104. Conner, B.C.; Schwartz, M.H.; Lerner, Z.F. Pilot Evaluation of Changes in Motor Control after Wearable Robotic Resistance Training in Children with Cerebral Palsy. *J. Biomech.* **2021**, *126*, 110601. [CrossRef]
105. Jonsdottir, J.; Lencioni, T.; Gervasoni, E.; Crippa, A.; Anastasi, D.; Carpinella, I.; Rovaris, M.; Cattaneo, D.; Ferrarin, M. Improved Gait of Persons with Multiple Sclerosis after Rehabilitation: Effects on Lower Limb Muscle Synergies, Push-Off, and Toe-Clearance. *Front. Neurol.* **2020**, *11*, 668. [CrossRef] [PubMed]
106. Kadone, H.; Kubota, S.; Abe, T.; Noguchi, H.; Miura, K.; Koda, M.; Shimizu, Y.; Hada, Y.; Sankai, Y.; Suzuki, K.; et al. Muscular Activity Modulation During Post-Operative Walking With Hybrid Assistive Limb (HAL) in a Patient With Thoracic Myelopathy Due to Ossification of Posterior Longitudinal Ligament: A Case Report. *Front. Neurol.* **2020**, *11*, 102. [CrossRef] [PubMed]
107. Kinugawa, K.; Mano, T.; Wada, H.; Ozaki, M.; Shirai, D.; Imura, T.; Kido, A. Improvement in Lower Extremity Hemiplegia in a Post-Operative Brain Tumor Patient by Applying an Integrated Volitional Control Electrical Stimulator. *J. Phys. Ther. Sci.* **2022**, *34*, 473–477. [CrossRef]
108. Calafiore, D.; Negrini, F.; Tottoli, N.; Ferraro, F.; Ozyemisci-Taskiran, O.; de Sire, A. Efficacy of Robotic Exoskeleton for Gait Rehabilitation in Patients with Subacute Stroke: A Systematic Review. *Eur. J. Phys. Rehabil. Med.* **2022**, *58*, 1–8. [CrossRef] [PubMed]

109. Chen, G.; Chan, C.K.; Guo, Z.; Yu, H. A Review of Lower Extremity Assistive Robotic Exoskeletons in Rehabilitation Therapy. *Crit. Rev. Biomed. Eng.* **2013**, *41*, 343–363. [CrossRef]
110. Kostov, A.; Stein, R.B.; Popović, D.; Armstrong, W.W. Improved Methods for Control of FES for Locomotion. *IFAC Proc. Vol.* **1994**, *27*, 445–450. [CrossRef]
111. Peri, E.; Ambrosini, E.; Pedrocchi, A.; Ferrigno, G.; Nava, C.; Longoni, V.; Monticone, M.; Ferrante, S. Can FES-Augmented Active Cycling Training Improve Locomotion in Post-Acute Elderly Stroke Patients? *Eur. J. Transl. Myol.* **2016**, *26*, 6063. [CrossRef] [PubMed]
112. Duncan, P.W.; Sullivan, K.J.; Behrman, A.L.; Azen, S.P.; Wu, S.S.; Nadeau, S.E.; Dobkin, B.H.; Rose, D.K.; Tilson, J.K.; Cen, S.; et al. Body-Weight–Supported Treadmill Rehabilitation after Stroke. *N. Engl. J. Med.* **2011**, *364*, 2026–2036. [CrossRef]
113. Ivey, F.M.; Hafer-Macko, C.E.; Macko, R.F. Exercise Rehabilitation after Stroke. *NeuroRX* **2006**, *3*, 439–450. [CrossRef]
114. Iaccarino, M.A.; Bhatnagar, S.; Zafonte, R. Chapter 26—Rehabilitation after Traumatic Brain Injury. In *Handbook of Clinical Neurology*; Grafman, J., Salazar, A.M., Eds.; Traumatic Brain Injury, Part I; Elsevier: Amsterdam, The Netherlands, 2015; Volume 127, pp. 411–422.
115. Mehrholz, J.; Wagner, K.; Rutte, K.; Meiβner, D.; Pohl, M. Predictive Validity and Responsiveness of the Functional Ambulation Category in Hemiparetic Patients After Stroke. *Arch. Phys. Med. Rehabil.* **2007**, *88*, 1314–1319. [CrossRef]
116. Demeurisse, G.; Demol, O.; Robaye, E. Motor Evaluation in Vascular Hemiplegia. *Eur. Neurol.* **2008**, *19*, 382–389. [CrossRef] [PubMed]
117. Fugl-Meyer, A.R.; Jääskö, L.; Norlin, V. The Post-Stroke Hemiplegic Patient. *Scand. J. Rehabil. Med.* **1975**, *7*, 73–83.
118. Mahoney, F.I.; Barthel, D.W. Functional Evaluation: The Barthel Index: A Simple Index of Independence Useful in Scoring Improvement in the Rehabilitation of the Chronically Ill. *Md. State Med. J.* **1965**, *14*, 61–65.
119. Wrisley, D.M.; Marchetti, G.F.; Kuharsky, D.K.; Whitney, S.L. Reliability, Internal Consistency, and Validity of Data Obtained with the Functional Gait Assessment. *Phys. Ther.* **2004**, *84*, 906–918. [CrossRef]
120. Berg, K.; Wood-Dauphinee, S.; Williams, J.I. The Balance Scale: Reliability Assessment with Elderly Residents and Patients with an Acute Stroke. *Scand. J. Rehabil. Med.* **1995**, *27*, 27–36.
121. Franchignoni, F.; Horak, F.; Godi, M.; Nardone, A.; Giordano, A. Using Psychometric Techniques to Improve the Balance Evaluation System's Test: The Mini-BESTest. *J. Rehabil. Med. Off. J. UEMS Eur. Board Phys. Rehabil. Med.* **2010**, *42*, 323–331. [CrossRef]
122. Klein, P.J.; Fiedler, R.C.; Rose, D.J. Rasch Analysis of the Fullerton Advanced Balance (FAB) Scale. *Physiother. Can.* **2011**, *63*, 115–125. [CrossRef]
123. Whitney, S.L.; Hudak, M.T.; Marchetti, G.F. The Dynamic Gait Index Relates to Self-Reported Fall History in Individuals with Vestibular Dysfunction. *J. Vestib. Res.* **2000**, *10*, 99–105. [CrossRef] [PubMed]
124. Butland, R.J.; Pang, J.; Gross, E.R.; Woodcock, A.A.; Geddes, D.M. Two-, Six-, and 12-Minute Walking Tests in Respiratory Disease. *Br. Med. J. Clin. Res. Ed* **1982**, *284*, 1607–1608. [CrossRef]
125. Peters, D.M.; Fritz, S.L.; Krotish, D.E. Assessing the Reliability and Validity of a Shorter Walk Test Compared With the 10-Meter Walk Test for Measurements of Gait Speed in Healthy, Older Adults. *J. Geriatr. Phys. Ther.* **2013**, *36*, 24–30. [CrossRef]
126. Podsiadlo, D.; Richardson, S. The Timed "Up & Go": A Test of Basic Functional Mobility for Frail Elderly Persons. *J. Am. Geriatr. Soc.* **1991**, *39*, 142–148. [CrossRef] [PubMed]
127. Tinetti, M.E. Performance-Oriented Assessment of Mobility Problems in Elderly Patients. *J. Am. Geriatr. Soc.* **1986**, *34*, 119–126. [CrossRef]
128. Brunnstrom, S. Motor Testing Procedures in Hemiplegia: Based on Sequential Recovery Stages. *Phys. Ther.* **1966**, *46*, 357–375. [CrossRef] [PubMed]
129. Collin, C.; Wade, D. Assessing Motor Impairment after Stroke: A Pilot Reliability Study. *J. Neurol. Neurosurg. Psychiatry* **1990**, *53*, 576–579. [CrossRef]
130. Verheyden, G.; Nieuwboer, A.; Mertin, J.; Preger, R.; Kiekens, C.; De Weerdt, W. The Trunk Impairment Scale: A New Tool to Measure Motor Impairment of the Trunk after Stroke. *Clin. Rehabil.* **2004**, *18*, 326–334. [CrossRef]
131. Palisano, R.; Rosenbaum, P.; Walter, S.; Russell, D.; Wood, E.; Galuppi, B. Development and Reliability of a System to Classify Gross Motor Function in Children with Cerebral Palsy. *Dev. Med. Child Neurol.* **1997**, *39*, 214–223. [CrossRef]
132. Meder, K.G.; LoJacono, C.T.; Rhea, C.K. A Systematic Review of Non-Pharmacological Interventions to Improve Gait Asymmetries in Neurological Populations. *Symmetry* **2022**, *14*, 281. [CrossRef]
133. Bernstein, N.A. *The Co-Ordination and Regulation of Movements*, 1st ed.; Pergamon Press: Oxford, NY, USA, 1967.
134. Latash, M.L. There Is No Motor Redundancy in Human Movements. There Is Motor Abundance. *Motor Control* **2000**, *4*, 259–261. [CrossRef] [PubMed]
135. Lee, D.D.; Seung, H.S. Learning the Parts of Objects by Non-Negative Matrix Factorization. *Nature* **1999**, *401*, 788–791. [CrossRef]
136. Li, Y.; Ngom, A. The Non-Negative Matrix Factorization Toolbox for Biological Data Mining. *Source Code Biol. Med.* **2013**, *8*, 10. [CrossRef]
137. Tresch, M.C.; Cheung, V.C.K.; d'Avella, A. Matrix Factorization Algorithms for the Identification of Muscle Synergies: Evaluation on Simulated and Experimental Data Sets. *J. Neurophysiol.* **2006**, *95*, 2199–2212. [CrossRef] [PubMed]
138. Olney, S.J.; Richards, C. Hemiparetic Gait Following Stroke. Part I: Characteristics. *Gait Posture* **1996**, *4*, 136–148. [CrossRef]

139. Coscia, M.; Monaco, V.; Martelloni, C.; Rossi, B.; Chisari, C.; Micera, S. Muscle Synergies and Spinal Maps Are Sensitive to the Asymmetry Induced by a Unilateral Stroke. *J. NeuroEng. Rehabil.* **2015**, *12*, 39. [CrossRef] [PubMed]
140. Bizzi, E.; Cheung, V.C.K.; d'Avella, A.; Saltiel, P.; Tresch, M. Combining Modules for Movement. *Brain Res. Rev.* **2008**, *57*, 125–133. [CrossRef] [PubMed]
141. d'Avella, A.; Saltiel, P.; Bizzi, E. Combinations of Muscle Synergies in the Construction of a Natural Motor Behavior. *Nat. Neurosci.* **2003**, *6*, 300–308. [CrossRef] [PubMed]
142. Tresch, M.C.; Saltiel, P.; Bizzi, E. The Construction of Movement by the Spinal Cord. *Nat. Neurosci.* **1999**, *2*, 162–167. [CrossRef]
143. Cheung, V.C.K.; Turolla, A.; Agostini, M.; Silvoni, S.; Bennis, C.; Kasi, P.; Paganoni, S.; Bonato, P.; Bizzi, E. Muscle Synergy Patterns as Physiological Markers of Motor Cortical Damage. *Proc. Natl. Acad. Sci. USA* **2012**, *109*, 14652–14656. [CrossRef]
144. Roh, J.; Rymer, W.Z.; Beer, R.F. Evidence for Altered Upper Extremity Muscle Synergies in Chronic Stroke Survivors with Mild and Moderate Impairment. *Front. Hum. Neurosci.* **2015**, *9*, 6. [CrossRef] [PubMed]
145. Steele, K.M.; Rozumalski, A.; Schwartz, M.H. Muscle Synergies and Complexity of Neuromuscular Control during Gait in Cerebral Palsy. *Dev. Med. Child Neurol.* **2015**, *57*, 1176–1182. [CrossRef] [PubMed]
146. Tang, L.; Li, F.; Cao, S.; Zhang, X.; Wu, D.; Chen, X. Muscle Synergy Analysis in Children with Cerebral Palsy. *J. Neural Eng.* **2015**, *12*, 046017. [CrossRef]
147. Singh, R.E.; Iqbal, K.; White, G.; Hutchinson, T.E. A Systematic Review on Muscle Synergies: From Building Blocks of Motor Behavior to a Neurorehabilitation Tool. *Appl. Bionics Biomech.* **2018**, *2018*, e3615368. [CrossRef] [PubMed]
148. Zhvansky, D.S.; Sylos-Labini, F.; Dewolf, A.; Cappellini, G.; d'Avella, A.; Lacquaniti, F.; Ivanenko, Y. Evaluation of Spatiotemporal Patterns of the Spinal Muscle Coordination Output during Walking in the Exoskeleton. *Sensors* **2022**, *22*, 5708. [CrossRef]

Disclaimer/Publisher's Note: The statements, opinions and data contained in all publications are solely those of the individual author(s) and contributor(s) and not of MDPI and/or the editor(s). MDPI and/or the editor(s) disclaim responsibility for any injury to people or property resulting from any ideas, methods, instructions or products referred to in the content.

bioengineering

MDPI

Review

The Use of Head-Mounted Display Systems for Upper Limb Kinematic Analysis in Post-Stroke Patients: A Perspective Review on Benefits, Challenges and Other Solutions

Paolo De Pasquale [1], Mirjam Bonanno [1,*], Sepehr Mojdehdehbaher [2], Angelo Quartarone [1] and Rocco Salvatore Calabrò [1]

1 IRCCS Centro Neurolesi Bonino-Pulejo, Cda Casazza, SS 113, 98124 Messina, Italy; paolo.depasquale@irccsme.it (P.D.P.); angelo.quartarone@irccsme.it (A.Q.); roccos.calabro@irccsme.it (R.S.C.)
2 Department of Mathematics, Computer Science, Physics and Earth Sciences (MIFT), University of Messina, Viale Ferdinando Stagno d'Alcontres, 31, 98166 Messina, Italy; sepehrbaher@gmail.com
* Correspondence: mirjam.bonanno@irccsme.it

Abstract: In recent years, there has been a notable increase in the clinical adoption of instrumental upper limb kinematic assessment. This trend aligns with the rising prevalence of cerebrovascular impairments, one of the most prevalent neurological disorders. Indeed, there is a growing need for more objective outcomes to facilitate tailored rehabilitation interventions following stroke. Emerging technologies, like head-mounted virtual reality (HMD-VR) platforms, have responded to this demand by integrating diverse tracking methodologies. Specifically, HMD-VR technology enables the comprehensive tracking of body posture, encompassing hand position and gesture, facilitated either through specific tracker placements or via integrated cameras coupled with sophisticated computer graphics algorithms embedded within the helmet. This review aims to present the state-of-the-art applications of HMD-VR platforms for kinematic analysis of the upper limb in post-stroke patients, comparing them with conventional tracking systems. Additionally, we address the potential benefits and challenges associated with these platforms. These systems might represent a promising avenue for safe, cost-effective, and portable objective motor assessment within the field of neurorehabilitation, although other systems, including robots, should be taken into consideration.

Keywords: motion capture; head-mounted display; virtual reality; upper limb kinematics; post-stroke; neurorehabilitation

Citation: De Pasquale, P.; Bonanno, M.; Mojdehdehbaher, S.; Quartarone, A.; Calabrò, R.S. The Use of Head-Mounted Display Systems for Upper Limb Kinematic Analysis in Post-Stroke Patients: A Perspective Review on Benefits, Challenges and Other Solutions. *Bioengineering* **2024**, *11*, 538. https://doi.org/10.3390/bioengineering11060538

Academic Editor: Carlo Albino Frigo

Received: 11 April 2024
Revised: 20 May 2024
Accepted: 21 May 2024
Published: 24 May 2024

Copyright: © 2024 by the authors. Licensee MDPI, Basel, Switzerland. This article is an open access article distributed under the terms and conditions of the Creative Commons Attribution (CC BY) license (https://creativecommons.org/licenses/by/4.0/).

1. Introduction

Human upper limb tasks require fine-tuned coordination of multiple joints and muscles for interaction with the surrounding environment. Upper limb functions are often impaired after a brain injury [1]. In 2017, it was estimated that approximately 1.12 million people in Europe were affected by cerebrovascular disorders, due to ischemic or haemorrhagic events [2]. Post-stroke patients can manifest altered motor patterns in the contralesional upper and lower limbs, due to paralysis and/or spasticity [3]. In detail, infarctions of the middle cerebral artery affecting the primary motor cortex and the integrity of the corticospinal tract have been associated with upper limb movement deficits (including weakness, decreased inter-joint coordination, and diminished finger dexterity) [3,4]. Over the last few decades, upper limb kinematic assessment has grown in popularity in a clinical context. Indeed, it has been increasingly used as a more objective outcome, especially in stroke patients. During robotic-assisted therapy, the use of kinematic assessment is fundamental to monitor the patient's progress and to tailor the rehabilitation process to their needs. It is noteworthy that some robotic devices can perform a complete kinematic assessment automatically, as suggested by Bonanno and Calabrò [5]. In detail, a new category of robotic devices, known as collaborative robots, is growing in popularity in the field of rehabilitation. Due to their inherent safety and reliability, collaborative robots are increasingly being explored in healthcare applications, including robotic rehabilitation and kinematic assessment, where they can provide physical interaction with patients driven by actuation systems [6].

The advantage of upper limb instrumental kinematic measurements, compared to standard clinical assessments (e.g., Fugl-Meyer, Box and Block, 9-Hole Peg Test), is that different aspects of motion can be tracked more objectively and continuously [4]. However, the diversity and heterogeneity of kinematic assessment techniques, instrumentation, and selected outcomes expanded along with this developing area, make it challenging to interpret results. Moreover, the upper limb tasks comprise a wide set of activities, including "non-contact" movements such as gestures, or "contact" movements such as grasping activities [7]. This assessment can be performed in dedicated environments, like camera-based motion laboratories, or using robot-based measurement systems [5,8,9].

In the literature, the most used motion capture (MoCap) systems are based on optoelectronic technology, which is often considered as the "gold standard" because it is the most accurate and reliable when compared to other MoCap technologies [9]. Specifically, optoelectronic systems detect light and estimate the 3D position of a marker by triangulation from multiple cameras. They involve two main categories of markers: passive and active [9]. The passive marker system uses markers that reflect infrared (IR) light to the cameras, e.g., the VICON (Oxford, UK) or Optitrack (Corvallis, OR, USA) systems, whereas active marker systems contain a source of light for the sensors (often IR), provided that they are powered. To reconstruct the motion of a skeletal segment, reflective skin markers are placed on anatomical landmarks, and kinematics are derived from the marker position by adapting them to a human body model [10,11]. Despite the accuracy and reliability of optoelectronic systems, the widespread implementation of such devices as large-scale screening tools is impeded by their costs. Consequently, their usage is confined only to specialised clinical laboratory settings [10]. Over the past decade, the advent of novel technologies for motion analysis utilising wearable inertial sensors has paved the way for new approaches in clinical practice for the delivery of patient-centred rehabilitative interventions [8,12]. Inertial measurement units (IMUs) are small devices that integrate different multiaxial sensors, like accelerometers, gyroscopes, and magnetometers, to detect movements. These systems are particularly popular since they have several advantages, being low-cost, wearable, easy-to-use, settings which allow continuous monitoring of movement both in the clinic and at home [8]. IMUs collect raw data from each integrated sensor through on-board sensor-fusion algorithms, based on Kalman filters [13], in order to obtain different motion parameters. Generally, these systems allow the recording of kinematics, providing data on angular acceleration, velocity, and spatial orientation, from different body parts, including upper limbs [8,10]. However, the main disadvantage is that IMUs are less accurate and reliable than optoelectronic systems. In fact, IMUs estimate the orientation of body segments which may be likely to cause drifting due to the integration of noisy measurements [10]. To overcome this issue, there are numerous strategies allowing the corrections of potential bias in the motion analysis. Another MoCap method consists of the image analysis captured by one or more cameras through computer vision algorithms, which extracts information to describe movement. These tracking systems exploit machine learning methodologies, facilitating the recognition of human body movements, hand gestures, and expression of the face [14].

Virtual Reality Head-Mounted Display (HMD-VR) Platforms

On the other hand, novel technologies have recently caught on in clinical settings, like virtual reality (VR) platforms which integrate MoCap systems. Virtual reality head-mounted displays (HMD-VR) technology offers various possibilities to track body posture. These include specific marker positions, or a camera integrated in the helmet combined with computer graphic algorithms to estimate hand position and gestures. VR employs real-time rendering to generate immersive visual experiences, constantly updating images based on the users' movements within the virtual space. Another method to provide virtual feedback is augmented reality, which combines the real world and computer-generated 3D content. Augmented reality adjusts the ongoing perception of a real-world environment, while VR entirely substitutes the user's real-world environment with a simulated one. These systems can incorporate visual, auditory, and haptic feedback (i.e., hand controller vibration) to enhance the sense of immersion. Visual and auditory feedback are widely integrated in VR setups, but haptic technology remains neglected, often limited to adjustable vibrating motors found in handheld controller or wearable tracker devices. These haptic systems can modulate intensity, frequency, and duration to enhance user experience [15]. Other devices can be used and integrated in HMD-VR platforms to provide a wider range of haptic vibration feedback, such as haptic gloves, vests, and whole-body suits [16,17], or haptic force magnitude feedback [18]. However, these devices can significantly increase the overall price of the HMD-VR system and are often prototypes and not available on the market yet. Thanks to the gaming world, the HMD-VR has made great strides in recent years in terms of system usability and reliability of movement estimation. With positional tracking, users can navigate within the virtual environment along the X, Y, and Z axes, enabling not only head rotation but also horizontal and vertical movement. In addition, hand controllers, as well as additional trackable devices, are equipped with sensors to monitor the users' interactions within a virtual space. The possibility of easily creating controlled virtual experimental environments integrating different types of sensors within such platforms, through opensource development libraries, has opened the way for complex rehabilitation applications [19,20].

Moreover, HMD-VR are easy-to-use and low-cost systems, which allow users to perform rehabilitation treatments for both inpatients and outpatients [21]. In the neurorehabilitation context, the HMD-VR platforms were mostly used to train upper limb functions in post-stroke patients. According to recent studies [22,23], these systems are promising tools used to improve upper limb functional outcomes and daily living activities in chronic post-stroke patients. HMD-VR platforms offer the possibility of performing task-oriented movements with high degrees of realism, which could enhance gesture relevance [21,24]. According to some studies, post-stroke patients reported high levels of satisfaction, engagement, and motivation during rehabilitation treatment with VR as well as HMD-VR, denoting an overall acceptance of the device. Despite their promising role in the rehabilitation context though, it is still not clear if they are reliable and advantageous systems when compared to the classical MoCap.

In this review, we aim to report the state of the art of HMD-VR used for the kinematic analysis of upper limbs in post-stroke patients, offering a comparison between the conventional MoCap and HMD-VR. Moreover, the emerging role of robots and collaborative robots with virtual and augmentative reality has also been investigated.

Given the narrative and perspective nature of this review, we have analysed the most relevant studies in the English language (such as studies with different designs, focusing on the kinematic evaluation of upper limb motor tasks in post-stroke patients, using HMD-VR compared with conventional MoCap). Finally, each article has been assessed by the authors considering its title, abstract, text, and scientific validity.

2. Kinematics of Upper Limb Movements in Post-Stroke Patients

Differently from lower limbs, upper limbs have many degrees of freedom, allowing subjects to perform fine movements of reaching and grasping for daily life activities (ADL). Hand and arm movements constantly influence human ADL and sensory functions, providing a constant interaction with the environment [7,25]. In terms of kinematics, the upper limbs are an open kinematic chain starting in the sternoclavicular joint and ending in finger joints [26]. Due to the specific structure of human proximal joints, only rotational motions in these joints are possible [26]. Upper limb movements also involve object manipulation, which entails controlling finger forces to execute specific grasp types. For instance, when reaching for an object, the hand adjusts to its location and shapes the fingers to grasp it securely [27,28]. During planar reaching tasks, where the aim is to reach a target at a set distance by moving the arm and/or forearm, healthy individuals swiftly move towards the target. If the target is far away and the task takes longer, they may make a secondary corrective movement based on sensory feedback (like visual cues) to adjust their trajectory. Healthy individuals not only achieve consistent accuracy in reaching the target, but they also adjust their movement speed to maintain a fairly constant duration [25].

Upper limb reaching or hand and arm grasping actions may be impaired in post-stroke patients [28,29]. These patients can manifest weakness related to the inability to activate specific upper limb muscles or segments with uncoordinated inter-joint movements [28,29]. This abnormality is caused by an irregular flexor synergy during reaching, resulting in reduced finger dexterity for grasping and an increase in trunk movements as compensation. Pathological flexor synergy entails abnormal muscle activation, resulting in an uncoordinated contraction of elbow flexors and shoulder abductors. This phenomenon becomes apparent during upper limb tasks necessitating fine motor control and finger manipulation [25,26,28,29]. After a stroke, patients typically have slower hand movements and may take longer to complete grasping actions. Their grip openings tend to be wider, and they may move their hand along a longer path while performing grasping tasks. Furthermore, they often experience reduced grip strength while grasping and lifting objects [26,30]. Stroke survivors employ various compensatory strategies to improve grasping, including narrowing finger spread and adjusting finger and hand joints. These adaptations aim to enhance their ability to grasp objects effectively after stroke [4,29].

In this sense, functional assessment of the upper limbs is fundamental to plan rehabilitation and treatment of post-stroke patients according to their needs. By or large, physiotherapists are used to administering clinical scales and tests to assess upper limb functionality. For example, the Fugl-Meyer Assessment (FMA) scale is often administered to patients with stroke to evaluate functional abilities of the shoulder, elbow, wrist, and hand [7]. Despite its large use in clinical practice, this scale does not allow for sufficiently discriminating physiological and the above-mentioned pathological movement behaviour. To this aim, new emerging technologies are growing in the field of neurorehabilitation. In recent years, kinematic analysis has increasingly been used to objectively assess upper limb motor function after stroke and evaluate therapy effectiveness [4,31]. Unlike gait analysis, upper limb kinematic assessment does not have specific evaluation protocols, which makes the kinematic assessment more difficult. Previous studies have used different tasks and MoCap systems to illustrate the kinematic features of upper limb motor impairments following stroke (see Table 1).

These tasks often simulate ADL such as drinking, carrying, reach-to-grasp motion, or playing games. The Second Stroke Recovery and Rehabilitation Roundtable (SRRR2) [32] suggested that post-stroke patients with moderate to severe upper limb impairments may benefit from simpler point-to-point tasks, also known as two-dimensional (2D) tasks [25,31]. Examples of these tasks include Hand-to-Mouth (HTM) [33], Finger-to-Nose [34], or LIGHT tasks [35]. This recommendation aimed to make kinematic analysis more accessible and valuable for these patients. For example, Huang et al. [36] evaluated the HTM in post-stroke patients. They analysed kinematic variables such as movement smoothness, movement velocity, movement trajectories, and the range of movement (RoM) of trunk and upper limb joints, using a 3D MoCap system. Similarly, Shwartz et al. [4] conducted a kinematic motion analysis of the upper limbs in post-stroke patients. They examined parameters including trunk movements (e.g., displacement), shoulder, forearm, and wrist movements, as well as movement time, peak velocity, the number of velocity peaks (NVP), and spectral arc length (SPARC).

Table 1. The main kinematic features extracted from motion analysis of the upper limb in post-stroke patients.

Kinematic Features	Description
Path length ratio	Path length ratio (PLR) in upper limb kinematic analysis is a quantitative measure used to assess movement efficiency. It is calculated by dividing the actual path length travelled by a specific point or segment of the upper limb during a movement task by the shortest possible path length for the same movement. Higher PLR indicates less efficient movement patterns.
Joint excursion	Joint excursion consists of the angular RoM experienced by a specific joint during a movement task. It is typically measured in degrees or radians and provides insight into the flexibility, mobility, and coordination of the joint during the execution of a motor task.
Smoothness	Smoothness in upper limb kinematic analysis is a quantitative measure used to assess movement quality during reaching. There are several methods to measure movement smoothness. One of these is calculated as the number of peaks detected in the velocity profile.
Movement time	Movement time refers to the duration taken to complete a specific movement task. It is a crucial measure in assessing motor control, coordination, and efficiency of upper limb movements.
Movement velocity	Movement velocity is calculated by dividing the displacement of the point or segment by the time taken to complete the movement. It provides information about the speed or pace of movement execution, and it can be used to understand motor performance, coordination dynamics, and task difficulty.
Peak velocity	Peak velocity refers to the maximum instantaneous velocity achieved by a specific point or segment of the upper limb during a movement task. It represents the highest speed attained during the movement.
Number of velocity peaks	The number of velocity peaks refers to the count of distinct instances where the velocity of a specific point or segment of the upper limb reaches a local maximum during a movement task. Each velocity peak corresponds to a moment of rapid acceleration or deceleration within the movement profile.

Choi et al. [37] measured upper limb parameters in the sagittal, coronal, and transverse planes. Indeed, the authors also considered the thoracic angles and RoM in tilt obliquity, and rotation at the four points between each phase. Also, angle deviations for each joint angle were calculated during the task to estimate the degree of movement deviation, comparing them with healthy controls. On the other hand, Guzik-Kopyto et al. [38] identified 16 parameters, encompassing maximum joint angles and motion ranges of the joints, spinal kinematics to detect compensatory movements, and the velocity of movements to assess upper limb physical efficiency and capability.

Aprile et al. [39] assessed the kinematics of drinking activity in post-stroke patients, examining parameters like arm elongation, trunk forward inclination, and trunk axial rotation to understand their impact on reaching movements. In addition, the authors calculated the RoM of the elbow during the reaching tasks, and measured mouth forward displacement during specific phases to determine trunk inclination.

Kinematic analysis is vital in post-stroke patients for quantitatively detecting movement alterations, including compensations and pathological synergies. It also enables the monitoring of progress following rehabilitation interventions. In addition, it could influence treatment decisions and maximise recovery. This may also improve the confidence of patients in the efficacy of the interventions during the chronic stage.

3. Technologies for Motion Capture: Optoelectronic, IMUs, and Vision-Based Motion Tracking and Other Solutions

In this paragraph, we aim to describe the most common MoCap systems used for kinematic motion analysis. We have explored different types: optoelectronic, IMU, and vision-based motion tracking, as well as robotic devices and HMD-VR, pointing out their technical features and differences (see Table 2).

Table 2. Shows a schematic description of the advantages and limitations of HMD-VR systems relative to conventional MoCap and robot-based systems used for upper limb kinematic motion analysis.

MoCap System	Portability	Markerless	Easy-to-Use	Tele-Monitoring	Untethered	Rehabilitation	Cost
Optoelectronic	No	No	No	No	No	No	High
IMU	Yes	No	Yes	Yes	No	No	Medium
Vision-based	Yes	Yes	Yes	Yes	No	No	Low
Robot	No	Yes	No	No	No	Yes	High
HMD-VR	Yes	Yes	Yes	Yes	Yes	Yes	Low

Optoelectronic devices represent a valid instrument for upper limb functional assessment providing a valid tool for 3D kinematic motion analysis [9,35]. These systems use light to estimate the 3D position of a marker by triangulation from multiple cameras, and they can use active and passive markers [9,10]. The latter are covered by photo-reflective materials that reflect the infrared light, while active markers emit IR light (see Figure 1A).

Figure 1. Conventional MoCap systems used for kinematic analysis of upper limb. (**A**) An example of a commercial optoelectronic MoCap system setup (Optitrack Flex 13 Motion Capture Camera [40]), which consists of several IR cameras where subjects are instrumented with reflective markers; (**B**) one of the subjects, who is wearing one of the commercially available IMU Mocap systems (Xsens Awinda [41]); (**C**) VBMA key points detection using an OpenPose pose estimation algorithm through a camera [42].

The 3D kinematic analysis of the upper limbs using optoelectronic devices can vary significantly in terms of marker sets, kinematic models, analysed functional movements, and reported kinematic outcomes. By or large, two sets of markers are used: anatomical or cluster marker sets. The former is placed in correspondence with body landmarks to build kinematics models, as anatomical landmarks are essential to define the local coordinate systems of a body segment (e.g., scapula, humerus, and thorax) in static conditions. On the other hand, cluster marker sets are placed in different body segments. These markers are calibrated to correspond to specific anatomical landmarks, facilitating accurate tracking of movement [43].

Motion recording systems based on inertial sensors allow the estimation of movement and orientation of specific body segments [8] (see Figure 1B). These systems typically rely on data from integrated accelerometers, gyroscopes, and magnetometers within compact, lightweight devices attached to the targeted body region. Accelerometers register velocity alterations along the sensor's three axes, enabling estimation of velocity and spatial displacement through integration processes. Gyroscopes measure angular velocity, enabling tracking of orientation changes, while magnetometers detect magnetic field strength and direction, providing absolute orientation relative to the Earth's magnetic field [8,10].

Position estimation based on inertial systems is not a direct process; indeed, it begins with estimating velocity from acceleration and then proceeds to estimating position from velocity through a double integration process.

This technique is known as dead reckoning (path integration), a navigation method used to estimate the current position based on a previously determined one, utilising speed, elapsed time, and course direction [44].

Commercial systems may comprise a single sensor, like the BTS G-WALK (BTS Bioengineering S.p.A., Milan, Italy), or multiple sensors like Xsens Link system (Movella Technologies, Enschede, The Netherlands), contingent on application requirements. For example, a single sensor positioned at the sacral level, combined with mathematical models, is adequate for evaluating walking, running, and jumping performances [10].

Conversely, complex kinematic analyses, like upper limb reaching tasks that involve numerous degrees of freedom, require the integration of multiple sensors and more sophisticated mathematical models [45].

Sensor placement typically involves specialised suits or elastic bands tailored to the target segment. Inertial systems are considered feasible also in post-stroke patients, as demonstrated by Held et al. [46]. To collect kinematic data, these authors used an Xsens full-body motion capture suit. In particular, IMUs registered kinematic parameters of movement during clinical assessment, both at the hospital and in home settings. This aspect is extremely advantageous because IMUs systems allow for the quantification of kinematic analysis outside a laboratory environment [8,28].

However, when comparing IMUs to optoelectronic systems, they are considered less accurate. This is because they do not directly record position (unlike optoelectronic systems); instead, they estimate it from acceleration [47].

However, when comparing IMUs to optoelectronic systems, they are considered less accurate. Secondly, IMU systems notably suffer from drift accumulation which leads to gradual change or offset in their output readings over time. This drift can be caused by various factors, including temperature changes, aging of internal components, imperfections in manufacturing, and electronic noise. Thirdly, metal objects in the evaluation environment may cause electromagnetic interference and significant distortions. Furthermore, conducting precise analyses involving multiple joints often necessitates the use of several sensors. However, due to the finite resolution and dynamic range of accelerometers, measurements of extremely small or large accelerations may not be reliable. [10,47]. On the other hand, IMU systems, as outlined in Table 2, provide significant advantages. They offer large capture volume capabilities and are portable, lightweight, and wireless. Additionally, they are adaptable to various environments, including outdoor conditions [48,49]. They boast cost-effectiveness and minimal recording latency, facilitating real-time applications. This is why they are often used for recording in real-time situations. In post-stroke patients, Nie et al. [50] analysed wrist reaching movements, comparing optoelectronic and IMU devices. The authors found that the mean error between the two systems was 0.09 ± 1.81 cm.

Moreover, another branch of MoCap is revolutionising the field of movement analysis, such as vision-based motion analysis systems (VBMA). VBMA is a type of markerless motion capture, which extracts information from subsequential images to describe movement [10]. Generally, this type of MoCap uses cameras with high resolution to ensure high accuracy. Hence, these tracking systems incorporate cameras or camera arrays to facilitate the recording of human body movements. Subsequent application of computer vision and machine learning techniques enables the capture of gestures of the hands and expressions of the face [14].

Significant implementations of these technologies include software libraries like MediaPipe [51], MoveNet [52], and OpenPose [53] (see Figure 1C). These libraries empower developers to cover a wide area of the image and track multiple individuals simultaneously, with numerous points, reaching up to 500, and including finger and facial features. Technical evaluations indicate that fast movements or considerable distances from the camera can cause a notable decline in the accuracy of segment recognition. This often leads to incorrect delineation of skeletal nodes. A significant limitation arises from the inability of a single camera to accurately establish the third coordinate of points (along the Z-axis), which makes it impossible to determine the distance between the camera and the object. Nonetheless, these limitations may be less impactful in certain musculoskeletal rehabilitation exercises, particularly those involving simple movements along one or two axes. Alternative solutions such as camera arrays, specialised cameras equipped with depth sensors or stereo capabilities provide opportunities for capturing the necessary data on the positioning of body parts [14,54]. According to Faity et al. [55], the Kinect, a device for VBMA, was reliable in assessing trunk compensations, hand range of motion, movement time and mean velocity with a moderate to excellent reliability, in post-stroke. In contrast, this system did not show the same accuracy in estimating elbow and shoulder range of motion, time to peak velocity, and path length ratio.

Other Technologies for Motion Capture: Focus on Collaborative Robots

Beyond conventional MoCap, other technologies can perform an objective and quantitative kinematic motion analysis, such as robotic devices. In fact, these technologies are commonly used for rehabilitation purposes, but they also can be translated to both analyse the patient's limb trajectory and accurately register spatial–temporal parameters of movement, collecting a large amount of data. These devices may be characterised by an end-effector attached to a set of arms connected by proper joints. Thanks to kinematic calculation methods and integrated measurement devices, the position of the end-effector or the arms can be measured [5]. In particular, robots integrate measurement devices, called encoders, which record joints angular deviations through optical, mechanical, or magnetic technology. These devices provide a reliable and objective motion analysis, also during rehabilitation training. For example, some authors [56–58] evaluated the range of joint upper limb movements, movement velocity, accuracy, and smoothness in active training, using robotic devices (i.e., Armeo Power, Armeo Spring).

Moreover, within rehabilitation, robotic devices can play pivotal role in upholding and overseeing the quality of users' movements, which is essential for customised treatment effectiveness. Consequently, robot-assisted therapy is gaining attention as a method for addressing motor function issues, particularly in the upper limbs, facilitating repetitive, high-intensity training while ensuring movement precision [59–61]. Incorporating robots can also diminish the necessity for continuous, hands-on supervision by therapists, as they can automate tasks and track the speed and accuracy of movements continuously [59–61]. Collaborative robots, called also cobots, are a small class of industrial robotic devices able to reduce human effort and minimise the risk of high-velocity impacts or injuries by collisions. In general, cobots can be differentiated from traditional robots by their flexible characteristics and safety features, which are provided by sensor-based control methodologies [62]. Additionally, these systems can use augmented reality, and may be customised for each patient.

Some authors [63,64] used cobots for both rehabilitation and motion evaluation purposes. In fact, all the data from the instrumented robot arm is available and the process can be monitored with accuracy. The system can also integrate decisions on increasing the difficulty of the training and its progress in order to motivate the patients. For example,

Kyrkjebø et al. [63] used an industrial cobot (UR5e) as a rehabilitation tool for upper limb training in post-stroke patients. The authors found that this device is feasible, since the robot can be customised to execute a wide range of movements suitable for the treatment process. In addition, the feasibility of cobots is enhanced when it is combined with accurate force and torque measurements.

Reinkensmeyer et al. [64] used a lightweight exoskeleton (Pneu-Wrex) to train upper limbs in post-stroke subjects, allowing a wide RoM of the arm in a 3D space by incorporating pneumatic actuators to generate active forces. In particular, the robot constantly monitored the assistance the patient needed to achieve the current task, and then provided slightly less assistance than the estimated amount. In this way, post-stroke subjects were encouraged to make more effort, improving their motor skills. Rodrigues et al. [65] combined cobots with augmented reality.

This work proposes grouping a collaborative robot with one specific augmented reality equipment to create a rehabilitation system where some gamification levels might be added to provide a better and more motivating experience to patients.

To date, studies on robotic devices have mostly focused on helping patients with neurological disorders improve their motor skills and functional recovery. However, the potential role of robots, regarding cobots, in the growing field of motion analysis deserves further investigation.

4. Benefits and Challenges of Virtual Reality Head-Mounted Display Platforms

Given the remarkable diversity observed in literature regarding the technology used, the upper limb motor tasks performed, and the kinematic metrics analysed, specific recommendations have been developed. These recommendations aim to establish standardised methods for assessing upper limb kinematics after stroke [33]. However, in recent years, an increasing number of motion analysis technologies have become commercially available, such as HMD-VR, which have different characteristics and features that have not yet been fully investigated in the field of rehabilitation [19]. In this paragraph, we have shed some light on the benefits and challenges of the different types of HMD-VR platforms compared with conventional marker-based MoCap systems, since they are considered as the "gold standard".

HMD-VR platforms, which are portable and low-cost, serve dual purposes: motor training and objective kinematic analysis of the upper limbs (see Table 2). HMD-VR platforms integrate validated technologies such as optoelectronic, IMUs, and VBMA for motion tracking to estimate the user's position in virtual space [20]. Obtaining a good estimate of the user's position is necessary to move and interact within VR environments. In HMD-VR platforms, the environment is displayed within the helmet, and the view of the physical world is completely obscured. Typically, these systems use head and hand position tracking for most applications. More complex applications may also require tracking of other body parts using multiple sensors or integrating data within biomechanical models [66]. In detail, HMD-VR incorporates various technical features to enhance the user's experience. These include high-resolution displays for detailed visuals, wide fields of view for immersion, and high refresh rates for smooth motion. Tracking technology monitors the user's head movements and position, while comfort features like adjustable straps and lightweight designs ensure prolonged use. Built-in audio systems or support for external headphones provide spatialised audio, and connectivity options enable interaction with external devices. Additionally, some HMDs come with handheld controllers and additional wearable trackers to interact with the virtual environment and record kinematics. Adjustable straps positioned on body segments enable recording of paretic limb motion without requiring the user to hold the controllers. These features collectively contribute to the performance, comfort, and immersion provided by HMDs in VR applications [67]. Moreover, these systems should possess adaptability for diverse settings and pathologies, as well as scalability features. Specifically, HMD-VR systems exhibit adaptability, providing portability, integration with other technologies (such as sensors and computers), interactivity via hand gestures, voice commands, and motion controllers, as well as comfort and ergonomic features (including adjustable headbands, cushioned padding, and lightweight materials). It is noteworthy that the adaptability of HMD-VR as a treatment device for other neurological patients has already been already demonstrated for use in treating Parkinson's disease [68] and in traumatic brain injury [69].

In addition, HMD-VRs offer scalability features, related to the following: (i) HMD-VR systems (e.g., these systems allow switching from one HMD-VR to another); (ii) degrees of VR (e.g., visual quality and latency which depend on the type of feedback and the desired performance); (iii) numbers of collaborators (e.g., numbers of users connected simultaneously in the same VR environment). However, it is not a common practice nowadays to use and interact with multiple users within VR, probably due to a lack of appropriate haptic feedback; (iv) the number of recorded objects/joints/limbs is based on the type of technology (e.g., in HMD-VR outside-in, the number of recorded objects depends on the number controllers/trackers presented in the scene and integrated within the device; however, in HMD-VR inside-out tracking, additional controllers (marker-based) need integration, or complex mathematical models (markerless) must be employed, ensuring they remain within the camera's field of view) [70].

A promising trajectory in motion analysis lies in the direction of a fully automated, non-intrusive method. Such an approach has the potential to represent a significant advancement for both research and practical applications within the domains of biomechanics and neurorehabilitation [19,21]. Nevertheless, most clinical studies have investigated the effects of the HMD-VR platform as a tool for rehabilitation intervention, due to its enriched and controlled environment [19,21]. Relatively little attention has been directed towards the potential utility of HMD-VR as kinematic measurement devices. Notably, only a handful of studies have examined the kinematic accuracy of HMD-VR in comparison to conventional MoCap systems [66,71], and even less research has investigated this issue in post-stroke patients [50,72].

Nowadays, there are different methods that rely on different technologies to track the user's position: outside-in (Figure 2A) and inside-out (Figure 2B) [10,66].

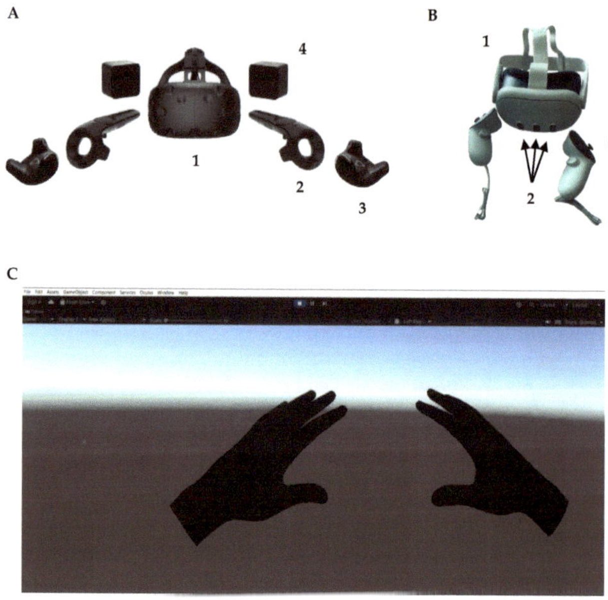

Figure 2. HMD-VR platforms. (**A**) An outside-in HMD-VR platform (HTC Vive, HTC Europe Co., Ltd., Slough, Berkshire, UK). This system consists of an HMD (1), 2 controllers (2), 2 trackers (3),

2 base stations or "lighthouses" (4). (**B**) An inside-out system (Meta Quest 3, Meta Technologies LLC, New York, NY, USA), which simply consists of an HDM (1) with integrated cameras (2) and 2 controllers (3). (**C**) Three-dimensional hand model representation developed with Unity [73] software (version 2023.2.1f1) and the device's software development kit (Meta XR Core SDK, version 65.0) [74].

The first systems commercially available have outside-in technology; that is, they allow the position of the HMD (Figure 2A, 1) and handheld controllers and wearable trackers (Figure 2A, 2, 3) to be estimated through external cameras or sensors (Figure 2A, 4). These systems, such as the HTC Vive (HTC Europe Co., Ltd., Slough, Berkshire, UK) or the Oculus Rift (Oculus, Irvine, CA, USA), integrate information from external cameras with IMUs placed inside the wearable devices, allowing reliable and accurate estimation even in large working environments [24]. The IR light-sensitive cameras rely on optical technology to estimate the position of a given set of markers that is placed on the HMD or controller. Typically, two "lighthouse" station bases emit flashes from an IR LED array at a fixed frequency, along with vertical and horizontal lasers sweeping in both horizontal and vertical directions across the room. By analysing the sequence in which the photosensors on the HMD and controllers receive these laser sweeps, the position of the tracked devices can be determined [66]. Since the external base stations are placed in stationary locations, the estimated position of the tracked objects depends on the lighthouses. Indeed, a calibration procedure is required to set system components and achieve reliable and accurate measurements. The tracking volume depends on the type and placement of the base stations. Considerable tracking spaces can be achieved up to a maximum size of 7 m $\times$ 7 m. For object tracking, it is only necessary that the objects are within the tracking volume of the base stations. The data stream from the HMD-VR platforms reaches the PC through a cable. Graphics processing is then carried out by the PC, which is responsible for updating the virtual scene [75].

On the other hand, inside-out systems have been gaining popularity in recent years, since their features are more suitable in the field of gaming and entertainment than the outside-in systems. First, the main difference lies in the fact that in the inside-out systems, the camera is not located externally but physically inside the HMD (Figure 2B, 2) [75].

Two different types (i.e., marker-based, and markerless) of tracking objects modalities are possible within inside-out systems. In particular, when the application requires a controller to interact with the VR environment, the same technology of outside-in systems, based on IR and IMU, is used. However, in contrast to outside-in systems, inside-out systems do not rely on external cameras or base stations to detect controller position. Instead, in these cases, the IR LED in the controllers is detected by HMD cameras, which integrate this information with data from multiple IMUs placed on the controllers [76]. The inside-out systems also allow the user to interact with the VR environment using hands and without any markers [77]. A combination of computer vision algorithms, including visual-inertial mapping, place recognition, geometry reconstruction, and cameras placed on the HMD, can be employed to estimate the position and gesture of the hands [72]. Position and gestures are extrapolated from the recorded images through the integration of stereoscopic information obtained from the cameras and VBMA [14], as shown in Figure 2C. For example, in Figure 2C, hand pose and position, extrapolated from recorded images, are shown through a 3D model representation. This 3D model representation was developed with Unity (Unity Technologies, San Francisco, CA, USA) [73] software, which is one of the most popular VR environments development software. In addition, Unity software allows 2D and 3D VR environment developments on different platforms, like desktop, mobile, augmented, and virtual reality. To aid and facilitate multi-platform content development, different tools such as the software development kit (SDK) and specific open source software, like SteamVR (Valve Corporation, WA, USA) [78], provide a wide set of libraries.

Regarding the tracking volume, significant differences are evident among the different types of HMD-VR systems. In inside-out devices, the tracking volume is limited by the visual space of the cameras that are on the helmet. Therefore, the volume is dynamic and restricted when it is compared to outside-in systems. Objects can therefore be tracked if they are within the field of view of the cameras [79]. If hands are beyond the camera's field of view, tracking becomes unfeasible, except in marker-based situations where position estimation is still achievable, albeit with less reliability and for a limited duration, solely relying on IMUs. (see Figure 3).

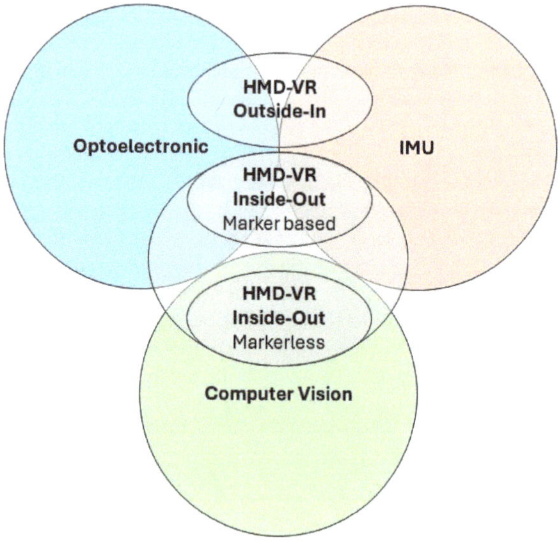

Figure 3. Tracking systems for HMD to estimate upper limb motion. Figure shows motion tracking technology exploited by HMDs to estimate upper limb (end-effector) position.

These systems allow untethered, i.e., unconstrained configuration [80]. All processing of the virtual scene and updating of object positions within the room as well as the viewpoint itself, are handled directly by the HMD. Alternatively, when high graphical or computational demands are present, this processing can be offloaded to a PC via cable. Given that, both HMD outside-in and inside-out present different characteristics and meeting points, which are all displayed in Table 3.

Table 3. Table shows the differences/meeting points between outside-in and inside-out, both marker-based and markerless.

Features	Outside-In	Inside-Out (Marker-Based)	Inside-Out (Markerless)
Portability	✗	✓	✓
Untethered	✗	✓	✓
Hand gesture	✗	✗	✓
HMD field of view tracking	✓	✓	✓
External field of view tracking	✓	✗	✗

Considering the aforementioned aspects, HMD-VR platform and conventional MoCap systems differ for tracking performance accuracy and precision. Some authors have compared these technologies in both healthy and post-stroke patients (see Table 4).

Table 4. Description of the most significant studies evaluating accuracy of HMD-VR for upper limb position estimation.

Study Reference	MoCap Systems Comparison	Performed Task	Parameters of Precision/Accuracy	Kinematic Assessment
		HMD Outside-In		
[50]	IMUs and HMD VR sensor (Vive) compared with Optoelectronic system (Vicon)	Wrist position, during reaching tasks, with respect to the shoulder	Compared to a traditional optical tracking system, both methods accurately tracked the wrist during reaching, with mean signed errors of 0.09 ± 1.81 cm and 0.48 ± 1.58 cm for the IMUs and Vive, respectively.	Normalised mean endpoint speed (Smoothness)
[66]	HTC Vive HMD and Vive tracker	Reaching tasks	HTC Vive headset and Vive Trackers showed that both can track joint rotation and position with reasonable accuracy and a very low end to latency of 6.71 ± 0.80 ms.	Joint rotation and position
		HMD Inside-Out Marker-based		
[81]	HMD (Oculus Quest 2) compared with Qualysis optical capture system	Upper limb rotational and translational movements	The results showed a mean absolute error of 13.52 ± 6.57 mm at a distance of 500 mm from the HMD along the x-direction. The maximum mean absolute error for rotational displacements was found to be 1.11 ± 0.37° for a rotation of 40° around the z-axis.	Translational and rotational movement
[71]	HMD (Oculus Touch v2) controller compared with IMU	Flexion–extension movement of the forearm.	The level of agreement between the measurements of these devices was 0.999 with a 95% confidence interval (ranged from 0.996 to 1.000). The accuracy degrades at flexion values between 70° and 110°, peaking at 90°.	Range of motion of elbow in the sagittal plane
		HMD Inside-Out Marker-less		
[72]	HMD (Oculus Quest 2, Meta) compared with Optoelectronic system (Optitrack).	Reaching	Maximum distance: mean slope = 0.94 ± 0.1; peak velocity: mean slope = 1.06 ± 0.12).	Peak velocity and hand position

For example, Nie et al. [50] evaluated the tracking accuracy (0.48 ± 1.58 cm) of the wrist during the reaching task in post-stroke optoelectronic, outside-in HMD-VR. On the other hand, some authors compared the tracking performance of inside-out marker-based HMD-VR with optoelectronic systems. For example, Carnevale et al. [81] tracked translational (mean absolute error of 1.35 ± 0.66 cm) and rotational (mean absolute error of 1.11 ± 0.37°) shoulder movements. Also, Jost et al. [82] analysed positional data to determine the translational and rotational accuracy of the HMD (0.17 ± 0.07 cm, 0.34 ± 0.38°) and controllers (0.44 ± 0.29 cm, 1.13 ± 1.23°). Similarly, Monica and Aleotti [75] found an average translation error for the HMD of about 1.83 cm and an average rotation error of 0.77°.

Casile et al. [72] evaluated upper limb movements in post-stroke patients, comparing the inside-out markerless HMD-VR with optoelectronic marker-based MoCap. In particular, the authors found a slope close to 1 (maximum distance: 0.94 ± 0.1; peak velocity: 1.06 ± 0.12) from a linear regression analysis of peak velocity and hand positional data.

Furthermore, other authors compared the tracking performances of HMD-VR inside-out markerless platforms with IMUs. For instance, Trinidad-Fernandez et al. [20] calculated the absolute error of HMD (0.48 ± 0.09°), while Rojo et al. [71] compared controllers' tracking accuracy to measure the elbow's motion in the sagittal plane (intra-rater reliability of 0.999, with a 95% confidence interval).

Moreover, there is another issue to be considered for the implementation of HMD-VR in current clinical practice that is referred to regulatory considerations. In fact, CE marking indicates that a product has been assessed by the manufacturer and deemed to meet European safety, health, and environmental protection requirements. While this process is relatively short worldwide, in Europe, new laws have complicated this pathway. Obtaining a CE marking is a long and complex path that initially requires the recognition of the device as a medical one, and the Medical Device Regulation must be applicable. So the path to CE marking is long, time-consuming, and money-consuming, and this could hinder investments in innovation systems, like HMD-VR, for rehabilitation [83].

On the other hand, the European Union is allocating sustained funding for rehabilitation technologies, displacing private investments while contributing to the reconstruction of a more environmentally friendly, technologically advanced Europe. The World Health Organization (WHO) acknowledges the significance of health financing as a vital tool in advancing objectives such as enhancing accessibility, shielding against financial burdens, and facilitating the delivery of high-quality services. The WHO proposes that health financing strategies can be leveraged to advance these objectives specifically for rehabilitation services, thus allowing access to new technological devices [83].

5. Discussion

In this perspective review, we primarily aim to highlight the benefits and challenges of the different types of HMD-VR for the motion analysis of upper limbs in post-stroke patients, as compared to conventional MoCap systems. We identified the most used tracking technologies in clinical practice, encompassing their main technical features and comparing them to the novel HMD-VR. Despite the wide range of literature about the use of optoelectronics and IMUs for the motion analysis of upper limbs [8,29,48,49], few studies have, to date, dealt with HMD-VR technologies [50,66,72,81,82]. It seems that HMD-VR has been largely used as a tool for rehabilitation of upper limbs [19,80], but its potential use for motion analysis is still overlooked. Some authors used HMD-VR to assess visual neglect and motor performance in patients with Parkinson's disease, demonstrating that HMD–VR allows instrumental quantitative and objective measurements [84].

In the context of upper-limb rehabilitation, reliable motion tracking is essential for effective rehabilitation [28]. In particular, hand tracking has a significant impact on the interaction between the patient and the virtual targets. Indeed, these systems should be able to recognise the individual differences among the different types of grips and range of movement of each individual patient [4,29,35]. Another point is that the VR rehabilitation systems must be able to record all data generated during the performance of the exercises. These data allow therapists to objectively assess the patient's progress over time.

5.1. Benefits and Challenges of HMD-VR in a Neurorehabilitation Context

Differently from conventional tracking systems, HMD-VR platforms exploit the benefits of VR. They offer simultaneous task-oriented training and assessment within a controlled and safe environment that closely resembles real-world scenarios. Therefore, patients can perform ADL (cooking, driving, etc.) without any risk, in the simulated scenarios, through different levels of difficulty for each task [19]. A key benefit of using HMD-VR for motor assessment is the motivation of the patients. As already stated by Saldana et al. [19], these systems increase the patients' motivation and engagement during rehabilitation tasks. Fostering patients' intrinsic motivation to actively participate in the therapeutic process is especially important, particularly when conventional methods may lead to boredom or lack of interest [85]. The available studies [22,85,86] on neurorehabilitation indicate that HMD-VR exhibits the potential to enhance tailored assessment and intervention methods by engaging patients within an immersive virtual environment. According to different authors [22,36,86,87], HMD-VR can be an effective tool to improve upper limb motor outcomes and kinematic factors, as well as ADLs, in post-stroke patients. In particular, De Giorgi et al. [88] used HMD-VR to perform virtual art therapy, which remarkably promotes the autonomy of post-stroke patients in their ADLs and upper limb functioning, in terms of muscle strength and pinching.

Furthermore, HMD-VR platforms were perceived as easy to use, and user-friendly by post-stroke patients. Some authors reported high levels of usability, and adherence to the treatment [87,89], which could have positively influenced their intention in using the VR exercise system. In this way, the authors enable multisensorial stimulation, involving auditory, visual, and tactile feedback, to promote patients' engagement.

To enhance active participation and adherence to the treatment, Rodrigues et al. [65] proposed a prototype of combined augmented reality interface with collaborative arm robot. This integrated method, combining augmented reality with a robotic arm, enables the creation of novel user experiences by harnessing gamification principles. These include introducing challenging scenarios, implementing reward systems, and even encouraging users to surpass their previous high scores. Collaborative robots are growing in popularity as translational devices, and they can be used as a valid and alternative to VR tools, especially in those patients suffering from cybersickness. In rehabilitation, they can be used in active or passive mode. In active mode, they can guide the patients in the early stages when they need to learn (relearning mechanism). Gradually, they can be used in passive mode where the patients are guiding the robot arm. In addition, they also provide kinematic information about a patient's performance [90].

Like any other therapy, rehabilitation with HMD-VR has its limitations and challenges. In fact, it should be also considered that these systems can elicit cybersickness (including dizziness, headaches, disorientation, and fatigue) as well as side-effects which may limit the usability of HMD-VR. Some studies have shown that side effects tend to get worse in full VR immersion, having a negative effect on static balance [91]. According to Caserman et al. [66], some participants reported cybersickness, such as dizziness, nausea, or a headache, while wearing an HMD for a longer time. Another side-effect that could be elicited by immersive VR is the loss of visual contact within the environment and rehabilitation setting (including physiotherapists), with potential negative consequences for patients' psychological state.

Despite the limited evidence on the side effects of HMD-VR, this must be considered when choosing the type and the duration of rehabilitation for a neurological patient. In addition, it should be considered that post-stroke patients may suffer from psychological conditions, like anxiety and depression, which could overload the sickness effects provided by the VR. This is the case when it is recommended to use other devices, including robots and cobots. However, a previous case study [92] reported that rehabilitation in an immersive VR environment, in a post-stroke subject, was effective in improving psychological symptoms. In this sense, the authors spelt out the use of immersive VR as a promising device for the of psychological post-stroke symptoms, including anxiety.

Another clinical challenge for the implementation of HMD-VR in post-stroke rehabilitation consists of some difficulties when the subject is managing the technology [24]. Indeed, interaction with HMD-VR requires a long press of the button on the controller, which can be demanding for post-stroke subjects. To this aim, patients could benefit from robotic-assisted upper limbs training, which supports the paretic limb against throughout the RoM required by the exercise, preventing the subject from becoming discouraged due to muscle weakness. In contrast to VR, robotic devices can allow the mobilisation of the hand and fingers, also giving a physiological trajectory of movement while avoiding abnormal muscle compensations due to spasticity.

However, robotic devices can require a long time for setup and specialised staff to operate with them. In fact, precise alignment between users and robotic devices is crucial to prevent adverse interaction forces that could lead to discomfort and safety concerns. To overcome this issue, soft exoskeletons, composed of flexible textiles or elastomers, present a promising solution by enhancing user compliance compared to rigid orthoses. However, the limited availability of robotic devices in rehabilitation centres due to cost, maintenance, and specialised staff requirements poses a significant obstacle to their integration into clinical practice at scale. These factors contribute to the current limited adoption of robot-based assessments in clinical settings.

On the other hand, HMD-VR exploits the conventional tracking methodology of these technologies by cutting down on costs and complexity, thanks to their link with the world of consumer electronics. The accessibility of these devices should help make future studies using HMD-VR with larger populations more feasible. Considering that HMD-VR represents a growing and potential field, it should be also analysed in terms of cost-effectiveness. The whole HMD-VR platform (HMD, computer, and/or controllers, and/or lighthouses) is relatively low-cost when compared with other VR devices [93]. In addition, the primary savings due to the use of HMD-VR for post-stroke patients rely on its utilisation for both evaluation and treatment with or without the presence of the therapist. In this latter condition, HMD-VR allows for the reduction in costs related to therapist's service and travel expenses for both the therapist (i.e., domiciliary therapy) and the caregiver (i.e., ambulatory care) in terms of money, time spent, and effort. To this aim, some authors [94] investigated the effects of an outside-in HMD-VR for telerehabilitation, in a post-stroke patient. Although this is only one case study, the authors have shown that HMD-VR can be an ecological, user-friendly, and adjunctive rehabilitation approach to conventional treatment, even for those subjects living in rural regions.

5.2. Reliability, Technical Features, and Limitations of HMD-VR

On the other hand, the analysis of motion in the neurorehabilitation context allows the instrumental registration and monitoring of patients' movement parameters (Table 1) in a more objective way than clinical scales (clinical purpose). The multitude of data produced from motion analysis can be used to build machine learning algorithms, detecting prognostic and predictive factors for better motor outcomes (research purpose) [5].

To this end, both accuracy and precision of tracking measurements are essential for obtaining reliable results. Indeed, many studies have focused on the tracking measurement quality of such systems compared with conventional MoCap systems. For instance, Nie et al. [50] and Caserman et al. [66] found that outside-in HMD-VR devices (HTC Vive) track the wrist during reaching, with mean signed errors of 0.48 ± 1.58 cm (36) and a latency of 6.71 ± 0.80 ms (42) when compared to a traditional optical tracking system. According to Nie et al. [50], the IMUs and HMD-VR (HTC Vive) necessitate minimal alignment and calibration steps compared to optical tracking systems like Vicon. These procedures take at most 10 min and can be performed by individuals, including those with stroke. In particular, Caserman et al. [66] reported encouraging results on the accuracy of the HMD (HTC Vive). Indeed, the system efficiently provided real-time joint configuration, enabling smooth and accurate movement tracking.

Regarding inside-out marker-based device (Oculus Quest 2), Carnevale et al. [81] measured the rotational and translational movements of the upper limbs with a mean absolute error of $1.11 \pm 0.37°$ and 13.52 ± 6.57 mm, respectively (51). These results were registered at 500 mm from the HMD along the x-direction. On the other hand, a recent study [82] investigating the accuracy of the same inside-out marker-based device reported a reduced absolute error. The researchers [82] pointed out that the best controller performance occurred when the HMD wearer observed it closely, which contrasts with Carnevale's et al. [81] study where they varied the distance between the headset and controller.

In contrast to previous studies, Rojo et al. [71] found a high agreement level of 0.999 (with a 95% confidence interval) between measurements of forearm flexion-extension using both Oculus Touch v2 controllers and IMU devices. However, the controllers exhibit inaccuracies, especially near a 90° angle in the sagittal rotation plane. Despite this, they adequately capture the full range of motion of the elbow joint in virtual environments. According to the authors [71], while minor misalignments may be acceptable in VR applications where precision is not critical, for tasks requiring high accuracy in orientation measurement, an IMU sensor is preferable.

Finally, comparing markerless inside-out HMD-VR with optoelectronic systems, Casile et al. [72] found a maximum distance of 0.94 ± 0.1 and a peak velocity of 1.06 ± 0.12. The authors suggested that the inside-out markerless HMD-VR (Oculus Quest) provided hand position and peak velocity estimates that closely matched those of a marker-based commercial system. In addition, Casile et al. found that Oculus Quest was also sensitive in distinguishing pathological from healthy upper limb movements.

From a technical point of view, HMDs present some differences and limitations among them. For example, the outside-in systems enable the estimation of upper limb during complex tasks, even when they are outside of the field of view of the HMD. In contrast, inside-out can only estimate the position of the upper limbs if they are within the field of view of the HMD cameras [75]. Additionally, outside-in systems are generally faster and more accurate than inside-out systems, and the accuracy of these systems can be easily improved by adding more cameras. Also, outside-in systems are capable of tracking even in dark settings, and controllers can be tracked even while they are behind the user's back [75]. Indeed, inside-out systems cannot estimate the position of the patient's paretic hand if it is outside the field of view of HMD cameras during a reaching task of an object, framed with the HMD. These systems are easier to set up than outside-in, since there is no need to install fixed cameras for the calibration. Regardless of whether these systems are inside-out or outside-in, marker-based systems are generally less reliable in estimating hand gestures [14]. In fact, these systems employ a controller, enabling the estimation of a few basic hand gestures decoded through the controller's buttons. Conversely, the HMD-VR inside-out markerless systems facilitate complex hand gesture recognition but necessitate computer vision algorithms. It is noteworthy that the HMD-VR platform is based on conventional and validated MoCap methods to estimate user's motion. Conventional methods are accurate although they present some disadvantages related to their expensiveness, which are time-consuming for instrumentation and data processing, requiring qualified staff to be used in some cases [10]. In addition, these systems have not yet been recognised as medical devices; in fact, they do not have the CE mark. However, their use in the world of gaming and education has been translated to the neurorehabilitation field. Overall, these aspects could explain why these devices are not always used in clinical settings, but they are largely diffused in the context of research.

5.3. Clinical Perspectives and Future Directions

From a clinical point of view, the motion analysis of hand gesture is fundamental to track those compensatory strategies that post-stroke patients use to perform a grasping function. Some examples refer to reduced finger abduction, proximal interphalangeal joint flexion, and metacarpophalangeal joint extension during object grasping [28,29,48]. To overcome these issues, rehabilitation modalities should ideally be affordable, easily transportable, and user-friendly for long-term use. Specifically, portability and ease of use are crucial for promoting treatment adherence and enabling continuous care at home. While full embodiment in a virtual avatar may not be necessary, the presence of the affected body part in the virtual environment is essential. Creating a feeling of presence by synchronising limb movements in real-time is crucial, and monitoring both the affected and intact limbs may be necessary for the task. Therefore, HMD-VR platforms meet most of the needs of post-stroke patients, both in terms of treatment and kinematic motion analysis. However, the dose of treatment and which patients to select is still to be understood. Researchers and clinicians must consider the risks faced by certain individuals when engaging in VR. Examples include those with epilepsy, seizure disorders, recent concussions, and elderly adults experiencing vision impairment due to age-related changes. These groups may encounter adverse effects during VR experiences, highlighting the importance of clinical patients' examination.

In the end, we could hypothesise that HMD-VR should be considered task-specific-based. This aspect has an important bearing on clinical practice, because the choice of HMD-VR is closely linked to the type of task that needs to be assessed. Clinicians need to consider their intended use and what they want to achieve, in terms of reliability and effectiveness. To this aim, the collaboration between clinicians bioengineering professional figures should be also considered. This aspect can promote the implementation of innovative technologies, like HMD-VR, in the neurorehabilitation setting. For example, physiotherapists and neurologists could collaborate more with bioengineers to share bidirectionally practical information, which is needed to plan specific motor training based on objective and quantitative findings.

Indeed, since these devices are being designed for the gaming/entertainment world, the newest models may not always be the best solution in the rehabilitation field, especially in neurological patients. Future research and clinical trials aiming to validate the effectiveness of HMD-VR technology in rehabilitation may involve investigating the accuracy of both outside-in and inside-out HMD-VR systems compared to conventional MoCap methods across various motor tasks. Another future consideration is the utilisation of these systems at home for both treatment and kinematic motion analysis, ensuring the continuity of care. Additionally, the extensive data collected from these devices could be leveraged to develop machine learning algorithms capable of predicting outcomes. Finally, we present some perspectives about the use of HMD-VR platform for upper limb motion estimation in neurorehabilitation context. From our literature analysis, we gather the following points:

- A comparison between the different types and brands of HMD-VR platforms is still missing, intended as measurement and rehabilitation tools. In addition, among the studies that we have included in this review, there is no homogeneity in results regarding accuracy and precision analysis.
- According to the authors (Table 4), these systems are suitably accurate and reliable to be used as rehabilitation tools and MoCap systems.
- The selection of one device over another depends on its intended use and on the severity degree of the disease. Therefore, clinicians must consider the reliability and effectiveness of the instrument.
- It is noteworthy that VR is not the only technology capable of providing an objective and quantitative assessment of movement as well as delivering rehabilitation treatment. In fact, robotic devices can be tailored to meet the patient's needs, offering both precise movement evaluation and intensive, repetitive, and task-oriented treatment options.

6. Conclusions

According to the reviewed literature, the HMD-VR platform can be a safe, cost-effective, and portable tool for upper limb kinematic motion analysis in the context of neurorehabilitation. The HMD-VR platforms not only provide the advantage of offering an objective and quantitative assessment of upper limb function but also support rehabilitation treatment. Particularly for post-stroke patients, VR presents a compelling option due to its immersive and safe environment enriched with multisensory stimuli. This immersive experience enhances patient attention, fostering increased engagement and adherence to treatment protocols. Yet, it is crucial to consider both the duration of treatment and the severity of patients' conditions when formulating rehabilitation plans, taking into consideration also other devices, including robots and cobots. Indeed, the use of VR is still far from being considered standard in clinical practice. Future studies should explore HMD–VR applications in neurological clinical settings and home-based training to better investigate whether and to which extent these promising tools can be used in the movement assessment of patients with neurological disorders, including stroke.

Author Contributions: Conceptualisation, P.D.P.; methodology, P.D.P. and M.B.; validation, all authors; investigation, M.B., P.D.P. and S.M.; resources, A.Q.; data curation, P.D.P. and M.B.; writing—original draft preparation, P.D.P. and M.B.; writing—review and editing, R.S.C.; visualisation, S.M.; supervision, R.S.C. and A.Q.; project administration, R.S.C.; funding acquisition, A.Q. All authors have read and agreed to the published version of the manuscript.

Funding: This research was supported by Current Research Funds, 2024, Italian Ministry of Health.

Institutional Review Board Statement: Not applicable.

Informed Consent Statement: Not applicable.

Data Availability Statement: Not applicable.

Acknowledgments: We thank Anthony Longo and Agata Grosso for the English editing.

Conflicts of Interest: The authors declare no conflicts of interest.

References

1. Anwer, S.; Waris, A.; Gilani, S.O.; Iqbal, J.; Shaikh, N.; Pujari, A.N.; Niazi, I.K. Rehabilitation of Upper Limb Motor Impairment in Stroke: A Narrative Review on the Prevalence, Risk Factors, and Economic Statistics of Stroke and State of the Art Therapies. *Healthcare* **2022**, *10*, 190. [CrossRef] [PubMed]
2. Wafa, H.A.; Wolfe, C.D.A.; Emmett, E.; Roth, G.A.; Johnson, C.O.; Wang, Y. Burden of Stroke in Europe: Thirty-Year Projections of Incidence, Prevalence, Deaths, and Disability-Adjusted Life Years. *Stroke* **2020**, *51*, 2418–2427. [CrossRef] [PubMed]
3. Murphy, S.J.X.; Werring, D.J. Stroke: Causes and Clinical Features. *Medicine* **2020**, *48*, 561–566. [CrossRef] [PubMed]
4. Schwarz, A.; Bhagubai, M.M.C.; Nies, S.H.G.; Held, J.P.O.; Veltink, P.H.; Buurke, J.H.; Luft, A.R. Characterization of Stroke-Related Upper Limb Motor Impairments across Various Upper Limb Activities by Use of Kinematic Core Set Measures. *J. NeuroEngineering Rehabil.* **2022**, *19*, 2. [CrossRef] [PubMed]
5. Bonanno, M.; Calabrò, R.S. Robot-Aided Motion Analysis in Neurorehabilitation: Benefits and Challenges. *Diagnostics* **2023**, *13*, 3561. [CrossRef] [PubMed]
6. Chiriatti, G.; Bottiglione, A.; Palmieri, G. Manipulability Optimization of a Rehabilitative Collaborative Robotic System. *Machines* **2022**, *10*, 452. [CrossRef]
7. Pan, B.; Huang, Z.; Jin, T.; Wu, J.; Zhang, Z.; Shen, Y. Motor Function Assessment of Upper Limb in Stroke Patients. *J. Healthc. Eng.* **2021**, *2021*, 6621950. [CrossRef] [PubMed]
8. Gu, C.; Lin, W.; He, X.; Zhang, L.; Zhang, M. IMU-Based Motion Capture System for Rehabilitation Applications: A Systematic Review. *Biomim. Intell. Robot.* **2023**, 3, 100097. [CrossRef]
9. Topley, M.; Richards, J.G. A Comparison of Currently Available Optoelectronic Motion Capture Systems. *J. Biomech.* **2020**, *106*, 109820. [CrossRef]
10. van Schaik, J.E.; Dominici, N. Chapter 5—Motion Tracking in Developmental Research: Methods, Considerations, and Applications. In *Progress in Brain Research*; Hunnius, S., Meyer, M., Eds.; New Perspectives on Early Social-cognitive Development; Elsevier: Amsterdam, The Netherlands, 2020; Volume 254, pp. 89–111.
11. Zago, M.; Luzzago, M.; Marangoni, T.; De Cecco, M.; Tarabini, M.; Galli, M. 3D Tracking of Human Motion Using Visual Skeletonization and Stereoscopic Vision. *Front. Bioeng. Biotechnol.* **2020**, *8*, 181. [CrossRef] [PubMed]
12. Aguilera-Rubio, Á.; Alguacil-Diego, I.M.; Mallo-López, A.; Cuesta-Gómez, A. Use of the Leap Motion Controller® System in the Rehabilitation of the Upper Limb in Stroke. A Systematic Review. *J. Stroke Cerebrovasc. Dis.* **2022**, *31*, 106174. [CrossRef]
13. Potter, M.V.; Cain, S.M.; Ojeda, L.V.; Gurchiek, R.D.; McGinnis, R.S.; Perkins, N.C. Evaluation of Error-State Kalman Filter Method for Estimating Human Lower-Limb Kinematics during Various Walking Gaits. *Sensors* **2022**, *22*, 8398. [CrossRef] [PubMed]
14. Colyer, S.L.; Evans, M.; Cosker, D.P.; Salo, A.I.T. A Review of the Evolution of Vision-Based Motion Analysis and the Integration of Advanced Computer Vision Methods Towards Developing a Markerless System. *Sports Med. Open* **2018**, *4*, 24. [CrossRef] [PubMed]
15. Gibbs, J.K.; Gillies, M.; Pan, X. A Comparison of the Effects of Haptic and Visual Feedback on Presence in Virtual Reality. *Int. J. Hum. Comput. Stud.* **2022**, *157*, 102717. [CrossRef]
16. Lindeman, R.W.; Page, R.; Yanagida, Y.; Sibert, J.L. Towards Full-Body Haptic Feedback: The Design and Deployment of a Spatialized Vibrotactile Feedback System. In Proceedings of the ACM Symposium on Virtual Reality Software and Technology, Osaka, Japan, 8–10 December 2021.
17. Pacchierotti, C.; Sinclair, S.; Solazzi, M.; Frisoli, A.; Hayward, V.; Prattichizzo, D. Wearable Haptic Systems for the Fingertip and the Hand: Taxonomy, Review, and Perspectives. *IEEE Trans. Haptics* **2017**, *10*, 580–600. [CrossRef] [PubMed]
18. Li, F.; Chen, J.; Ye, G.; Dong, S.; Gao, Z.; Zhou, Y. Soft Robotic Glove with Sensing and Force Feedback for Rehabilitation in Virtual Reality. *Biomimetics* **2023**, *8*, 83. [CrossRef] [PubMed]

19. Saldana, D.; Neureither, M.; Schmiesing, A.; Jahng, E.; Kysh, L.; Roll, S.C.; Liew, S.-L. Applications of Head-Mounted Displays for Virtual Reality in Adult Physical Rehabilitation: A Scoping Review. *Am. J. Occup. Ther.* **2020**, *74*, 7405205060p1–7405205060p1 [CrossRef] [PubMed]
20. Trinidad-Fernández, M.; Bossavit, B.; Salgado-Fernández, J.; Abbate-Chica, S.; Fernández-Leiva, A.J.; Cuesta-Vargas, A.I. Head-Mounted Display for Clinical Evaluation of Neck Movement Validation with Meta Quest 2. *Sensors* **2023**, *23*, 3077. [CrossRef] [PubMed]
21. Fregna, G.; Paoluzzi, C.; Baroni, A.; Cano-de-la-Cuerda, R.; Casile, A.; Straudi, S. Head-Mounted Displays for Upper Limb Stroke Rehabilitation: A Scoping Review. *J. Clin. Med.* **2023**, *12*, 7444. [CrossRef] [PubMed]
22. Kiper, P.; Godart, N.; Cavalier, M.; Berard, C.; Cieślik, B.; Federico, S.; Kiper, A.; Pellicciari, L.; Meroni, R. Effects of Immersive Virtual Reality on Upper-Extremity Stroke Rehabilitation: A Systematic Review with Meta-Analysis. *J. Clin. Med.* **2024**, *13*, 146. [CrossRef]
23. Amini Gougeh, R.; Falk, T.H. Head-Mounted Display-Based Virtual Reality and Physiological Computing for Stroke Rehabilitation: A Systematic Review. *Front. Virtual Real.* **2022**, 3. [CrossRef]
24. Marek, K.; Zubrycki, I.; Miller, E. Immersion Therapy with Head-Mounted Display for Rehabilitation of the Upper Limb after Stroke—Review. *Sensors* **2022**, *22*, 9962. [CrossRef] [PubMed]
25. Kwakkel, G.; Van Wegen, E.; Burridge, J.H.; Winstein, C.J.; van Dokkum, L.; Alt Murphy, M.; Levin, M.F.; Krakauer, J.W. Standardized Measurement of Quality of Upper Limb Movement after Stroke: Consensus-Based Core Recommendations from the Second Stroke Recovery and Rehabilitation Roundtable. *Int. J. Stroke* **2019**, *14*, 783–791. [CrossRef] [PubMed]
26. Duprey, S.; Naaim, A.; Moissenet, F.; Begon, M.; Chèze, L. Kinematic Models of the Upper Limb Joints for Multibody Kinematics Optimisation: An Overview. *J. Biomech.* **2017**, *62*, 87–94. [CrossRef] [PubMed]
27. De Baets, L.; Vanbrabant, S.; Dierickx, C.; van der Straaten, R.; Timmermans, A. Assessment of Scapulothoracic, Glenohumeral, and Elbow Motion in Adhesive Capsulitis by Means of Inertial Sensor Technology: A Within-Session, Intra-Operator and Inter-Operator Reliability and Agreement Study. *Sensors* **2020**, *20*, 876. [CrossRef] [PubMed]
28. Schwarz, A.; Bhagubai, M.M.C.; Wolterink, G.; Held, J.P.O.; Luft, A.R.; Veltink, P.H. Assessment of Upper Limb Movement Impairments after Stroke Using Wearable Inertial Sensing. *Sensors* **2020**, *20*, 4770. [CrossRef] [PubMed]
29. Schwarz, A.; Kanzler, C.M.; Lambercy, O.; Luft, A.R.; Veerbeek, J.M. Systematic Review on Kinematic Assessments of Upper Limb Movements After Stroke. *Stroke* **2019**, *50*, 718–727. [CrossRef] [PubMed]
30. Valtin, M.; Salchow, C.; Seel, T.; Laidig, D.; Schauer, T. Modular Finger and Hand Motion Capturing System Based on Inertial and Magnetic Sensors. *Curr. Dir. Biomed. Eng.* **2017**, *3*, 19–23. [CrossRef]
31. Mesquita, I.A.; da Fonseca, P.F.P.; Pinheiro, A.R.V.; Velhote Correia, M.F.P.; da Silva, C.I.C. Methodological Considerations for Kinematic Analysis of Upper Limbs in Healthy and Poststroke Adults Part II: A Systematic Review of Motion Capture Systems and Kinematic Metrics. *Top. Stroke Rehabil.* **2019**, *26*, 464–472. [CrossRef] [PubMed]
32. Bernhardt, J.; Hayward, K.S.; Dancause, N.; Lannin, N.A.; Ward, N.S.; Nudo, R.J.; Farrin, A.; Churilov, L.; Boyd, L.A.; Jones, T.A.; et al. A Stroke Recovery Trial Development Framework: Consensus-Based Core Recommendations from the Second Stroke Recovery and Rehabilitation Roundtable. *Int. J. Stroke* **2019**, *14*, 792–802. [CrossRef] [PubMed]
33. Prange-Lasonder, G.B.; Alt Murphy, M.; Lamers, I.; Hughes, A.-M.; Buurke, J.H.; Feys, P.; Keller, T.; Klamroth-Marganska, V.; Tarkka, I.M.; Timmermans, A.; et al. European Evidence-Based Recommendations for Clinical Assessment of Upper Limb in Neurorehabilitation (CAULIN): Data Synthesis from Systematic Reviews, Clinical Practice Guidelines and Expert Consensus. *J. NeuroEngineering Rehabil.* **2021**, *18*, 162. [CrossRef] [PubMed]
34. Johansson, G.M.; Grip, H.; Levin, M.F.; Häger, C.K. The Added Value of Kinematic Evaluation of the Timed Finger-to-Nose Test in Persons Post-Stroke. *J. NeuroEngineering Rehabil.* **2017**, *14*, 11. [CrossRef] [PubMed]
35. van der Vliet, R.; Selles, R.W.; Andrinopoulou, E.-R.; Nijland, R.; Ribbers, G.M.; Frens, M.A.; Meskers, C.; Kwakkel, G. Predicting Upper Limb Motor Impairment Recovery after Stroke: A Mixture Model. *Ann. Neurol.* **2020**, *87*, 383–393. [CrossRef] [PubMed]
36. Huang, X.; Liao, O.; Jiang, S.; Li, J.; Ma, X. Kinematic Analysis in Post-Stroke Patients with Moderate to Severe Upper Limb Paresis and Non-Disabled Controls. *Clin. Biomech.* **2024**, *113*, 106206. [CrossRef] [PubMed]
37. Choi, H.; Park, D.; Rha, D.-W.; Nam, H.S.; Jo, Y.J.; Kim, D.Y. Kinematic Analysis of Movement Patterns during a Reach-and-Grasp Task in Stroke Patients. *Front. Neurol.* **2023**, *14*, 1225425. [CrossRef] [PubMed]
38. Guzik-Kopyto, A.; Nowakowska-Lipiec, K.; Krysiak, M.; Jochymczyk-Woźniak, K.; Jurkojć, J.; Wodarski, P.; Gzik, M.; Michnik, R. Selection of Kinematic and Temporal Input Parameters to Define a Novel Upper Body Index Indicator for the Evaluation of Upper Limb Pathology. *Appl. Sci.* **2022**, *12*, 11634. [CrossRef]
39. Aprile, I.; Rabuffetti, M.; Padua, L.; Di Sipio, E.; Simbolotti, C.; Ferrarin, M. Kinematic Analysis of the Upper Limb Motor Strategies in Stroke Patients as a Tool towards Advanced Neurorehabilitation Strategies: A Preliminary Study. *Biomed. Res. Int.* **2014**, *2014*, 636123. [CrossRef] [PubMed]
40. Flex 13-In Depth. Available online: http://optitrack.com/cameras/flex-13/index.html (accessed on 21 May 2024).
41. MVN Awinda | Movella.Com. Available online: https://www.movella.com/products/motion-capture/xsens-mvn-awinda (accessed on 28 April 2024).
42. Wade, L.; Needham, L.; McGuigan, P.; Bilzon, J. Applications and Limitations of Current Markerless Motion Capture Methods for Clinical Gait Biomechanics. *PeerJ* **2022**, *10*, e12995. [CrossRef] [PubMed]

43. Boser, Q.A.; Valevicius, A.M.; Lavoie, E.B.; Chapman, C.S.; Pilarski, P.M.; Hebert, J.S.; Vette, A.H. Cluster-Based Upper Body Marker Models for Three-Dimensional Kinematic Analysis: Comparison with an Anatomical Model and Reliability Analysis. *J. Biomech.* **2018**, *72*, 228–234. [CrossRef] [PubMed]
44. Bodily, K.D.; Daniel, T.A.; Sturz, B.R. The Roles of Beaconing and Dead Reckoning in Human Virtual Navigation. *Learn. Motiv.* **2012**, *43*, 14–23. [CrossRef]
45. Ponsiglione, A.M.; Ricciardi, C.; Amato, F.; Cesarelli, M.; Cesarelli, G.; D'Addio, G. Statistical Analysis and Kinematic Assessment of Upper Limb Reaching Task in Parkinson's Disease. *Sensors* **2022**, *22*, 1708. [CrossRef] [PubMed]
46. Held, J.P.O.; Klaassen, B.; Eenhoorn, A.; van Beijnum, B.-J.F.; Buurke, J.H.; Veltink, P.H.; Luft, A.R. Inertial Sensor Measurements of Upper-Limb Kinematics in Stroke Patients in Clinic and Home Environment. *Front. Bioeng. Biotechnol.* **2018**, *6*, 27. [CrossRef] [PubMed]
47. Zhou, L.; Fischer, E.; Tunca, C.; Brahms, C.M.; Ersoy, C.; Granacher, U.; Arnrich, B. How We Found Our IMU: Guidelines to IMU Selection and a Comparison of Seven IMUs for Pervasive Healthcare Applications. *Sensors* **2020**, *20*, 4090. [CrossRef] [PubMed]
48. Merlau, B.; Cormier, C.; Alaux, A.; Morin, M.; Montané, E.; Amarantini, D.; Gasq, D. Assessing Spatiotemporal and Quality Alterations in Paretic Upper Limb Movements after Stroke in Routine Care: Proposal and Validation of a Protocol Using IMUs versus MoCap. *Sensors* **2023**, *23*, 7427. [CrossRef]
49. Unger, T.; de Sousa Ribeiro, R.; Mokni, M.; Weikert, T.; Pohl, J.; Schwarz, A.; Held, J.P.O.; Sauerzopf, L.; Kühnis, B.; Gavagnin, E.; et al. Upper Limb Movement Quality Measures: Comparing IMUs and Optical Motion Capture in Stroke Patients Performing a Drinking Task. *Front. Digit. Health* **2024**, *6*, 1359776. [CrossRef] [PubMed]
50. Nie, J.Z.; Nie, J.W.; Hung, N.-T.; Cotton, R.J.; Slutzky, M.W. Portable, Open-Source Solutions for Estimating Wrist Position during Reaching in People with Stroke. *Sci. Rep.* **2021**, *11*, 22491. [CrossRef] [PubMed]
51. GitHub-Google-Ai-Edge/Mediapipe: Cross-Platform, Customizable ML Solutions for Live and Streaming Media. Available online: https://github.com/google-ai-edge/mediapipe (accessed on 21 May 2024).
52. MoveNet: Modello di Rilevamento Della Posa Ultra Veloce e Preciso. TensorFlow Hub. Available online: https://www.tensorflow.org/hub/tutorials/movenet?hl=it (accessed on 10 April 2024).
53. CMU-Perceptual-Computing-Lab/Openpose: OpenPose: Real-Time Multi-Person Keypoint Detection Library for Body, Face, Hands, and Foot Estimation. Available online: https://github.com/CMU-Perceptual-Computing-Lab/openpose (accessed on 10 April 2024).
54. Ceseracciu, E.; Sawacha, Z.; Cobelli, C. Comparison of Markerless and Marker-Based Motion Capture Technologies through Simultaneous Data Collection during Gait: Proof of Concept. *PLoS ONE* **2014**, *9*, e87640. [CrossRef]
55. Faity, G.; Mottet, D.; Froger, J. Validity and Reliability of Kinect v2 for Quantifying Upper Body Kinematics during Seated Reaching. *Sensors* **2022**, *22*, 2735. [CrossRef] [PubMed]
56. Galeoto, G.; Berardi, A.; Mangone, M.; Tufo, L.; Silvani, M.; González-Bernal, J.; Seco-Calvo, J. Assessment Capacity of the Armeo® Power: Cross-Sectional Study. *Technologies* **2023**, *11*, 125. [CrossRef]
57. Merlo, A.; Longhi, M.; Giannotti, E.; Prati, P.; Giacobbi, M.; Ruscelli, E.; Mancini, A.; Ottaviani, M.; Montanari, L.; Mazzoli, D. Upper Limb Evaluation with Robotic Exoskeleton. Normative Values for Indices of Accuracy, Speed and Smoothness. *NeuroRehabilitation* **2013**, *33*, 523–530. [CrossRef] [PubMed]
58. Longhi, M.; Merlo, A.; Prati, P.; Giacobbi, M.; Mazzoli, D. Instrumental Indices for Upper Limb Function Assessment in Stroke Patients: A Validation Study. *J. NeuroEng. Rehabil.* **2016**, *13*, 52. [CrossRef] [PubMed]
59. Schweighofer, N.; Choi, Y.; Winstein, C.; Gordon, J. Task-Oriented Rehabilitation Robotics. *Am. J. Phys. Med. Rehabil.* **2012**, *91*, S270–S279. [CrossRef] [PubMed]
60. Maciejasz, P.; Eschweiler, J.; Gerlach-Hahn, K.; Jansen-Troy, A.; Leonhardt, S. A Survey on Robotic Devices for Upper Limb Rehabilitation. *J. NeuroEngineering Rehabil.* **2014**, *11*, 3. [CrossRef] [PubMed]
61. Wagner, T.H.; Lo, A.C.; Peduzzi, P.; Bravata, D.M.; Huang, G.D.; Krebs, H.I.; Ringer, R.J.; Federman, D.G.; Richards, L.G.; Haselkorn, J.K.; et al. An Economic Analysis of Robot-Assisted Therapy for Long-Term Upper-Limb Impairment after Stroke. *Stroke* **2011**, *42*, 2630–2632. [CrossRef]
62. Cherubini, A.; Navarro-Alarcon, D. Sensor-Based Control for Collaborative Robots: Fundamentals, Challenges, and Opportunities. *Front. Neurorobot.* **2021**, *14*, 576846. [CrossRef] [PubMed]
63. Kyrkjebø, E.; Johan Laastad, M.; Stavdahl, Ø. Feasibility of the UR5 Industrial Robot for Robotic Rehabilitation of the Upper Limbs After Stroke. In Proceedings of the 2018 IEEE/RSJ International Conference on Intelligent Robots and Systems (IROS), Madrid, Spain, 1–5 October 2018; pp. 1–6.
64. Reinkensmeyer, D.J.; Wolbrecht, E.T.; Chan, V.; Chou, C.; Cramer, S.C.; Bobrow, J.E. Comparison of 3D, Assist-as-Needed Robotic Arm/Hand Movement Training Provided with Pneu-WREX to Conventional Table Top Therapy Following Chronic Stroke. *Am. J. Phys. Med. Rehabil.* **2012**, *91*, S232–S241. [CrossRef] [PubMed]
65. Rodrigues, J.C.; Menezes, P.; Restivo, M.T. An Augmented Reality Interface to Control a Collaborative Robot in Rehab: A Preliminary Usability Evaluation. *Front. Digit. Health* **2023**, *5*, 1078511. [CrossRef] [PubMed]
66. Caserman, P.; Garcia-Agundez, A.; Konrad, R.; Göbel, S.; Steinmetz, R. Real-Time Body Tracking in Virtual Reality Using a Vive Tracker. *Virtual Real.* **2019**, *23*, 155–168. [CrossRef]

67. Pandita, S.; Stevenson Won, A. Chapter 7—Clinical Applications of Virtual Reality in Patient-Centered Care. In *Technology and Health*; Kim, J., Song, H., Eds.; Academic Press: Cambridge, MA, USA, 2020; pp. 129–148. ISBN 978-0-12-816958-2.
68. Campo-Prieto, P.; Cancela-Carral, J.M.; Rodríguez-Fuentes, G. Wearable Immersive Virtual Reality Device for Promoting Physical Activity in Parkinson's Disease Patients. *Sensors* **2022**, *22*, 3302. [CrossRef] [PubMed]
69. Varela-Aldás, J.; Palacios-Navarro, G.; Amariglio, R.; García-Magariño, I. Head-Mounted Display-Based Application for Cognitive Training. *Sensors* **2020**, *20*, 6552. [CrossRef] [PubMed]
70. Memmesheimer, V.M.; Ebert, A. Scalable Extended Reality: A Future Research Agenda. *Big Data Cogn. Comput.* **2022**, *6*, 12. [CrossRef]
71. Rojo, A.; Cortina, J.; Sánchez, C.; Urendes, E.; García-Carmona, R.; Raya, R. Accuracy Study of the Oculus Touch v2 versus Inerti Sensor for a Single-Axis Rotation Simulating the Elbow's Range of Motion. *Virtual Real.* **2022**, *26*, 1651–1662. [CrossRef]
72. Casile, A.; Fregna, G.; Boarini, V.; Paoluzzi, C.; Manfredini, F.; Lamberti, N.; Baroni, A.; Straudi, S. Quantitative Comparison of Hand Kinematics Measured with a Markerless Commercial Head-Mounted Display and a Marker-Based Motion Capture System in Stroke Survivors. *Sensors* **2023**, *23*, 7906. [CrossRef] [PubMed]
73. Unity Real-Time Development Platform | 3D, 2D, VR & AR Engine. Available online: https://unity.com/ (accessed on 20 May 2024).
74. Downloads-Meta XR Core SDK (UPM). Available online: https://developer.oculus.com/downloads/package/meta-xr-core-sdk/ (accessed on 20 May 2024).
75. Monica, R.; Aleotti, J. Evaluation of the Oculus Rift S Tracking System in Room Scale Virtual Reality. *Virtual Real.* **2022**, *26*, 1335–1345. [CrossRef]
76. Pereira, D.; Oliveira, V.; Vilaça, J.L.; Carvalho, V.; Duque, D. Measuring the Precision of the Oculus Quest 2's Handheld Controllers. *Actuators* **2023**, *12*, 257. [CrossRef]
77. Kharoub, H.; Lataifeh, M.; Ahmed, N. 3D User Interface Design and Usability for Immersive VR. *Appl. Sci.* **2019**, *9*, 4861. [CrossRef]
78. SteamVR su Steam. Available online: https://store.steampowered.com/app/250820/SteamVR/ (accessed on 20 May 2024).
79. Mirzaei, B.; Nezamabadi-pour, H.; Raoof, A.; Derakhshani, R. Small Object Detection and Tracking: A Comprehensive Review. *Sensors* **2023**, *23*, 6887. [CrossRef] [PubMed]
80. Fregna, G.; Schincaglia, N.; Baroni, A.; Straudi, S.; Casile, A. A Novel Immersive Virtual Reality Environment for the Motor Rehabilitation of Stroke Patients: A Feasibility Study. *Front. Robot. AI* **2022**, *9*, 906424. [CrossRef] [PubMed]
81. Carnevale, A.; Mannocchi, I.; Sassi, M.S.H.; Carli, M.; De Luca, G.; Longo, U.G.; Denaro, V.; Schena, E. Virtual Reality for Shoulder Rehabilitation: Accuracy Evaluation of Oculus Quest 2. *Sensors* **2022**, *22*, 5511. [CrossRef] [PubMed]
82. Jost, T.A.; Nelson, B.; Rylander, J. Quantitative Analysis of the Oculus Rift S in Controlled Movement. *Disabil. Rehabil. Assist. Technol.* **2021**, *16*, 632–636. [CrossRef]
83. Calabrò, R.S.; Müller-Eising, C.; Diliberti, M.L.; Manuli, A.; Parrinello, F.; Rao, G.; Barone, V.; Civello, T. Who Will Pay for Robotic Rehabilitation? The Growing Need for a Cost-Effectiveness Analysis. *Innov. Clin. Neurosci.* **2020**, *17*, 14–16. [PubMed]
84. Arias, P.; Robles-García, V.; Sanmartín, G.; Flores, J.; Cudeiro, J. Virtual Reality as a Tool for Evaluation of Repetitive Rhythmic Movements in the Elderly and Parkinson's Disease Patients. *PLoS ONE* **2012**, *7*, e30021. [CrossRef] [PubMed]
85. Aderinto, N.; Olatunji, G.; Abdulbasit, M.O.; Edun, M.; Aboderin, G.; Egbunu, E. Exploring the Efficacy of Virtual Reality-Based Rehabilitation in Stroke: A Narrative Review of Current Evidence. *Ann. Med.* **2023**, *55*, 2285907. [CrossRef] [PubMed]
86. Demeco, A.; Zola, L.; Frizziero, A.; Martini, C.; Palumbo, A.; Foresti, R.; Buccino, G.; Costantino, C. Immersive Virtual Reality in Post-Stroke Rehabilitation: A Systematic Review. *Sensors* **2023**, *23*, 1712. [CrossRef] [PubMed]
87. Chen, X.; Liu, F.; Lin, S.; Yu, L.; Lin, R. Effects of Virtual Reality Rehabilitation Training on Cognitive Function and Activities of Daily Living of Patients With Poststroke Cognitive Impairment: A Systematic Review and Meta-Analysis. *Arch. Phys. Med. Rehabil.* **2022**, *103*, 1422–1435. [CrossRef] [PubMed]
88. De Giorgi, R.; Fortini, A.; Aghilarre, F.; Gentili, F.; Morone, G.; Antonucci, G.; Vetrano, M.; Tieri, G.; Iosa, M. Virtual Art Therapy: Application of Michelangelo Effect to Neurorehabilitation of Patients with Stroke. *J. Clin. Med.* **2023**, *12*, 2590. [CrossRef] [PubMed]
89. Castillo, J.F.V.; Vega, M.F.M.; Cardona, J.E.M.; Lopez, D.; Quiñones, L.; Gallo, O.A.H.; Lopez, J.F. Design of Virtual Reality Exergames for Upper Limb Stroke Rehabilitation Following Iterative Design Methods: Usability Study. *JMIR Serious Games* **2024**, *12*, e48900. [CrossRef] [PubMed]
90. Pilacinski, A.; Pinto, A.; Oliveira, S.; Araújo, E.; Carvalho, C.; Silva, P.A.; Matias, R.; Menezes, P.; Sousa, S. The Robot Eyes Don't Have It. The Presence of Eyes on Collaborative Robots Yields Marginally Higher User Trust but Lower Performance. *Heliyon* **2023**, *9*, e18164. [CrossRef] [PubMed]
91. Park, S.; Lee, G. Full-Immersion Virtual Reality: Adverse Effects Related to Static Balance. *Neurosci. Lett.* **2020**, *733*, 134974. [CrossRef] [PubMed]
92. De Luca, R.; Manuli, A.; De Domenico, C.; Lo Voi, E.; Buda, A.; Maresca, G.; Bramanti, A.; Calabrò, R.S. Improving Neuropsychiatric Symptoms Following Stroke Using Virtual Reality. *Medicine* **2019**, *98*, e15236. [CrossRef] [PubMed]

93. Calabrò, R.S.; Bonanno, M.; Torregrossa, W.; Cacciante, L.; Celesti, A.; Rifici, C.; Tonin, P.; De Luca, R.; Quartarone, A. Benefits of Telerehabilitation for Patients With Severe Acquired Brain Injury: Promising Results From a Multicenter Randomized Controlled Trial Using Nonimmersive Virtual Reality. *J. Med. Internet Res.* **2023**, *25*, e45458. [CrossRef] [PubMed]
94. Choukou, M.-A.; He, E.; Moslenko, K. Feasibility of a Virtual-Reality-Enabled At-Home Telerehabilitation Program for Stroke Survivors: A Case Study. *J. Pers. Med.* **2023**, *13*, 1230. [CrossRef] [PubMed]

Disclaimer/Publisher's Note: The statements, opinions and data contained in all publications are solely those of the individual author(s) and contributor(s) and not of MDPI and/or the editor(s). MDPI and/or the editor(s) disclaim responsibility for any injury to people or property resulting from any ideas, methods, instructions or products referred to in the content.

MDPI AG
Grosspeteranlage 5
4052 Basel
Switzerland
Tel.: +41 61 683 77 34

Bioengineering Editorial Office
E-mail: bioengineering@mdpi.com
www.mdpi.com/journal/bioengineering

Disclaimer/Publisher's Note: The title and front matter of this reprint are at the discretion of the Guest Editor. The publisher is not responsible for their content or any associated concerns. The statements, opinions and data contained in all individual articles are solely those of the individual Editor and contributors and not of MDPI. MDPI disclaims responsibility for any injury to people or property resulting from any ideas, methods, instructions or products referred to in the content.

www.ingramcontent.com/pod-product-compliance
Lightning Source LLC
Chambersburg PA
CBHW041358210926
51850CB00013B/3116
9783725838431